ACS SYMPOSIUM SERIES **544**

Silk Polymers

Materials Science and Biotechnology

David Kaplan, EDITOR
Natick Research, Development, and Engineering Center, U.S. Army

W. Wade Adams, EDITOR
Materials Directorate, Wright Laboratory, U.S. Air Force

Barry Farmer, EDITOR
University of Virginia

Christopher Viney, EDITOR
University of Washington

Developed from the Workshop on Silks: Biology, Structure, Properties, Genetics sponsored by the Division of Polymer Chemistry, Inc., Charlottesville, Virginia, January 28–29, 1993

American Chemical Society, Washington, DC 1994

Library of Cong[obscured] Data

Workshop on Silks: Biology, Structure, Properties, Genetics (1993: Charlottesville, Va.)

Silk polymers: materials science and biotechnology / David Kaplan . . . [et al.].

p. cm.—(ACS symposium series, ISSN 0097–6156; 544)

"Developed from the Workshop on Silks: Biology, Structure, Properties, Genetics, sponsored by the Division of Polymer Chemistry, Inc., Charlottesville, Virginia, January 28–29, 1993."

Includes bibliographical references and index.

ISBN 0–8412–2743–8

1. Silk—Congresses. 2. Polymers—Congresses. 3. Molecular biology—Congresses. 4. Polymeric composites—Congresses.

I. Kaplan, David, 1953– . II. Title. III. Series.

TS1546.W67 1993
620.1′97—dc20 93–37155
CIP

The paper used in this publication meets the minimum requirements of American National Standard for Information Sciences—Permanence of Paper for Printed Library Materials, ANSI Z39.48–1984. ∞

PRINTED IN THE UNITED STATES OF AMERICA

1993 Advisory Board

ACS Symposium Series

M. Joan Comstock, *Series Editor*

Foreword

THE ACS SYMPOSIUM SERIES was first published in 1974 to provide a mechanism for publishing symposia quickly in book form. The purpose of this series is to publish comprehensive books developed from symposia, which are usually "snapshots in time" of the current research being done on a topic, plus some review material on the topic. For this reason, it is necessary that the papers be published as quickly as possible.

Before a symposium-based book is put under contract, the proposed table of contents is reviewed for appropriateness to the topic and for comprehensiveness of the collection. Some papers are excluded at this point, and others are added to round out the scope of the volume. In addition, a draft of each paper is peer-reviewed prior to final acceptance or rejection. This anonymous review process is supervised by the organizer(s) of the symposium, who become the editor(s) of the book. The authors then revise their papers according to the recommendations of both the reviewers and the editors, prepare camera-ready copy, and submit the final papers to the editors, who check that all necessary revisions have been made.

As a rule, only original research papers and original review papers are included in the volumes. Verbatim reproductions of previously published papers are not accepted.

M. Joan Comstock
Series Editor

Contents

Preface

SILKS HAVE BEEN A SCIENTIFIC CURIOSITY FOR CENTURIES, and new insights about these polymers are surfacing with improved analytical methods and the tools of molecular biology. Silks include a broad range of primarily protein-based high-molecular-weight polymers often associated with insects, silkworms, and orb-weaving spiders. Spider silks are becoming as well studied as silkworm silks because of newly developed methods of study, their interesting mechanical properties, and the availability of a family of spider silks with functional properties tailored for a range of applications.

Researchers in such diverse fields as biosynthesis, processing, and mechanical properties are interested in silks, and a great deal can be learned and mimicked from complex biological systems involved in silk production that will have broad implications for polymer science and materials science. The genetic control and tailorability of silk polymers during biosynthesis provides an unusual system for the study of structure–function relationships as well as tightly regulated gene expression. Processing of silk polymers in an aqueous phase under ambient conditions results in water-insoluble high-strength fibrous products, a process that is of interest from the standpoint of environmental compatibility. Mechanical properties of silk fibers have a unique combination of toughness and stiffness, and the fibers are environmentally stable yet biodegradable products.

This book reflects the work presented at the 1993 International Workshop on Silk. The symposium was organized by the editors to provide a forum for the latest advances in the field and to promote interchange among researchers. The symposium and the book together represent the first compilation of information on the biology, structure, processing, properties, and genetics of silks.

What is known about silk, and, just as important, what is not known is described clearly in this volume. The first section presents issues of general biology, genetics, and synthesis of native and synthetic silks and silklike materials. The second section deals with how these polymers are characterized and modeled and gives some of their properties. Finally, the third section describes the processing of silks, followed by current and potential applications.

This interdisciplinary book will be of interest to many—from graduate students getting started in their research to museum curators concerned about preserving silk articles. Materials scientists and engineers, biochemists, biologists, chemists, molecular biologists, and protein chemists will also find it useful. The information on silks provided here will give newcomers and experienced researchers alike the latest information in the field.

We thank all of the participants in the symposium. The interchange and discussion held in the oval room of the Jefferson Rotunda resulted in one of the most enjoyable and productive scientific sessions with which we have been involved. We also acknowledge the contributions of Marlene Jewer, who took care of much of the correspondence; E. P. Socci, J. A. Young, E. K. KariKari, and J. S. Morris for help with the logistics of the symposium; Anne Wilson for coordinating the publication of the book; the University of Virginia Department of Materials Science for hosting and sponsoring the symposium; and the Air Force Wright Laboratory Materials Directorate, through a grant to Lawrence Associates, Inc., and the U.S. Army Research Office for co-funding the symposium.

We dedicate this book to two of the inspirational pioneers and leaders in the field, Bob Work and Yoshiaki Suzuki.

DAVID KAPLAN
Natick Research, Development,
and Engineering Center
U.S. Army
Natick, MA 01760–5020

W. WADE ADAMS
Materials Directorate, Wright
Laboratory
U.S. Air Force
Wright-Patterson AFB, OH 45433–7702

BARRY FARMER
University of Virginia
Charlottesville, VA 22903–2442

CHRISTOPHER VINEY
University of Washington
Seattle, WA 98195

July 15, 1993

THE FAR SIDE By GARY LARSON

Tough spiders

Biology, Genetics, and Synthesis

Chapter 1

Silk: Biology, Structure, Properties, and Genetics

David Kaplan[1], W. Wade Adams[2], Barry Farmer[3], and Christopher Viney[4]

[1]Biotechnology Division, Natick Research, Development, and Engineering Center, U.S. Army, Natick, MA 01760–5020
[2]Materials Directorate, Wright Laboratory, WL/MLPJ, Wright-Patterson AFB, OH 45433–7702
[3]Center for Bioengineering, University of Washington, Seattle, WA 98195
[4]Department of Materials Science and Engineering, University of Virginia, Charlottesville, VA 22903–2442

General Biology

Silks are produced by the more than 30,000 known species of spider [1], and by most of the 113,000 species in the insect order *Lepidoptera* [2]. In addition, silks are produced by members of several other insect orders [3]. For every silk that has been characterized in any detail, over 1000 uncharacterized silks are known to exist.

This book describes only a very small number of spider, silkworm and aquatic midge larva silks. The reader will discover (or be reminded) that silks are truly remarkable materials in many respects. On statistical grounds alone, it is highly unlikely that the silks characterized to date have shown us the limits of what Nature can achieve in regard to the mechanical properties of fibroin proteins. Concern about the extinction of (especially uncharacterized) species therefore must be maintained far beyond the agenda of specialized ecology: materials technology also stands to lose when Nature's lessons go untaught.

The most carefully characterized silks are the cocoon silk of the domesticated silkworm *Bombyx mori*, and the dragline and capture threads of large, orb-weaving spiders such as *Nephila clavipes*. *B. mori* has been studied extensively because of its long (at least 5000 year) history of domestication and use as a source of textile grade fiber [4]. Domestication of *B. mori* is so advanced that this insect now depends entirely on humans for feeding and protection, in which respect it is unique among insects. The contribution by **Goldsmith & Shi** points out that highly inbred silkworm strains have resulted, that differ with respect to characteristics such as number of generations produced per year, larval growth rates, climate tolerance, disease resistance, silk yield, and silk mechanical properties. Indeed, *B. mori* is second only to the fruit fly, *Drosophila melanogaster,* as an insect model used for genetic studies. **Goldsmith & Shi** report on progress towards constructing a complete map on which *B. mori* characteristics are genetically linked - that is, which characteristics are transferred together during cross-breeding. Such a map will expedite optimization of breeding strategies and genetic engineering, thus facilitating efforts to tailor strains that exhibit predetermined combinations of desirable characteristics.

0097–6156/94/0544–0002$06.00/0

N. clavipes and spiders of the genus *Araneus* are studied because they are relatively large and easy to obtain. Until mid-century, spider silk found widespread use as reticules in optical instruments, because the fibers are fine, strong, and resistant to extremes of temperature. Fine platinum wire or etched glass is preferred in more modern instruments, because silk will sag in humid weather. Unprocessed *Nephila* silk still supplies some New Guinea, Solomon Island and New Hebrides natives with gill nets, dip nets, fishing lures, and silk for weaving bags and headdresses. Historically, spider silks have not found widespread use in processed textiles, because spiders are solitary and predatory (and so are more difficult than silkworms to raise in large numbers), and they do not produce a convenient large batch of silk in a cocoon. In 1709, Bon de Saint-Hilaire demonstrated that spider egg sac silk could be used for fabrics (stockings and gloves) in the same way that silkworm silk is used [5]. However, it was concluded by the French Academy that several hundred thousand spiders would be needed for a pound of silk, so that a silk industry based on spiders could never be profitable. Nearly three centuries later, the advent of biotechnology points the way to obtaining economically useful quantities of spider silk protein from genetically modified bacteria or other suitable host systems.

Several aspects of the biology and ecology of silk-producing organisms are relevant to the goal of processing silk from bacterial culture. The effect of many impurities on the color of silk, and the way in which impurities are bound to the fibroin molecules and distributed in the microstructure, are not understood fully.

The paper by **Tillinghast** addresses various physiological aspects of silk synthesis and transformation to fiber. It is expected that all morphological structures associated with the natural silk spinning process are assignable to some function, and familiarity with such detail may be useful to the materials processor. For example, the observed multiplicity of secretory granules in representative silk glands suggests that spider silks do not consist of just one type of protein. Light microscopy and scanning electron microscopy suggest that the dragline silk from many species of spider has a skin-core microstructure; this too has been interpreted as evidence for the heterogeneity of silk [6]. However, the contribution by **Mahoney et al.** describes detailed microscopic studies on the dragline silk from *Nephila clavipes* which do not indicate such a core, even with abraded fibers. Heterogeneity could be consistent with some genetic evidence [7,8]. The contribution by **Hinman et al**. identifies two distinct genes encoding two proteins (Spidroin 1 and Spidroin 2) in *N. clavipes* dragline silk; while the contribution by **Mello et al**. reports only one protein. Many of these contradictory findings reflect the dearth of detailed biochemical and morphological information on silk fibers, a gap that is beginning to be addressed by taking advantage of many new characterization methods as reflected in the contributions to this book.

In the field of biomimetics, one must constantly be mindful that Nature's optimization processes occur over very long periods of time and may not be directed towards a single or obvious goal such as maximizing strength. The optimization can be significantly more subtle, and may seek to minimize the total energy cost by favoring cheaper molecules and more expensive microstructures [9]. One cannot presume to understand Nature's individual microstructure or process design without an appreciation for the complex interaction between the organism and its environment. Again, ecological studies become relevant to the materials sciences. The choice of amino acids used in a particular silk may be governed not only by what gives the greatest strength, or extensibility, or toughness, but must also take account of what amino acid sources are available locally. The recycled web of an orb weaver may enable partial conservation of essential amino acids. If a spider is *not* an orb weaver, its choice of dietary species has to optimize essential amino acid intake [10]. Much of the water in the glue on orb webs is drawn from the atmosphere, and so ingestion of the web provides a significant contribution to the daily intake of water by the web-building spider [11].

Silk is such an evolutionarily specialized material that its mechanical properties and spectral reflectivity may dictate the habitat in which a spider species can live, and thus what prey it can capture, and thus its reproductive success [12]. This complex relationship between spider ecology and silk constitutes an unusually clear example of interaction between molecular and organismal evolution, and is described in more detail in the contribution by **Craig**. Evolutionary questions are also addressed in the paper by **Vollrath**, in the course of assessing the relative success of cribellate spiders (capture threads without glue) and cribellate spiders (capture threads with glue). It requires more energy to spin cribellate silk, but cribellate silk is also more effective at capturing and retaining prey. However it is not clear whether the cribellate silk is intrinsically a more effective trap, or whether it simply appears to be so because flying insects have evolved to the point where they can better escape the glue on the more prevalent cribellate webs.

Two recurring themes in materials mechanical property optimization are the formation of composites, and microstructural hierarchy [13]. Silk is no exception [14]. The paper by **Vollrath** discusses how entire webs succeed as composite energy absorbing structures that have been optimized to dissipate energy associated with sudden out-of-plane loads. The stiff radial threads and compliant capture threads work together to absorb energy while simultaneously maintaining support. Individual silk fibers from the *Araneus aurantia* egg sac outer cover are a composite consisting of round fibrils (sub-micron diameter) in an amorphous matrix [15]. Individual fibrils of *Nephila* dragline silk have a composite microstructure of crystalline regions (30% of dry silk volume; length-to-thickness ratio approximately 5:1) in an amorphous matrix (70% of dry silk volume; chain segments are approximately 15 amino acid residues long) [16]. The stated attributes of the crystalline and amorphous domains are extrapolated from dragline mechanical properties and from X-ray diffraction; the size and connectivity of individual domains are unknown.

As a result of their composite and hierarchical microstructure, spider silks have several mechanisms for undergoing reversible deformation needed for prey capture:

* Tensile deformation may convert amorphous regions to helices in dragline. The helices revert to disordered conformations when stress is released [17]. Under ambient conditions, the amorphous regions behave like a rubber with a low T_g or sub-ambient T_g, preserving appreciable stiffness while simultaneously exhibiting a significant capacity to absorb energy [18].
* The water-based glue droplets on capture thread function like miniature windlasses, collecting around balls of loose core fiber. The coiled thread within each droplet unwinds under load, and retains a coating of glue. The coat's surface tension creates a small force that helps to maintain tautness and winds up loose thread when the load is removed [19].
* NMR indicates that water enables reversible mobility at the molecular level in capture thread [20]. This mechanism appears to be unavailable in frame silk, which is more highly crystalline.
* Cribellate silks are an interwoven network of silks, including crimped threads that act as springs.

Individual mechanisms receive more detailed treatment in chapters by **Vollrath**, by **Hinman et al.**, and by **Gosline et al.**

Genetics

Silk protein synthesis in silkworms and spiders is a highly specialized process for protein expression under tight regulatory control in the cell. Genetic regulation in the epithelial cells lining the posterior region of silk glands is under tight transcriptional control. In the case of the silkworm, activation starts at the anterior part of the posterior region of the silk gland beginning with the 5th instar, and spreads towards the posterior end of the gland.

Early work on silk genetics focused entirely on the silkworm. Through the 1970s the heavy chain fibroin-encoding gene was partially characterized by a number of researchers and found to consist of a repetitive motif [21-24]. The fibroin heavy chain gene is ≈16 kb (kilobases) with a 5' adenine/thymine rich region, a 970 bp (base pair) intron, a nonrepetitive coding region of 414 bp, a core repeat region of ≈15 kb (consisting of 10 crystalline-encoding domains of 1-2 kb interspersed with amorphous-encoding domains of ≈220 bp each), a polyadenine tail to the 3' end, and a flanking region at the 3' end. Only about 2 kb of the fibroin gene has been sequenced, including the 5' flanking region [21,22].

Codon preference (triplet nucleotide base sequence) has been addressed in the silkworm with the glycine codon (GGA) not used before a serine codon but often used before an alanine codon, and GGU and GGA are prefered codons for glycine, GCU for alanine, and UCA for serine [23,24]. This preference relates to the maintenance of stability and compatability with host translation systems [23,24]. Restriction maps were generated for the fibroin gene, although most restriction enzymes studied were unable to cleave in the repetitive coding region to due to the high guanine and cytosine content. The fibroin gene was present as a single copy per haploid genome [25]. Cell-free systems supplemented by insect tRNAs were capable of forming full length fibroin [26]. Discontinuous translation was observed in the cell free system and in major ampullate glands maintained in culture. This discontinuous translation was presumed to correspond to the different time required for ribosomal recognition and binding at each codon and may be due to sub-optimal concentrations of specific tRNAs due to the preponderance of glycine, alanine and serine in the proteins.

The fibroin light chain was sequenced by Yamaguchi et al. [27]. Three cysteine residues were identified, two involved in an intramolecular disulfide linkage and the other located near the C-terminus and potentially available for linkage to the fibroin heavy chain. From the genetic sequence the encoded fibroin light chain protein was determined to have a molecular weight of 25.8 kDa. The synthesis of heavy and light fibroin chains is co-regulated such that equimolar amounts of the two proteins are synthesized [28]. The two genes are encoded on different chromosomes. The family of sericins are encoded by at least two genes with different gene splicing events giving rise to different mRNAs ranging in size from 2.8 kb to 10.5 kb [29-31].

Natural "wild" populations of silkworms have also been screened for gene variability. It has been found that significant differences exist in the fibroin gene size. This variation is assumed to be tolerable due to the repetitive nature of the encoded protein [23,24,32]. Studies on homologous recombination in natural populations of silkworms found signficant variations in crystalline-encoded domains, which result in a high degree of polymorphism in length and organization of the fibroin gene locus, and lead to a 15% size variation in the encoded proteins. Since the changes identified would not alter the beta-sheet secondary structure, it is presumed that these variations are tolerable.

In the late 1970s, Suzuki and Oshima cloned a silkworm fibroin gene into a bacterial host, *Eschericia coli* [33,34]. Problems were encountered with gene rearrangements and stability (in part due to the repetitive gene structure), incompatibility with prokaryotic host system tRNA ratios, and mRNA secondary structure. Issues of protein stability, toxicity in foreign hosts and full length processing were not addressed due to the inability to stabilize the gene.

There is strong interest in silkworm genetics due to the unusual translation system wherein large quantities of protein are produced at a very specific stage in the life cycle of the silkworm. The regulation of this process is of great interest to molecular biologists and developmental biologists. The process is under strong translational controls and results in high levels of protein production (≈300 μg of fibroin per cell) [35]. In addition, the fibroin mRNA is very stable with a half life of days [25]. *B. mori* is also being studied as a host system to express recombinant proteins due to this high level of protein production and strong regulatory control.

In recent years there has been some focus on spider silk genetics [36,37]. Candelas et al. [38-40] identified translational pauses using excised glands from *N. clavipes* which displayed discontinuous extension of the protein in a similar process to that described for the silkworm. Lombardi and Kaplan [41,42] reported on cloning a dragline silk encoding gene from *N. clavipes* from a cDNA library. Problems were encountered with gene stability in the prokaryotic host. Some of these limitations continue to be addressed both with native genes as well as synthetic gene constructs. Xu and Lewis [7] and Hinman and Lewis [8] have reported partial sequence data for cDNA clones from *N. clavipes* major ampullate gland which encode for spidroin I and spidroin II (see contribution from **Hinman et al.**). A total of 2.3 kilobases of a clone from the 3' end of the gene for spidroin I and 2 kilobases of a clone from the 3' end of of the gene for spidroin II were sequenced. The data illustrate the repetitive nature of the gene in the coding region. Codon preference was assessed and cytosine or guanine preference was significantly reduced in the third position for glycine, alanine or glutamine in both spidroin I and II clones.

There remains a great deal to resolve about silk genes from spiders and the silkworm. None of the genes has been fully sequenced and characterized and they remain poorly understood in terms of introns/exons, upstream regions, copy numbers, similarity to other silk-producing genes, etc. Part of the problem relates to the difficulty in characterizing highly repetitive gene structures. Further, they contain unusual secondary structure at the mRNA level which is responsible for the stability of the mRNA. Difficulties in sequencing, stability in host systems, and difficulty in using gene amplification techniques are obstacles to resolving these problems. The genetics of the aquatic midge have also been studied and are described by **Case and Smith** in a chapter in this volume. A synthetic gene construct encoding some of the known protein sequence data was expressed.

The focus of much of the recent work on silk genetics has been on the construction and expression of synthetic silk-like genes. This has resulted in part because of lack of full understanding of the native silk genes and in part because of difficulties encountered in stabilizing silk genes from silkworms and spiders in foreign hosts. With the synthetic gene approach, the latter issue can be readily overcome by the judicious choice of codons to match that of the host system chosen for expression. Two chapters in this volume (**Cantor et al.** and **O'Brien et al.**) summarize some of the current work on synthetic silk-like gene constructs.

Cappello and Crissman [43] and Cappello *et al.* [44] report on synthetic silk-like genes expressed in *E. coli* in combinations with other coding domains to tailor the functional properties of the silk-like polymers produced. For example, elastin-like domains are interspersed with the silk domains to provide block copolymers with rigid and elastomeric regions. In addition, fibronectin domains for surface binding have also been added to to coding domains. Some of these recombinant proteins have been commercialized. An approach to spinning fibers and assessing their properties is described in the chapter by **Cappello and McGrath**. **Anderson et al.** discuss the characterization of some of these recombinant proteins and the relationship between processing conditions and morphology. A ternary phase diagram was developed for one of these synthetic silk-like proteins. The importance of elucidating these types of relationships in developing higher order structures in materials science is addressed. McGrath *et al.* [45,46] report on the use of synthetic gene technology to produce chain folded lamellae with surface residues tailored to specific functions. An overview of this work in terms of the types of proteins produced and their detailed structural characterization is presented in the contribution by **Cantor et al**. Finally, Hoess *et al.* [47] reported on the use of synthetic gene technology to produce recombinant proteins containing 15-mer repeats which form cross-beta silk-like polymers useful for fibers with interesting film and nonlinear optical applications. Repetitive genes have been constructed and recombinant proteins 40 kDa in size were produced [47]. The proteins have been expressed in *E. coli* and consist of 9.7% of the total cell protein. A solution of 95% protein was anisotropic and exhibited a cross-beta secondary structure by X-ray analysis. The details of this work are presented in the contribution by **O'Brien et al.**

Processing / Physical Characterization

Researchers involved in materials characterization must be concerned that their selected samples are representative of *in situ* conditions, and to what extent specimen preparation causes changes in composition, conformation or microstructure. There is no guarantee that the molecular conformation or microstructure of silk recovered from a freshly dissected gland is identical to that of silk stored inside a gland in a live animal.

The NMR spectroscopic studies of **Asakura et al.** include ^{13}C solution spectra obtained from live silkworms. These data yield amino acid sequence information for the fibroin molecules. The same authors have used solid state NMR to characterize molecular conformation (measure torsional angles) in silk. The latter technique is necessarily limited to solid material exhibiting a significant degree of uniform molecular alignment. Thus, NMR enables the molecular conformation of silk II to be measured, but the conformation of silk I remains a subject of debate.

Two techniques are presented that will be useful for studying real-time molecular order changes *in situ* as silk is drawn from a live spider. Raman spectroscopy is described in the chapter by **Gillespie et al.** The principal limitation to such work is the considerable fluorescence that masks the Raman signal; possible fixes include changing the source wavelength (e.g., FT-Raman), or maintaining the spiders under conditions that eliminate incorporation of the fluorescent impurity into the silk. *In situ* characterization by synchrotron X-ray studies are described in the contribution by **McNamee et al**. The principal limitations here are the small scattering volume presented by a single fiber, together with the low atomic weight elements that constitute the sample. It should also be noted from the paper by **Vollrath** that fiber obtained by forcible silking may have inferior properties compared to natural fiber –so the two materials cannot be presumed to have identical microstructures.

Several papers describe characterization studies performed on solubilized silk, silk-like proteins, or solid products prepared from such solutions (**Muller et al.**, **Mello et al.**, **Anderson et al.**, **Asakura et al**. and **Viney et al.**). In each case, a protocol for preparing the protein solution is described. The contributions by **Anderson et al.** and **Viney et al.** also show how ternary phase diagrams can be useful aids for comparing the relative amounts of protein, water, and chaotropic salt in solutions from which solid products are processed, or in samples used for characterizing molecular order in the fluid state. Studies by **Anderson et al.** were carried out with synthetic silk-like proteins generated from recombinant DNA processes. **Viney et al.** study native silks (silkworm, spider, aquatic midge, and regenerated native silks). Kinetic parameters – i.e. the rate at which fibers are spun from solution, and the rate at which water evaporates from silk solutions – are shown to have a significant effect on microstructure (**Magoshi et al.** and **Viney et al.**).

Magoshi et al. describe a variety of energy-conserving spinning processes used by the silkworm to control molecular orientation and generate cocoon fibers, and highlight the need to mimic these processes in artifical systems. **Viney et al.** present preliminary evidence for aggregates of globular molecules interacting noncovalently as a key element in the formation of rod structures comprising the liquid crystalline phase for these types of materials. It is clear that additional characterization work is needed to fully understand the structures present during the different stages of processing, particularly if we are to successfully duplicate this process outside the organism.

Muller et al. describe the formation of thin films from solubilized silk. This technique is of particular interest because it provides an opportunity to control the secondary structure. Initial evidence based on Fourier Transform Infrared Analysis supports the presence of a silk II (water-insoluble) conformation under the initial conditions examined. The goal is to manipulate the structure of silk by controlling factors such as pH in the subphase (in a Langmuir trough), temperature, and pressure on the surface of the trough. **Magoshi et al.** in another chapter describe the thermal properties of silkworm silks using Differential Scanning Calorimetry and highlight the transition from random coil to beta sheet above 180°C. **Cunniff et al.** present related data for spider silk.

Structure and Modeling

The essence of the crystalline structure of silk has been known since 1955 when Marsh et al. [48] described the antiparallel, hydrogen bonded ß-sheet arrangement in the silk II structure of *B. mori* fibroin. As experimental and computational techniques advanced, refinements to that structure were made by Fraser et al. [49] and by Colonna-Cesari et al. [50]. Much of the refinement has dealt with the apportionment of the c-axis of the unit cell to the relative distances between adjacent ß-sheets. The sheets in the generally accepted structure of Marsh et al. are asymmetric, having one surface comprised primarily of the alanyl methyl groups and the other surface comprised of the glycyl hydrogen atoms. In the silk II structure, the sheets are arranged back to back in alternation - e.g., with two sheets having their glycyl faces juxtaposed followed by an alanyl-alanyl juxtaposition. Clearly, the spacings between the sheets differ, with the former being about 3.79Å and the latter being about 5.27Å.

More recently an alternate structure has been proposed by Takahashi et al. [51]. While their refined structure still consists of antiparallel hydrogen bonded sheets, the structure within the sheets is different. Adjacent strands are oriented in opposite directions, yielding a sheet with two identical surfaces comprised of both alanyl and glycyl components. The inter-sheet distance for packing these symmetric sheets is half of the c-axis spacing. In a contribution to this volume, **Takahashi** describes this proposed structure. In particular, the data set from which the structure was refined contained a considerably larger number of reflections than has been previously analyzed. In addition, the structure reflects an improved analysis of the data in terms of anisotropic temperature factors, random replacements of 1/3 of the alanine residues by serine, and statistical disorder in the orientation of adjacent sheets. The NMR data from Nicholson et al. [52] confirm the silk II structure described by Marsh et al. [48] and more recently by Takahashi et al. [51].

The silk II structure from *B. mori* is obviously not the only silk structure observed experimentally. The structures of spider silks differ from silkworm silks, though both are basically antiparallel ß-sheet structures. Other species produce silks having α-helical structures, while still others have cross-ß-sheet structures. Even among the ß-sheet materials, different amino acid compositions (by virtue of diet and genetics) alter the structure slightly. The most studied, and still not fully resolved, other structure is that identified as silk I. This structure is unstable to mechanical deformation, transforming to the more common silk II structure when subjected to drawing, shearing, rolling or spinning. Its importance lies in its possible relation to the structure of the mesophase material collected directly from the silk gland - e.g., unprocessed by the spider's or silkworm's extrusion and drawing mechanisms. The mechanical instability of this structure largely precludes obtaining an oriented specimen (and a fiber-diffraction pattern) from which to determine the structure.

Two models are currently favored for the structure of silk I. One proposed by Lotz and Keith [53] consists of antiparallel, hydrogen-bonded sheets comprised of molecules having a "crankshaft" conformation. The crankshaft conformation consists of alternating alanyl units in a ß conformation (ϕ,ψ) of (-123°, 122°), and glycyl units in a left-handed α–helical conformation, (ϕ,ψ) of (57°, 72°). Though unique in proteins, this conformation agrees reasonably well and quantitatively with the (rather sparse) X-ray data available. More recently, Fossey et al. [54] have proposed a structure comprised of chains having alanyl residues with (ϕ,ψ) of (-80°, 150°) and glycyl residues in a conformation with (ϕ,ψ) of (-150, 80°). In a contribution to this volume, **Fossey and Kaplan** describe the molecular modeling studies from which they identified their proposed structure, and compare its features with the available diffraction data.

Molecular modeling studies have proceeded hand-in-hand with diffraction studies in the determination of the structures of silk. The relative simplicity of the composition of the protein sequence found in the crystal lends itself naturally to such studies. In fact, the explorations of the silk structure could be said to have focused on developing the methodology of modeling as much as on determining the structure. Though model building has always been a part of diffraction analysis, computerization of the modeling allowed it to become quantitative and rapid enough to be predictive rather than just corroborative. The paucity of experimental diffraction data available from silk (as well as many other fibrous proteins) required a multi-faceted approach to solve the structural problem, and the concurrent development of computational methods (and for that matter computers themselves) offered a potentially powerful new tool, one in need of refinement and suitable test cases.

For the most part, computational studies have used an alternating copolymer of L-alanine and glycine as a model for the crystalline component of silk. Crystals of alanine-glycine and silk have been shown to be isomorphous with only the intersheet spacing varying. The presence of serine, which by composition replaces approximately one alanine in three, has largely been ignored. Structural calculations have proved valid, nonetheless, probably owing to the easy accommodation of the occasional hydroxyl group in the region between sheets normally populated by the alanyl methyl groups. As noted above, modeling studies of ß-sheet structures of poly-L-alanine and poly(L-alanine-glycine) led to the refinement of early structure work on silk II by examining the distances between different faces of the asymmetric ß-sheets. Further, the calculations of various polypeptide structures, including that of polyglycine I, helped resolve apparent disparities between the silk II structure and those of the homopolymers, polyglycine and poly-L-alanine, which were initially supposed to reflect the different sheet spacings.

Computational modeling was also used by Lotz and Colonna-Cesari [55] to explore chain conformations and packing arrangements appropriate to the silk I structure. Conformations having alanyl residues in a ß conformation and glycyl residues in a left handed α-helical conformation were found to be most stable. Chain conformations with the torsion angles of the residues interchanged, an arrangement that could not be ruled out on the basis of the diffraction data alone, were found to be inconsistent with either the observed intersheet distances or low energy packing arrangements. For the β_{ala}-α_{gly} conformations, a parallel chain arrangement was found to be most energetically stable, again a possibility that could not be ruled out experimentally, although chain folding considerations would favor an antiparallel arrangement. The calculations also revealed several structures of comparable energy, differing by the orientations of adjacent sheets as well as positional disordering. No quantitative comparison with the diffraction data was undertaken.

An extensive conformational analysis for single chains of poly(alanine-glycine) was carried out by Oka et al. [56]. Several low energy conformations were identified. Of those which were consistent with the experimentally determined chain repeat distance, the crankshaft model was found to be a little higher in energy than a helix having (ϕ,ψ) values of (-70°, 103°) and (-75°, 42°) for the alanyl and glycyl residues, respectively.

Fossey and Kaplan (as reported in this volume) have carried out a considerably more extensive modeling analysis, examining and refining a spectrum of conformations for the polypeptide chains, packed within sheets with parallel or antiparallel directions, and arranged in polar or antipolar orientations (e.g., with all alanyl residues oriented to the same face of the ß-sheet or half to each face). They also considered the different ways of stacking the sheets. One of the lowest energy structures corresponded closely to the silk II structure. Another of the structures that emerged from the energy minimization was found to have dimensions comparable to those identified for silk I, and a rough correspondence between observed and calculated d-spacings. No intensity calculation or comparison was undertaken. As noted above, the structure is comprised of chains having (ϕ, ψ) of values (-80°, 150°) and (-150°, 80°) for alanyl and glycyl residues, respectively. Other conformations and structures proposed for silk I were all found to have higher total (intramolecular + intermolecular) energies.

Characterization of Silk Proteins

Early studies demonstrated that silkworm cocoons from *B. mori*, designed for encapsulation of the molting larvae, consist almost entirely of protein. The silk proteins isolated from the cocoon have been characterized more extensively than any of the other silks. Initially one fibroin protein was identified (fibrous main structural element), along with a family of sericin proteins which serve to bind the fibroin chains to form the cocoon composite structure. In more recent studies, a second, lower molecular weight fibroin chain has also been identified [27,57]. The higher molecular weight fibroin, termed heavy chain, is generally reported to have a molecular weight of 325 kDa and the fibroin light chain is 25 kDa. The size of the light chain has been confirmed by isolation and sequencing of the gene encoding this protein [27]. The molecular weight of the heavy chain remains an estimate since the full genetic element encoding this protein has not been sequenced. An intermolecular disulfide bond has been identified between the light and heavy fibroin chains and is postulated to play a role in maintaining solubility and preventing premature formation of certain secondary structures prior to silk fiber spinning. Analysis of protein compositions from wild-type silkworms indicate variability in composition and in the size of the fibroin heavy chain compared to the domestic silkworm [23,24].

A number of groups have reported the amino acid composition of silkworm fibroin [3,58-61]. Generally, domestic silkworm silk heavy chain fibroin contains approximately 87% short chain amino acids (45% glycine, 30% alanine, 12% serine). Some fibroins contain up to 96% of these three amino acids. The fibroin light chain contains a very different amino acid composition of 15% aspartate, 14% alanine, 11% glycine, 11% serine and a trace of cysteine. The consensus peptide crystalline repeat in the heavy chain fibroin is: GAGAGSGAAG$[SG(AG)_2]_8$Y (G=glycine, A=alanine, S=serine, Y=tyrosine) [62]. The amorphous domains are less well defined but generally include some of the bulkier side chain amino acids.

The family of proteins termed sericins, which comprise approximately 25% of the silkworm cocoon by weight, vary in molecular weight from 20 kDa to 310 kDa [29,32] and contain glycine, serine and aspartic acid totaling over 60%. The hydrophilic nature of sericin facilitates its relatively straightforward removal in hot water during cocoon processing to purify the fibroin for silk fiber and textile applications.

Spider silk proteins, designed for catching prey, are a more complex family of proteins than the those of the silkworm, primarily due to the specialization of different types of silks for different functions. Each silk is synthesized in different pairs of glands in the abdomen of the spider, and most have been only poorly characterized to date. Kaplan *et al.* [36] recently reviewed the compositions of silks reported in the literature. The dragline silk from the spider *N. clavipes* generally contains about 37% glycine, 21% alanine and 5% serine [63].

A major focus of spider silks has been on the dragline fiber, both for its mechanical properties and its intricate engineering into the orb web. Candelas and coworkers [38-40] published information on the protein composition of dragline fibers from *N. clavipes*. They reported one protein with a molecular weight of 320 kDa as determined by electrophoresis. The chapter by **Mello et al.** presents preliminary evidence for an estimated molecular weight of 275 kDa for the dragline silk protein. Upon chemical reduction, the major ampullate gland protein provides a double band at lower molecular weights, both of which contain the same amino acid composition. Initial evidence for disulfide bonds within the dragline protein is also presented in this chapter. It should be recognized, based on the recent work of Beavis *et al.* [64] that apparent molecular weights of silk-like proteins containing a significant percentage of glycine-alanine repeats determined by polyacrylamide gel electrophoresis can be significantly different than weights determined by mass spectroscopy.

Xu and Lewis [7] and Hinman and Lewis [8] reported the presence of two proteins, spidroin I and II, in the dragline silk from *N. clavipes*. The evidence for the two proteins is derived from the analysis of partial cDNA clones generated from the major ampullate gland of the spider. In the chapter by **Hinman et al.** a description of the two proteins is provided. Spidroin I contains a repeat unit that is not rigidly conserved and contains three regions: a six amino acid domain, a 13 amino acid domain consisting of polyalanine runs of six to nine residues, and a 15 amino acid conserved region containing GGx repeats where x = alanine, tyrosine, leucine or glutamine [7]. Spidroin II also consists of a three region repeat motif, with a highly conserved 15 amino acid domain containing the repeat GPGxy where xy is EE or GT, a heptapeptide repeat of GPSGPGS coupled to a polyalanine run of six to 10 residues, and a region with pentapeptide repeats with various deletions [8]. **Mello et al.** report on the repeat motif AAGGAGQGG (A=alanine, G=glycine, Q=glutamine) identified from major ampullate gland protein which appears 14 times in the spidroin I sequence reported earlier by Xu and Lewis [7], and the repeat motif YGGLGSQGAGRGG (Y=tyrosine, G=glycine, L=leucine, S=serine, Q=glutamine, A=alanine, R=arginine) identified from the dragline fiber which appears 10 times in the spidroin I sequence.

It is clear from the literature that a consensus has not yet been reached on the content of the dragline fiber. Apart from the dragline silk, few of the other silks have been well characterized in terms of their protein content and sequence. Andersen [65] reported on the amino acid composition of many of the different types of silks from the spider, *Araneus diadematus*. However, the number of different proteins contained in these silks, the sequences of these proteins, and the relationship between primary structure and functions of the silks are unknown.

The proteins from the aquatic midge and the caddis fly, capable of producing silks underwater to form structural tubes for shelter, foraging and filter-feeding, have also been studied [66]. Characterization of some of these proteins is summarized in the chapter by **Case et al.** The midge produces a family of proteins which range in size from 38 kDa to over a million. In general, these silk proteins include tandem repeats in a fashion similar to those of the silkworm and spider, although the composition and sequences are very different. For example, the unusally high cysteine content in the midge (7.3 mole percent) and relatively low percentage of small side chain amino acids in the midge and caddis fly, 27.8% and 41.1%, respectively, clearly distinguish these types of silks from the silkworm and spider silks.

In order to obtain proteins for analysis in many of the above cited studies, silkworm cocoons or spider silk obtained from webs or under controlled silking conditions [67] must be solubilized. Procedures for solubilizing silkworm silk are described in the chapter by **Muller et al.**, and procedures for solubilizing spider silks are described in the chapter by **Mello et al.** Often lithium bromide salts, propionic/hydrochoric acid mixtures, or 85% formic acid are used followed by dialysis in water or suitable buffers.

Becker and Tuross report on the degradation of *B. mori* silk by irradiation with UV and visible light. They report selective degradation of tyrosine, acidic and ionizable amino acids during the initial process, leading to peptide bond cleavage. With additional exposure the degradation process appears to be less selective for specific amino acids. The issue of environmental stability of silks is of particular interest for the preservation of materials in museums, since many ceremonial clothing and related items on display or in storage contain silks.

Properties of Silk

A book such as this, in which the contributors' expertise and the referenced material cross many boundaries, is destined to use technical terms that have multiple definitions. One of these is the term "elastic". A mechanical or materials engineer will describe a material as elastic if a load causes deflections that can be recovered more or less instantly and reversibly once the load is removed. Biologists may instead describe a material as elastic if it can be extended to very large (but not necessarily recoverable) strains before breaking. Thus, for different reasons, both spider dragline and capture spiral may be thought of as being elastic! If one is at least aware of this potential ambiguity, the intended meaning can usually be inferred from the particular context of use.

Mechanical property measurement is open to ambiguity too. A comparison of the strength, or stiffness, or toughness, or ductility of different materials strictly is valid only if all the samples are tested under identical loading conditions. Tensile modulus (stiffness, or resistance to deformation) comparisons of dragline and synthetic polymers, for example, should be made for samples of equal length and diameter (and machine compliance corrected), gripped in the same way, and strained at the same rate. The stiffness measurements should be made from a consistent part of the stress-strain curve, which in turn should be plotted consistently (e.g., as *nominal* stress (load/initial area) versus *nominal* strain (extension/original length)).

Most of the interest in silk fibers is due to their unusual combination of mechanical properties, which distinguishes them from other natural or synthetic fibers (see chapters by **Cunniff et al.** and **Bunning et al.**). Spider dragline and silkworm cocoon silks can exhibit up to 35% elongation, with tensile strengths approaching those of high performance synthetic fibers such as Kevlar 29 [37], while the energy absorbed before breaking can exceed that of Kevlar or even steel. Some theoretical calculations have suggested that the tensile modulus of a polypetide unit could exceed that of polyethylene [68], but even more intriguing is the prediction that the compressive properties of these biopolymers could be the highest yet seen for an organic fiber (see paper by **Pachter et al**. Besides these mechanical properties, silk's visual appearance (translucency, luster, and ultraviolet reflectance) and engineering aspects of web and cocoon construction and function are also of great interest (see paper by **Gosline et al.**).

Most fibers, which have some viscoelasticity, suffer a reduction in elongation to break with increased loading rates, even as their ultimate tensile strength and modulus increase. Silk, however, is rather novel in that, while the expected increases in strength and stiffness occur, elongation actually increases with faster deformation. **Cunniff et al.** report a comparison between conventional tensile testing of spider silk at 10%/s rates and very high speed elongation (ca 1,000,000%/s), under conditions mimicking ballistic impact. Although they obtained actual tensile properties somewhat lower than previously reported for dragline silk (1.1 GPa strength versus 1.5 for Work [69] and 22 GPa modulus versus 30 GPa for Zemlin [62]), their data for the ballistic resistance of silk did show a 1% elongation to break increase (from 9 to 10%) from static to dynamic loading. Zemlin [62] previously reported 30 GPa, 1.9 Gpa and 17.4% for strength, modulus, and extensibility to break, respectively, although the errors were reported to be as high as 64%. More significantly, **Cunniff et al.**, hypothesize that tensile strengths of about 3 GPa, a modulus of 60 GPa, and elongations of 11% should be attainable in genetically engineered silk. These predicted values are sufficient to provide impetus to further research.

It has been reported that mechanically silking a spider results in mechanical properties inferior to those obtained from fibers removed from webs and directly tested (**Vollrath**). As the silking rate increases, the fiber diameter increases. This probably indicates a relaxation of the spider's ability to control the process, since higher draw rates normally correlate with smaller fiber diameters in conventional fiber spinning processes.

Becker et al. report a comparison of X-ray determined tensile moduli of spider and silkworm silk fibers. Following the reported crystal moduli measurements of Nakamae et al. [70], these authors refine the data for silkworm fibers to indicate a tensile modulus of the crystallites of 29 GPa, while they report the first measurements ever on spider silk, with a crystal modulus of 17 GPa. The rather large difference is attributed partly to the difference in the bulky side groups in the silk sequences for spider silk, with a commensurate difference in the *a* axis dimensions of the unit cell. Morphology differences including lower crystallinity in the spider fibers and possibly different crystallite arrangements (connectivity) probably account for the rest. Since the X-ray measurement requires an assumption of uniform stress in the fibers, differences in the arrangements of crystalline and amorphous (or more properly non-crystalline) regions in the fibers can greatly affect the X-ray results. Measurements on artificially aged or heat-treated fibers, also reported by **Becker et al.,** show lower X-ray moduli as well, casting further doubt on the validity of the iso-stress model. The next major question to be answered is the disparity of the X-ray measured modulus to the theoretical predictions, which range from 150 GPa reported by **Becker et al.** for a classical mechanics calculation involving a cluster of Ala chains, and 155 GPa calculated from a simple summation of bond force constants [71], to the semiempirical, quantum mechanical calculation (ca 60 GPA) of **Pachter et al.** in this volume. Clearly, this is an area ripe for further research.

A potential highly significant property of silk fibers is their apparent resistance to axial compressive deformation. Scanning electron microscope observations of spider silk fibers revealed that even when tied into a tight knot, the fibers suffer no kink-band failure on the compressive side of a bend. Since all synthetic high performance fibers fail by kinking in compression at relatively low stress levels, silk appears to be unique in this behavior. **Mahoney et al.** discuss the possible morphological basis for this property and report the first examination of spider silk fibers by atomic force microscopy. They show no evidence for an inherent fibrillar microstructure, although more research will be needed to verify their observed morphology of spider silk. The modeling by **Pachter et al**. provides a possible molecular basis for this resistance to kink formation.

Applications

Silkworm silks have found utility in textiles for thousands of years, and many new applications for these types of proteins are targeted (see contribution by **Bunning et al**. Silks are considered biomaterials or biopolymers. To many, biomaterial refers to any material, of whatever origin, that is used as an implant in the human body [72]. However, the term also has frequent use for what can more descriptively be called a biological material. It is the latter sense that it is most used in this book. Silkworm silk, therefore, can be seen as a biomaterial in both senses, it is a material of biological origin and it can also be used in surgical sutures. The contribution by **Bunning et al.** also highlights the "lessons to be learned from nature" issue. Silkworms and spiders process high molecular weight silk polymers under ambient environmental conditions in an aqueous phase. The resulting mechanical properties are remarkable when viewed in the context of the processing conditions used, particularly when compared to synthetic high performance materials. The ability to tailor silk structure to function, a task the spider has already accomplished though genetics and evolution, also can lead to new directions in the design of polymers targeted for specific functions.

Literature Cited:

1. Foelix, R. F. Biology of Spiders, **1992**, Harvard University Press, Cambridge, MA
2. Academic American Encyclopedia, **1993**, Grolier Electronic Publishing, Inc.
3. Lucas, F.; Rudall, K. M. In: Florkin, M.; Stotz, E. H. (Eds.) Comprehensive Biochem.: Extracellular and Supporting Structures **1968**, Elsevier, Amsterdam, p.475-558
4. Hyde, N. *National Geographic* **1984**, 165(1), 3.
5. Gertsch, W. J. American Spiders, **1949**, D. Van Nostrand Company, Inc., Princeton, NJ
6. Work, R. W. *Trans. American Microscopy Soc.* **1984**, 103: 113-121
7. Xu, M.; Lewis, R. V. *Proc. Natl. Acad. Sci.* **1990**, 87, 7120.
8. Hinman, M. B.; Lewis, R. V. *J. Biol. Chem.*, **1992**, 267, 19320.
9. Vincent, J. F. V. In: Viney, C.; Case, S. T.; Waite, J. H. (Eds.) Biomolecular Materials **1993**, Materials Research Society, Pittsburgh, PA, p.35-43..
10. Greenstone, M. H. *Nature* **1979**, 282, 501.
11. Edmonds, D. T.; Vollrath, F. *Proc. Roy. Soc. Lond.* **1992**, B248, 145.
12. Craig, C. L. *Trends in Ecol. Evoln.* **1992**, 7, 270.
13. Aksay I. A.; Baer E.; Sarikaya M.; Tirrell D. A. (Eds.), Hierarchically Structured Materials, **1992**, Materials Research Society, Pittsburgh, PA.
14. Kaplan, D. L.; Fossey, S.; Viney, C.; Muller, W. In: Aksay, I. A.; Baer, E.; Sarikaya, M.; Tirrell, D. A. (Eds.) Hierarchically Structured Materials **1992**, Materials Research Society, Pittsburgh, PA p.19-29.
15. Stubbs, D. G.; Tillinghast, E. K.; Townley, M. A.; Cherim, N. A. *Naturwissenschaften* **1992**, 79, 231.
16. Gosline, J. M.; DeMont, M. E.; Denny, M. W. *Endeavour* **1986**, 10, 37.
17. Dong, Z.; Lewis, R. V.; Middaugh, C. R. *Arch. Biochem. Biophy.* **1991**, 284, 53.
18. Gosline, J. M.; Denny, M., W.; DeMont, M. E. *Nature* **1984**, 309, 551.
19. Vollrath, F.; Edmonds, D. T. *Nature* **1989**, 340, 305.

20. Bonthrone K. M.; Vollrath F.; Hunter B. K.; Sanders J. K. M. *Proc. Roy. Soc. Lond.* **1992**, B248, 141.
21. Tsuijimoto, Y.; Suzuki, Y. *Cell* **1979**, 16, 425.
22. Tsuijimoto, Y.; Suzuki, Y. *Cell* **1979**, 18, 591.
23. Gage, L. P.; Manning, R. F. *J. Biol. Chem.* **1980a**, 255, 9444.
24. Gage, L. P.; Manning, R. F. *J. Biol. Chem.* **1980b**, 255, 9451.
25. Suzuki, Y.; Gage, L. P., Brown, D. D. *J. Mol. Biol.* **1972**, 70, 637.
26. Lizardi, P. M. *Proc. Natl. Acad. Sci.* **1979**, 76, 6211.
27. Yamaguchi, K., Kikuchi, Y., Takagai, T., Kikuchi, A., Oyama, F.; Shimura, K.; Mizuno, S. *J. Mol. Biol.*, **1989**, 210, 127.
28. Couble, P. M.; Chvillard, M., Moine, A.; Ravel-Chapuis, P., Prudhomme, J.-C. *Nucleic Acid Res.* **1985**, 13, 1801.
29. Gamo, T.; Inokuchi, T.; Laufer, H. *Insect Biochem.* **1977**, 7, 285.
30. Okamoto, H.; Ishikawa, E.; Suzuki, Y. *J. Biol. Chem.* **1982**, 257, 15192.
31. Ishikawa, E.; Suzuki, Y. *Dev. Growth Different.* **1985**, 27, 73.
32. Sprague, K. U. *Biochem.* **1975**, 14, 925.
33. Oshima, Y.; Suzuki, Y. *Proc. Natl. Acad. Sci.* **1977**, 74, 5363.
34. Suzuki, Y.; Oshima, Y. *Cold Spring Harbor Symp. Quant. Biol.* **1977**, 42, 147.
35. Tashiro, Y.; Morimoto, T.; Matsura, S.; Nagata, S. *J. Cell. Biol.* **1968**, 38, 574.
36. Kaplan, D. L.; Lombardi, S. L.; Muller, W. S.; Fossey, S. A. In *Biomaterials: Novel Materials From Biological Sources*; Ed., D. Byrom; Stockton Press, New York, **1991**; 1-53.
37. Kaplan, D. L.; Fossey, S.; Mello, C. M.; Arcidiacono, S.; Senecal, K.; Muller, W.; Stockwell, S.; Beckwitt, R.; Viney, C.; Kerkam, K. *MRS Bulletin* **1992**, Oct., 41.
38. Candelas, G. C., Candelas, T.; Ortiz, A; Rodriguez, O. *Biochem. Biophys. Res. Commun.* **1983**, 116, 1033.
39. Candelas, G. C.; Cintron, J. J. *Exptl. Zool.* **1981**, 216, 1.
40. Candelas, G. C.; Lopez, F. *Comp. Biochem. Physiol.* **1983**, 74, 637.
41. Lombardi, S. J.; Kaplan, D. L. *Acta Zool. Fennica* **1991**, 190, 243.
42. Lombardi, S. J.; Kaplan, D. L. *Polymer Preprints* **1990b**, 31(1), 195.
43. Cappello, J.; Crissman, J. *Polymer Preprints* **1990**, 31(1), 193.
44. Cappello, J.; Crissman, J.; Dorman, M.; Mikolajczak, M.; Textor, G.; Marquet, M.; Ferrari, F. A. *Biotech. Prog.* **1990**, 6, 198.
45. McGrath, K.; Tirrell, D. A.; Kawai, M.; Mason, T. L.; Fournier, M. J. *Biotech. Prog.* **1990**, 6, 188.
46. McGrath, K.; Fournier, M. J.; Mason, T. L.; Tirrell, D. A. *J. Am. Chem. Soc.* **1992**, 114, 727.
47. Hoess, R. H.; O'Brien, J. P.; Salemme, F. R. *PCT Intl. Appl.* WO 92 09,695, **1990**.
48. Marsh, R. E.; Corey, R. B.; Pauling, L. *Biochim. Biophys. Acta*, **1955**, 16, 1.
49. Fraser, R. D. B.; MacRae, T. P.; Stewart, F. H. C.; Suzuki, E. *J. Mol. Biol.* **1965**, 11, 706.
50. Colonna-Cesari, F.; Premilat, S.; Lotz, B. *J. Mol. Biol.* **1975**, 95, 71.
51. Takahashi, Y.; Gehoh, M.; Yuzuriha, K. *J. Poly. Sci. Pt. B: Poly. Phys.* **1991**, 29, 889.
52. Nicholson, L.K.; Asakura, T.; Demura, M.; Cross, T. A. *Biopolymers* **1993**, 33, 847.
53. Lotz, B.; Keith, H. D. *J. Mol. Biol.* **1971**, 61, 201.
54. Fossey, S. A., Nemthy, G., Gibson, K. D., Scheraga, H. A. *Biopolym.* **1991**, 3d1 1529.
55. Lotz, B.; Colonna-Cesari, F. *Biochimie* **1979**, 61, 205.
56. Oka, M.; Baba, Y.; Kagemoto, A.; Nakajima, A. *Polymer J.* **1990**, 222, 416.
57. Komatsu, K. Proc. *7th Intl. Wool Tex. Res. Conf.* **1985**, 1, 373.
58. Lucas, F.; Shaw, J. T. B.; Smith, S. G. *Biochem. J.*, **1962**, 83, 164.
59. Lucas, F.; Shaw, J. T. B.; Smith, S. G. *J. Mol. Biol.*, **1960**, 2, 339.

60. Lucas, F.; Shaw, J. T. B.; Smith, S. G. In *Advances in Protein Chemistry*, editors Anfinsen, C. B., Anson, M. L.; Bailey, K.; Edsall, J. T.; Academic Press, New York, **1958**, 13, 107.
61. Zemlin, J. C. Report 69-29-CM (AD684333), U. S. Army Natick Laboratories, Natick, Massachusetts, **1968**.
62. Strydom, D. J.; Haylett, T.; Stead, R. H. *Biochem. Biophys. Res. Commun.* **1977**, 79, 932.
63. Lombardi, S. J.; Kaplan, D. L. *J. Arachnol.* **1990a**, 18, 297.
64. Beavis, R. C.; Chait, B. T.; Creel, H. S.; Fournier, M. J.; Mason, T. L.; Tirrell, D. A. *J. Am. Chem. Soc.* **1992**, 114, 7584.
65. Andersen, S. O. *Comp. Biochem. Physiol.* **1970**, 35, 705.
66. Case, S. T.; Wieslander, L. In Structure, Cellular Synthesis and Assembly of Biopolymers, Ed. by Case, S. T. Springer-Verlag, Berlin, **1992**, 187.
67. Work, R. W.; Emerson, P. D. *J. Arachnol.* **1982**, 10, 1.
68. Hong, S. Y.; Kertesz, M. *Phys. Rev. B.* **1990**, 41, 11368.
69. Work, R. W., *J. Exper. Biol.* **1985**, 118, 379.
70. Nakamae, K.; Nishino, T.; Ohkubo, H. *Polymer* **1989**, 80, 1243.
71. Iizuka, E. *Biorheology* **1965**, 3, 1.
72. National Research Council, Materials Science and Engineering for the 1990s, **1989**, National Academy Press, Washington, D.C.

RECEIVED July 15, 1993

Chapter 2

General Properties of Some Spider Silks

Fritz Vollrath

Zoologisches Institut, Rheinsprung 9, CH 4051 Basel, Switzerland

Three silks are discussed to exemplify the extreme variability of spider silks. Orb weaver radial silk is rather similar to many other typical silks whether found in spiders or insects. The capture silks of the ecribellate and cribellate orb webs are highly atypical. Indeed they are not so much materials as mechanical systems on a microscopic scale.

Spiders, unlike insects, use their silks for many different purposes. Accordingly spiders have evolved a wide range of silks with surprisingly different mechanical properties. To date we have little hard data on the properties of silks other than the drag line and web silks of a handful of common spiders. However, these already give a good indication of the high degree of variability inherent in spider silks. They also point to the amazing degree to which phylogenetic design constraints of common silks (e.g. plasticisation by water) have been subverted by clever modifications; a fine case of ingenious tinkering by mother nature.

The performance of an orb web has been selected during evolution to take out-of-plane loads in maximum deflection. This can be greatly enhanced by combining the mechanical properties of two types of silk: tightly strung, stiff and relatively inelastic radial threads provide support and transmit vibrations whereas the soft and highly visco-elastic spiral threads stick to prey. The sticky capture threads are also able to maintain tension when stretched and relaxed in rapid succession (e.g. buffeting by wind) to prevent strands from agglutinating. Initial softness but great ultimate strength is crucial for prey capture in a 2D structure for it provides quick inelastic absorption of high kinetic energy coupled with lack of purchase for struggling legs. Ecribellate orb weavers have solved this mechanical problem by incorporating into their capture threads an inexpensive windlass system powered by the surface tension of water. Cribellate spiders in contrast use in these threads a complex and energy costly mechanical system that uses a combination of threads with different mechanical qualities.

Spider webs and silks

Spider webs range from the fortification of terrestrial burrows in mygalomorphs to the

0097–6156/94/0544–0017$06.00/0

aerial orb webs of the uloborids and araneids (Figure 1). These architecturally rather complicated modern orb webs are possible because these spiders have evolved silk glands that produce a wide range of different silks *(1-3)*. In stark contrast to the mygalomorphs which produce only a few types of silk *(4, 5)*, the highly derived orb weavers and their descendants use a fair number of different silk types, each with its own production and extrusion system. The female of the common garden spider, for example, produces at least 7 different silks *(1, 2, 6-9)*: (1) drag line and structural silk (major ampullate glands); (2) structural thread and auxiliary spiral (minor ampullate glands), (3) core fibres of capture spiral (fili- or flagelliform glands); (4) aqueous coating and glycoprotein glue for capture spiral (aggregate glands); (5) tough outer silk of egg sac (cylindrical glands); (6) soft inner silk of egg sac and silk for swathing prey (aciniform a/b glands); (7) cement silk for joints and attachment (piriform glands). Cribellate spiders in addition also produce in the cribellum hundreds of superfine threads that are combed into the hackled bands on capture threads *(10-15)*.

Orb web-architecture

Spider orb webs are high energy absorbing nets. They are typically planar in structure and have been designed by evolution to absorb primarily out-of-plane aerodynamic and impact loads *(16, 17)*. Their ability to function well depends not only on the material properties of their silks but also on the structural properties of web design. Although spider webs excel in both qualities, only a few studies e.g. *(17-21)* have been devoted to a closer examination of why this should be so. The only detailed study of orb webs linking structure and ecology published so far *(20)* concludes that "no one feature of web architecture characterises the amount of energy webs can absorb". This is only too true.

The structural hierarchy of an orb web contains a wide variety of structural elements. This includes different junction types *(22)*, including junctions that may allow appreciable sliding *(23)* and that thus may function as breaks contributing to the effectiveness of the web architecture *(24)*. The radial support threads and the inter-radial capture threads can show large differences in their mechanical properties, both between individual radials *(25)* and between radial and spiral threads *(25, 26)*. Moreover, there can exist pre-stress differences along each radius *(21)*, created during the construction of the auxiliary spiral and maintained by tension in the capture threads (be it powered by the cribellate spring or the ecribellate windlass). Both radial and spiral threads typically show marked time dependent responses to stress at very different strains *(26)*. It is likely that these differences in time dependence contribute to the dissipation of impact energy and the securing of captured insects.

All orb webs incorporate a variety of different silks, connected into a polar network with both firm and soft sections. The firm sections, the radii, (i) transmit vibrations that signal the presence of prey and (ii) form pathways on which the spider traverses its web. The radii also support the capture spiral which is viscous, sticky and extraordinarily elastic. This softness prevents the capture spiral from interfering with vibration transfer along the radii which would be dampened by a firmly strung connecting spiral. In addition, its very softness is necessary to arrest the insect's flight without catapulting it back out, trampoline-fashion, and to prevent purchase for the trapped prey struggling to escape.

The orb weaving uloborids (like other cribellate spiders), use for spiral threads a complicated mechanical system *(3, 22)* composed of different silks with different diameters and fracture strengths intimately interwoven to form cable networks incorporating crimped threads as springs . These cribellate spiders have to combine these silks into strands by hackling *(11, 14)* which is costly in terms of time and energy *(27)*. The windlass system of the ecribellate orb web *(26)* is 'made' by taking advantage of simple physical forces *(28)*, and thus such threads are much more

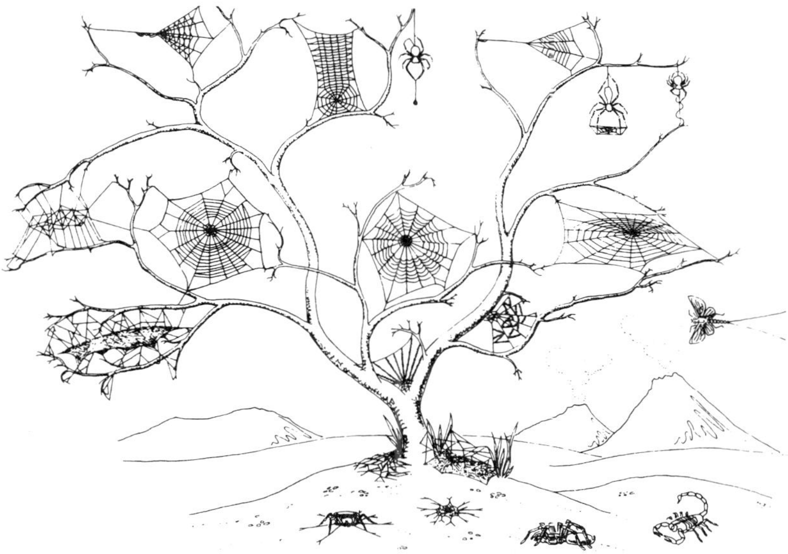

FIGURE 1 A selection of spider web types. On the ground, left of scorpion : cursorial and web-building mygalomorphs, amaurobiid, eresid and agelenid. In the tree: dictynid web; right-hand branch 2-D webs of : *Stegodyphus, Uloborus, Hyptiotes,* casting net of *Deinopis and* single line net of *Miagrammopes.*; center branch: *Araneus* orb web; left-hand branch : *Meta* orb web, 3-D knock-down webs of *Theridion* (above) and *Linyphia* (below), 3-D orb web of *Theridiosoma*, ladder orb web of *Scoloderus*. and droplet web of *Mastophora*. (Adapted from reference 69.)

economical to construct for the spider. This favourable cost-benefit ratio may be responsible for the immense adaptive radiation as well as numerical advantage of ecribellate orb weavers when compared to cribellate orb weavers.

Properties of ecribellate radial and capture threads

The radial threads in the typical orb web are dry, stiff and non-sticky. In the *Araneus* web they are slightly pre-tensed *(21, 25)*, but being made of a visco-elastic material they creep. This helps to balance out tensions between different structural members in the web. Radial threads can be taken as model spider silk having similar chemical and mechanical properties to spider drag line, the only other dry silk that has been studied extensively *(8, 19, 29-38)*. Such drag line silks are often forcibly drawn out using small motors *(39-41)*. This method of collecting may not matter for biochemical measurements. However, in pilot studies we found that such 'silked' threads can have mechanical properties inferior to radial threads of the same spiders, especially in breaking strain *(25)*. On the other hand, radial threads taken from orb webs always carry joints and residues of the capture spiral which might affect the mechanical behaviour of radials. For *Araneus diadematus*, however, these additions have no significant effect; this can be shown when 'impure' radials are compared with 'pure' radials of the same spiders but taken before spiral construction *(25)*.

On the whole, radial threads are typical silk that displays the typical behavior of other dry silks (Figure 2a). However, radial silk (like dragline silk) is of exceptionally high quality when compared with typical insect silks *(19)*. In contrast, the capture threads of ecribellate spiders show under normal conditions a mechanical behavior not shown by any other silk (Figure 2b). This atypical behaviour depends to an very large degree on the presence of the aqueous coating of these threads *(26)*.

The windlass mechanism of ecribellate capture threads

When radial threads are contracted even a small amount from their length in the web they sag *(26)* for in the web they are only slightly pre-tensed *(21, 25)*. The capture thread on the other hand can be contracted by a large amount, to a small faction of its web-length, without sagging *(26)*. It also extends to several hundred percent web-length without breaking, and returns immediately to web-length without obvious sag. This extraordinary behavior is not due to active pre-tension by the spider but to the properties inherent in the 'clever design' of the ecribellate capture thread.

Such a capture thread consists of a pair of core fibres each originating from a cylindrical gland evenly coated on extrusion with a viscous liquid originating from its two accompanying aggregate glands *(42-44)*. As such a liquid cylinder supported along its axis is unstable, it spontaneously flows into droplets regularly spaced along the core fibres (Figure 3). The coating liquid contains amino acids, glycoproteins, lipids and salts, as well as 80% water *(2, 45-48)*. The glycoproteins provide the web's adhesive, in the form of microscopic rings or nodules *(49)*. The nodules are distributed along the core fibres and are correlated to the original droplet distribution. It appears that they can slide on the thread, thus they avoid exerting localised force on the core threads and at the same time they can accumulate and strengthen any point of attachment to the prey (Vollrath unpubl.). The amino acids are extremely hygroscopic *(50)* and by drawing additional water from the atmosphere *(28)* they are directly responsible for the performance of these threads.

Water is crucial for the mechanical behavior of ecribellate capture silk *(26)*. Moreover, it provides ecribellate orb weavers with drinking water when the web is ingested for recycling *(28)*. Water is important on a molecular level: at intermediate contraction the plasticising effect of the absorbed water leads to super contraction and reversible elasticity where it plasticises and aids super contraction *(30, 51-53)*. These

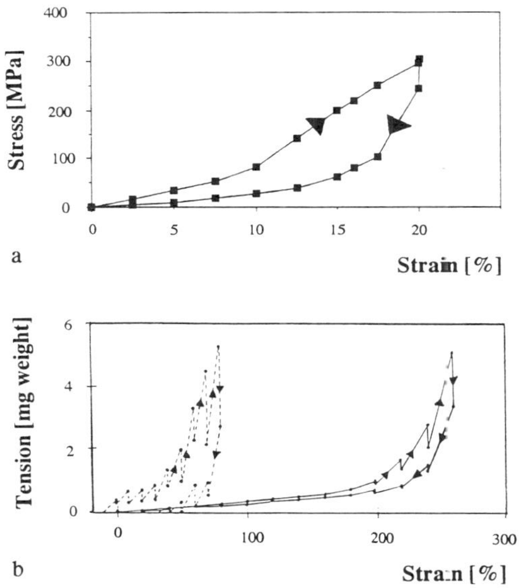

FIGURE 2 Stress-strain curves of *Araneus* orb web-threads. **(a)** Loading-unloading cycle from one sample of radial thread stretched to 60% of its potential. Loading was done continuously at 3.6 cm/min. (adaped from ref. *25*). **(b)** Comparison between the dynamic behavior of naturally wet and dry capture silk. For each step the imposed change in length was rapid and the system was then allowed to relax prior to the next change. Shown are coated (full line) and uncoated (broken line) sections of the same spiral capture thread. (Adapted from reference 26.)

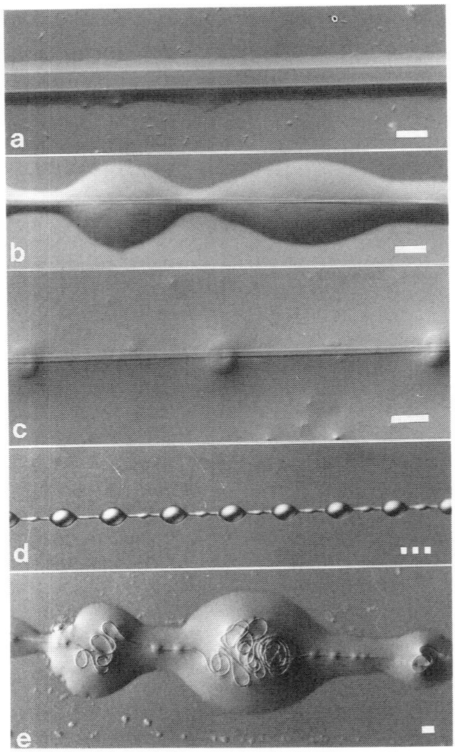

FIGURE 3 Capture thread of *Araneus diadematus* . **(a)** A section fixed 5 seconds after being laid down in the web. **(b)** Section fixed 60 seconds later. **(c)** Section fixed 240 seconds later no longer showing the coat; alternating small and large glycoprotein nodules have now formed underneath the droplets. **(d)** View of an unstretched capture thread suspended in air. **(e)** View of a similar thread contracted to 50% web-length and then laid onto a glass slide to show the loose core fibres gathered within the larger coalesced glue droplets. (Reproduced with permission from references 26,49.)

core threads are typical silk showing birefringence (Vollrath unpubl.); however, this can only be observed after the aqueous coat has been washed off.

In addition to acting on the molecular level, water also acts on a different structural level by providing a windlass system *(26)*. When a capture thread is contracted, overall tautness is always maintained. The surface tension of the aqueous coat takes up the hysteretic slack of the highly visco-elastic core fibres. Examination under a microscope reveals that the glue droplets begin to merge and the larger droplets act like tiny windlasses, collecting inside balls of loose core fibre. Stretching the thread unwinds the core fibres again. The coat's surface tension creates a very small force but is sufficient to maintain tautness even at large contractions.

Wetting also has an effect at large extension *(26)*. This is shown by comparing the elastic characteristics of the same capture thread, coated (wet) and uncoated (dry). For wet capture thread and extensions of up to about 200% web-length, the resisting force is small and reversible. At larger extensions this force rises dramatically and shows some hysteresis. The dry capture thread shows much greater stiffness and hysteresis even at low extensions, resembling very much a dry radial thread, albeit with a smaller force which is due partly to the smaller diameter of the core fibres *(26)* and partly to their molecular fine structure which seems to incorporate mobile elements *(54)*.

The distinction between the kinetic behavior of dry and coated threads can be observed in the time dependence of the tensile characteristics in a thread following small and rapid changes in length *(26)*. The coated thread immediately assumes the new equilibrium value while the dry thread on extension exhibits an initially large force which decays in time. Vollrath and Edmonds *(26)*demonstrated that it is the water which determines the changes in thread behaviour, by measuring the time characteristics of forces in a naturally dry radial thread first in air (dry) and then under water (wet). Immersion in water lowers the tension for a given extension. It also ensures the rapid attainment of the new equilibrium. A radial thread submersed under water behaves very much like a coated capture thread in air. The inverse is found when coated capture threads are measured in normal and dehydrated conditions. Removing just the water from a coated thread (by drying the environment with phosphorus pentoxide) leads to large tensions for small extensions, and a dramatic loss of reversibility. Restoring the dehydrated thread to ambient air immediately restores its original characteristics, including the rapidly reversible elasticity.

The crimped spring mechanism of cribellate capture threads

Ecribellate and cribellate capture threads are very different in form and function (Figures 4 and 5). We know much more about the mechanisms (for both adhesion and energy up-take) by which ecribellate capture threads work than those of cribellate capture threads. It is possible that the stickiness of the cribellate threads derives from electrostatic forces *(3, 4)*. Charging might either occur passively by the fine threads rubbing against each other or else done actively by the spider's combing motion during hackling. Experiments testing the stickiness of *Hyptiotes paradox* and *Uloborus walckenaerius* threads in strong electrostatic fields seem to indicate that electrostatic forces are of no significance for adhesion (Vollrath unpubl.). Thus it may be deduced by default that the operand forces could be van der Waals forces, analogous to the adhesion of gecko or fly foot pads *(55)*. This would be possible because hackled threads are extremely thin (0.03 µm diameter) and thus can come into very close contact with prey surfaces.

Thus the cribellate capture thread actually is a multi-fibre strand *(3, 10)*. Typically it consists of two axial support threads combined with two or more coiled threads that on extrusion or by combing are crimped to form a warped epi-spiral. These four threads are covered by the hackled silk in such a way that it surrounds the

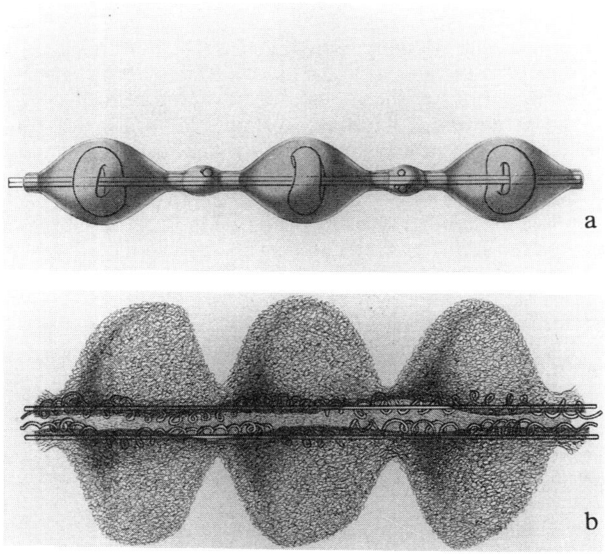

FIGURE 4 Interpretation of sticky capture threads. (**a**) Cribellate thread: a composite picture for a hypothetical spider based on SEM microscopy of *Deinopis* and *Uloborus* capture threads. The axial threads are generally thinner than the crimped threads. (**b**) Ecribellate thread: a composite picture based on Nomarsky and SEM microscopy of threads from *Araneus diadematus*, *Zygiella x-notata* and *Meta segmentata.* The coat is mostly water, the 'life savers' consist of glycoproteins.

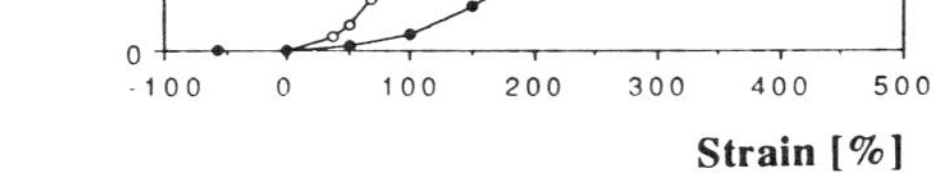

FIGURE 5 Stress-strain curves for typical cribellate and ecribellate capture threads stretched until they broke. The "droplet" strain is represented by the negative strain values. (Adapted from Köhler and Vollrath *in preparation.*)

support and coil threads like a tube slit open along its side. The hackled silk is combed out in alternating sweeps of the spider's especially modified hind legs. Thus the hackled coat forms alternating regions of puffed-up and drawn-out covering. It may be assumed that the reasonable elasticity of this rope system is due partly to the coiled threads which are assisted by the puffed regions of the hackled tube. Kinetic energy of prey would be absorbed by hackled threads first rearranging themselves in the woolly coating and then by breaking according to their stretch. In addition the support threads are stretched and the coil threads pulled against their crimp. The successive breaking of minor and major threads would explain the dips in the stress/strain diagrams of *Uloborus* capture threads (Figure 5)

The cribellate, like the ecribellate, capture thread is a complex mechanical system. But the cribellate thread is much more costly to construct. For the ecribellate thread to do its complex job the spider only has to provide core fibres and hygroscopic coating; physical laws do the rest. The cribellate thread relies on the spider laying down two types of core threads with different mechanical properties as well as hackling out the adhesive coating of the thread. This costs a fair amount of energy as well as material, and it is no wonder that cribellate orb weavers can take four times longer than their ecribellate colleagues to produce the same length of thread (Zschokke and Vollrath unpublished observations). Thus overall their webs are more costly to build *(27, 56)*. This should have an effect on their competitive ability.

It is sometimes claimed that cribellate webs may be more sticky than ecribellate webs. They are certainly very sticky *(57, 58)* as one can easily see when one feeds spiders by throwing flies at their webs: flies stick much better to cribellate than to ecribellate stands. Could this then be the factor that renders the cribellate orb web of *Uloborus* again competitive with its ecribellate look-alikes? If its web took four times more energy to build than the ecribellate *Araneus*' web, then it would only have to be four times more effective in prey capture to be equally cost efficient. A study of relative stickiness of the two web types indicates however (Vollrath unpublished) that cribellate silk was not inherently more sticky; at least not to any surface. It was significantly more sticky for fly wings but significantly less sticky for an equal area of plastic. Should we perhaps see the problem the other way round? Could it be that the relative preponderance of ecribellate webs has led flies to employ counter measures against these webs by evolving wings whose surfaces stick poorly to ecribellate glue? The analogy to the observation that moth and butterfly wing scales have evolved in an arms race with web spiders *(59)* is obvious.

Conclusions and outlook

We have only just begun to study the silks of spiders in their natural state, and to approach them without too many preconceptions inherited from studying other silks. Such pre- and misconceptions can lead to misinterpretations of observations. Spider silk are highly adapted materials and they are adapted to a wide range of tasks requiring a wide range of material properties. This includes complex mechanical systems like the cribellate and, even more, the ecribellate capture thread. These types of composite 'silks' far transcend mere materials. Thus statements about the ability of spider silk to absorb inordinate amounts of energy, if they are based on data from ecribellate or even cribellate capture thread, are wrought with danger of being misunderstood. In these cases energy is absorbed not so much by a silk thread but by a complex mechanical system, albeit made of silks. Thus the mechanism of a single araneid capture thread resembles the arresting unit of an aircraft carrier (albeit with a vast array of hydraulic dash pots in sequence); and the mechanism of the uloborid thread resembles a sprung towing rope.

Spider silks have evolved a wide range of biochemical and mechanical properties. We only have to look at the aqueous coating of the ecribellate capture

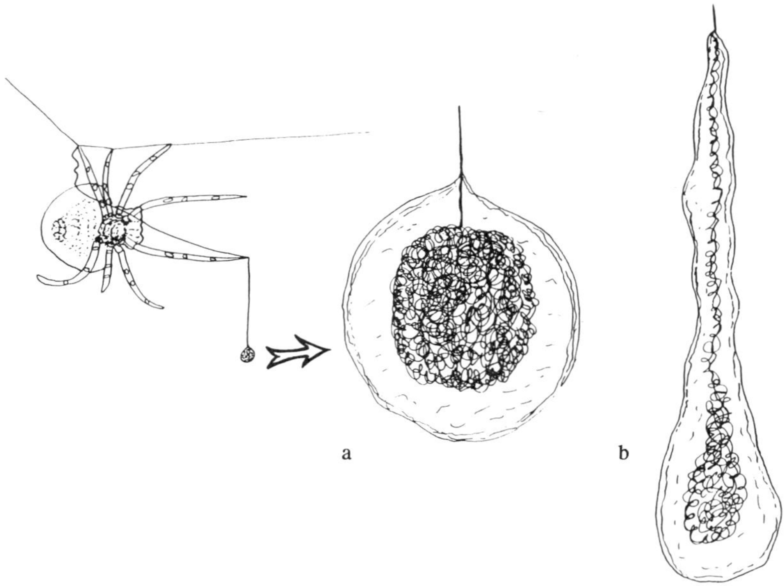

FIGURE 6 *Mastophora* and its bolas. The spider hangs on its support line (a frame/radius equivalent) and swings its sticky ball that incorporates in one huge droplet the plasticised core threads of the ancestral capture spiral. (Adapted from reference 70.)

thread to see the truth of this statement. The evolution of the aqueous coat has not only allowed the reduction of web building activity until the web consists of just the stick ball of the bolas spider (Figure 6). The viscid coat also contributes to the stickiness of the linyphiid tangle web *(60)* and lays at the origin of the glue cannons used during theridiid attack wrapping *(61)*. The viscid material of the aggregate glands is considered a silk although one would not think so simply from studying neither its material properties (which are those of a liquid) nor its amino acid composition (which is highly atypical). If it were a silk, and homologous to other spider silks, then it must have evolved from the 'typical' silk material of its ancestors. On the other hand, one might argue that the aggregate gland is a novel character evolved *de novo* and not by modification of other ancestral silk glands. The evolution of other cribellate and ecribellate silks pose similar questions *(61-66)*. These questions can only be solved by phylogenetic and comparative studies of silks and silk production systems *(61, 64, 67, 68)*. Therefore such studies will continue to be important not only for the ecologist interested in silks and webs but also for the silk-biochemist or silk-engineer. But no less important for everybody concerned will be further in-depth studies of spider ecology, morphology and biology. Only interdisciplinary collaboration will be able to solve the many mysteries that still lie at the heart of the spider's silks.

Literature Cited

1. Kovoor, J. In *Ecophysiology of Spiders* Nentwig, W. Ed. Springer: Berlin-Heidelberg-New York, 1987; pp. 160-186.
2. Tillinghast, E.K.; Townley, M. In *Ecophysiology of Spiders* Nentwig, W. Ed. Springer: Heidelberg, 1987; pp. 203-210.
3. Peters, H.M. In *Ecophysiology of Spiders* Nentwig, W. Ed. Springer: Berlin, 1987; pp. 187-202.
4. Coyle, J.E. In *Spiders: Webs, Behavior and Evolution* Shear, W.A. Ed. Stanford University Press: Stanford, Cal., 1986; pp. 259-305.
5. Palmer, J. *J. Morphol* **1985,** *186,* 195-207.
6. Lucas, F. *Discovery* **1964,** *25,* 20-26.
7. Peakall, D. *Journal of Experimental Zoology* **1964,** *156,* 345-350.
8. Denny, M.W. In *The Mechanical Properties of Biological Materials* Vincent, J.F.V. & Currey, J.D. Ed. Soc. Exp. Biol. Symp. 34., Cambridge University Press: Cambridge UK, 1980; pp. 245-271.
9. Vollrath, F. *Sci. Am.* **1992,** *266,* 70-76.
10. Lehmensick, R.; Kullmann, E. *Zool. Anz.* **1956,** 123-129.
11. Eberhard, W.G. *Bull. Br. arachnol. Soc.* **1988,** *7,* 247-251.
12. Friedrich, V.L.; Langer, R.M. *Amer. Zool.* **1969,** *9,* 91-96.
13. Opell, B. *Bull. Br. Arachnol. Soc* **1982,** *5,* 338-343.
14. Opell, B.D.; Roth, G.; Cushing, P.E. *J. Arachnol.* **1990,** *18,* 238-240.
15. Peters, H.M. *Zoomorph.* **1984,** *104,* 96-104.
16. Langer, R. *American Zoologist* **1969,** *9,* 81-89.
17. Craig, C.L.; Okubo, A.; Andreasen, V. *J. theor. Biol* **1985,** *115,* 201-211.
18. Denny, M. *J. Exp. Biol.* **1976,** *65,* 483-506.
19. Gosline, J.M.; DeMont, M.E.; Denny, M.W. *Endeavour* **1986,** *N.S.10,* 37-43.
20. Craig, C.L. *Biol. J. Linn. Soc.* **1987,** *30,* 135-162.
21. Wirth, E.; Barth, F.G. *J. Comp. Physiol. A* **1992,** *171,* 359-371.
22. Kullmann, E.; Frei, O.; Braun, T.; Raccanello, R. *Netze in Natur und Technik* Universität: Stuttgart, 1975.
23. Jackson, R.R. *Psyche* **1971,** *78,* 12-31.
24. Eberhard, W.G. *J. Nat. Hist.* **1976,** *10,* 481-488.
25. Köhler, T. *The mechanical properties of different threads of the orb webs of Araneus diadematus and Uloborus walkenaerius* Basel, 1992.
26. Vollrath, F.; Edmonds, D. *Nature* **1989,** *340,* 305-307.
27. Lubin, Y.D. In *Spiders: Webs, Behavior and Evolution* Shear, W.A. Ed. Stanford University Press: Stanford, 1986; pp. 132-171.
28. Edmonds, D.; Vollrath, F. *Proc. Roy. Soc. Lond.* **1992,** *248,* 145-148.
29. Zemlin, J.C. *A study of the mechanical behavior of spider silks.* Clothing and Organic Materials Laboratory, U.S. Army Natick Laboratories, 1968.
30. Work, R.W. *Text Res J* **1977,** *47,* 650-662.
31. Work, R.W. *Trans. Am. microscop. Soc.* **1978,** *97,* 180-191.
32. Work, R.W. *Trans Am Microsc Soc* **1981,** *100,* 1-20.
33. Work, R.W. *Trans Am Microsc Soc* **1984,** *103,* 113-121.
34. Work, R.W.; Young, C.T. *J. Arachnol.* **1987,** *15,* 65-80.
35. Griffiths, J.R.; Salanitri, V.R. *J. Mat. Sci.* **1980,** *15,* 491-496.
36. Iizuka, E. *J. appl. Polymer Sci.* **1985,** *41,* 163-171.
37. Dong, Z.; Lewis, R.V.; Middaugh, C.R. *Arch. Biochem. Biophys.* **1991,** *284,* 53-57.
38. Lewis, R.V. *Act. Chem. Res.* **1992,** *25,* 392-398.

39. Witt, P.N.; Reed, C.F.; Peakall, D.B. *A Spider's Web: Problems in Regulatory Biology.* Springer Heidelberg: 1968.
40. Work, R.W. *Text. Res. J.* **1976,** *46*, 485-492.
41. Work, R.W.; Emerson, P.D. *J Arachnol* **1982,** *10*, 1-10.
42. Warburton, C. *Quarterly Journal of Microscopical Science* **1890,** *31*, 29-39.
43. Sekiguchi, K. *Annot. Zool. Jap.* **1952,** *25*, 394-399.
44. Peters, H.M. **1955,** *10b*, 395-404.
45. Fischer, F.; Brander, J. *Hoppe-Seyler's Z. Physiol Chem* **1960,** *320*, 92-102.
46. Schildknecht, H.; Munzelmann, P.; Krauss, D.; Kuhn, C. *Naturwissenschaften* **1972,** *59*, 98-99.
47. Sinohara, H.; Tillinghast, E. *Biochem. Int.* **1984,** *9*, 315-317.
48. Vollrath, F.; Fairbrother, W.J.; Williams, R.J.P.; Tillinghast, E.K.; Bernstein, D.T.; Gallagher, K.S.; Townley, M.A. *Nature* **1990,** *345*, 526-528.
49. Vollrath, F.; Tillinghast, E.K. *Naturwissenschaften* **1991,** *78*, 557-559.
50. Townley, M.A.; Bernstein, D.T.; Gallagher, K.S.; Tillinghast, E.K. *J. exp. Zool.* **1991,** *259*, 154-165.
51. Work, R.W. *J. Arachnol* **1981,** *9*, 299-308.
52. Work, R.W. *J. Exp Biol* **1985,** *118*, 379-404.
53. Gosline, J.; Denny, M.; DeMont, M. *Nature (London)* **1984,** *309*, 551-552.
54. Bonthrone, K.M.; Vollrath, F.; Hunter, B.K.; Sanders, J.K.M. *Proc. Roy. Soc. Lond.* **1992,** *248*, 141-144.
55. McMahon, T.A.; Bonner, J.T. *On size and life* Freemann: New York, 1983.
56. Opell, B.D. *Behav. Ecol. Sociobiol.* **1990,** *26*, 375-383.
57. Eberhard, W.G. *J. Arachnol.* **1980,** *8*, 283.
58. Opell, B.D. *J. Arachnol.* **1989,** *17*, 112-113.
59. Eisner, T.; Alsop, R.; Ettershank, G. *Science* **1964,** *146*, 1058-1061.
60. Peters, H.M.; Kovoor, J. *Zoomorphol.* **1991,** *111*, 1-17.
61. Coddington, J. *J. Arachnol.* **1989,** *17*, 71-95.
62. Glatz, L. *Z. Morph. Tiere* **1972,** *72*, 1-26.
63. Glatz, L. *Z. Morph. Tiere* **1973,** *75*, 1-50.
64. Coddington, J. In *Spiders, Webs, Behaviour and Evolution.* Shear, W.A. Ed. Stanford University Press: 1986 ; pp. 319-363 .
65. Shear, W.A.; Palmer, J.M.; Coddington, J.A.; Bonamo, P.M. *Science* **1989,** *246*, 479-481.
66. Selden, P.A. *Nature* **1989,** *340*, 711-712.
67. Platnick, N.I. *Am. Mus. Nov.* **1990,** *2978*, 1-42.
68. Platnick, N.I. *Am. Mus. Nov.* **1991,** *3016*, 1-73.
69. Vollrath, F. *Trends Ecol. Evol.* **1988**, *3*, 331–335.
70. Eberhard, W. G. *Psyche* **1980**, *87*, 143–169.

RECEIVED June 29, 1993

Chapter 3

Silk Glands of Araneid Spiders

Selected Morphological and Physiological Aspects

Edward K. Tillinghast and Mark A. Townley

Department of Zoology, University of New Hampshire, Durham, NH 03824

Regionalization within the glandular epithelium of araneid silk glands has been revealed by several authors using various experimental methods. The relationship of this regionalization to the potential complexity of silk is discussed. We also review studies on the regulation of secretory protein synthesis in araneid silk glands, with an unavoidable emphasis on the major ampullate glands, the most frequently studied spider silk glands. A brief discussion of the relationship between web composition and the nutritional requirements of araneid spiders follows. Lastly, we describe a mechanism which allows juvenile araneid spiders to produce ampullate fibers during proecdysis (the preparatory period before ecdysis), a time when the primary major and minor ampullate glands are being remodeled and are temporarily nonfunctional. Two sets of secondary major and minor ampullate glands function alternately in successive proecdyses. (Only one set is functional in each juvenile stadium.) Both sets are nonfunctional and atrophied in adults (i.e. after the final molt).

The silk glands of araneid spiders are of several types (Figure 1). Of these, the major ampullate glands have received the most attention. They give rise to nonsticky silk fibers which are used in draglines and various elements of orb webs, including mooring lines, framelines, temporary and hub spirals, and radii. The functions of fibers produced by the minor ampullate glands are uncertain. They commonly accompany major ampullate fibers, especially in radii (*1,2*), and may serve to reinforce the elements in which they occur. However, Work (*2*) has argued against this interpretation. Certainly, their infrequent or seemingly random occurrence in some structural elements suggests that they are not critical to the functioning of such elements. To date, the use of minor ampullate fibers, without concurrent use of major ampullate fibers, has only been observed when bridging lines are being produced (*3*). This provides the first convincing evidence for a specific function for minor ampullate fibers. The flagelliform glands, which resemble the ampullate glands superficially, produce the core fibers of the orb web's adhesive spiral, while the aggregate glands produce the viscid, aqueous solution that envelops these core fibers (*4,5*). Pyriform glands give rise to attachment disks, the cements that secure ampullate fibers to substrates, and probably also to the deposits that are used, e.g. during web construction, to cement intersecting ampullate

0097–6156/94/0544–0029$06.00/0

fibers to one another (e.g. *1,2,5-7*). The aciniform glands are involved in various activities, including swathing prey (along with fibers from pyriform and ampullate glands) (e.g. *5,7,8*), producing 'molting threads' at ecdysis, and, probably at least in some species, eggcase construction. The principal fibers used in constructing the eggcase are products of the cylindrical (=tubuliform) glands, which are present only in females (e.g. *9-11*).

Spiders of the subfamily Nephilinae, as well as spiders in the genus *Zygiella*, were formerly classified as members of the family Araneidae (see *12*), but are presently included in the family Tetragnathidae (see *13*). (Both families are included in the superfamily Araneoidea.) *Nephila clavipes*, a member of the Nephilinae, has been and is often used in research on silk glands and their products. In this chapter, devoted primarily to discussing araneid silk glands, we have not hesitated to include the results of research performed using species in either of these transferred taxa. Where mention is made of other non-araneid spiders, the family to which the spiders belong is stated. Otherwise, it can be assumed that particular species referred to are members of the family Araneidae.

Regionalization in Silk Glands

From a gross morphological perspective, regionalization is immediately apparent in all silk glands in the form of a clear distinction between the gland's duct and the secretory epithelium of the body of the gland. The latter synthesizes and secretes into the gland's lumen proteins (in general) destined for extracorporeal use, while the former provides, in part, a conduit for passage of luminal contents to the spider's exterior. The duct is also undoubtedly involved in the final, critical processing of the material passing through its lumen and has been the subject of detailed ultrastructural studies in several types of silk glands from *Araneus diadematus* and *Nephila clavata* (*14-22*). In the major ampullate and minor ampullate glands of araneids, and to a lesser extent in the flagelliform glands, two morphological regions composing the body of the gland are recognized, the ampulla and the tail. The distinction between the two is essentially based on width, the tail being a relatively narrow tube and the ampulla a sac with an expanded lumen in which the secretory products of the epithelial cells are amassed (see Figure 4).

In many types of silk glands different and well segregated cell types can also be distinguished in the body of a gland (reviewed in *23,24*). In araneids, the secretory epithelia of minor ampullate, flagelliform and pyriform glands are composed of two, sharply delineated regions (cell types) which exocytose histochemically and ultrastructurally distinct secretory granules into the lumen, while major ampullate glands contain either two or three regions, depending on the genus (*23-25*). The border between ampulla and tail is not coincident with the border between cell types. For example, in the major ampullate glands of *Araneus* and *Argiope*, about three-fourths of the ampulla, that portion closest to the junction with the duct and furthest from the tail, is composed of one cell type while the remainder of the ampulla and the entire tail are composed of a second cell type (*26*). It has been observed that uniformity, with respect to secretory granules and other cellular components (e.g. glycogen particles, microtubules, Golgi apparatus, microvilli), does not necessarily exist even within one region (*14,26*). In the ampulla-specific cell type of the major ampullate glands of *Argiope bruennichi*, for example, gradual changes in both the size and histochemical affinities of the secretory granules are reportedly evident over the length of this region (*26*). With regard to the other cellular components mentioned, nonuniformity generally takes the form of gradual changes in abundance over the length of the region.

The secretory epithelia of aggregate, cylindrical and aciniform glands are composed of only one cell type in araneids (*23,24*). Nevertheless, regionalization of a sort may occur in such glands. Kovoor (*26*) has described certain secretory granules in the cylindrical glands of *Zygiella x-notata* which, based on histochemical evidence, are present only in the most distal third of the body of the gland. And it is evident just from dissecting gravid *Argiope aurantia* that their cylindrical glands are not uniform in all respects from end to end. Examining the length of the contorted body of one of these glands one encounters a dazzling array of colors, produced by the luminal contents, with segments of different color merging into one another. To our knowledge the nature of this phenomenon has not been investigated. In the aggregate glands of *Araneus diadematus*, Kovoor (*26*) has observed a narrow region in the body of the gland, occupying a position between the rest of the glandular body and the duct, which differs from the rest of the glandular epithelium. The cells in this region are narrower than elsewhere and histochemical tests reveal abundant glycogen deposits and granules rich in protein, unlike the results obtained for the bulk of the glandular epithelium.

Evidence from Fibers, Luminal Contents, Nucleic Acid Sequencing. The histochemical, histological and ultrastructural evidence available, not only from araneids but from a variety of representatives of the liphistiomorph, mygalomorph and araneomorph suborders (reviewed in *23,24*, see also *22,27-33*), argues strongly for many silks being composed of more than one protein. Evidence obtained by other methods from luminal contents, the silks themselves and, recently, from nucleic acid sequence data, lead to the same conclusion. In major ampullate silk fibers from several araneid species (as well as *Nephila* and *Nephilengys*) a sheath-core morphology was commonly observed after fibers had been supercontracted in water and then re-extended (*34*). Work and Young (*35*) have noted that this behavior certainly does not prove that sheath and core differ in chemical composition, but they cite the ease with which the two can be made to separate and the intraspecific and intraindividual variability in amino acid composition of major ampullate silk as evidence which suggests that such is the case. In silk glands of several mygalomorphs (*24,36,37*) and in the pyriform glands of *A. diadematus* (*38*) it has been reported that the products of the different cell types remain separate and distinct even within the lumen of the duct, giving rise to fibers with sheath-core morphologies. If the same description applies to araneid ampullate glands, as Kovoor and Zylberberg (*38*) have indicated, then it is possible that the sheath and core observed by Work (*34*) are products of the ampulla-specific and tail cell types, respectively.

Two reports have provided additional evidence that some silks have a substructure. Using transmission electron microscopy, structural heterogeneity has been observed in cylindrical, pyriform and aciniform type A fibers from two species in the family Uloboridae (*39*), as well as in cylindrical gland fibers from *A. aurantia* (*40*). Moreover, treatment of *A. aurantia* cylindrical gland fibers with trypsin and a solution of urea, SDS and 2-mercaptoethanol revealed these fibers to be composed of fibrils embedded within a matrix (*40*) (Figure 2). Even greater heterogeneity in the composition of cylindrical fibers from this species is indicated by the results of SDS-PAGE. When extracts made from eggcases and cylindrical gland luminal contents were electrophoresed more than ten proteins were resolved (*40*). That cylindrical gland fibers can be compositionally heterogeneous is not unexpected as it has been shown that two histochemically distinct types of secretion granules are synthesized by the (single cell type of) cylindrical glands in some species (*23,24*). In addition, TEM has revealed that individual secretory granules in the cylindrical glands of *Nephila clavata* can be heterogeneous (*20*).

Finally, in regard to heterogeneity in silk fibers, sequence data from partial cDNA clones has provided evidence of two, and only two, proteins in major ampullate silk from *Nephila clavipes* (*41-44*).

Phosphatases. High alkaline phosphatase activity was observed in the silk glands of certain insects and spiders more than forty years ago, with pronounced regional distributions reported in some glands (*45,46*). In Figure 3 we present data for the major ampullate gland alkaline and acid phosphatases of *Araneus cavaticus*, expressed in U/L. [One unit of alkaline or acid phosphatase activity is defined as that amount of enzyme which will hydrolyze 1 μmol of *p*-nitrophenyl phosphate or α-naphthyl phosphate,

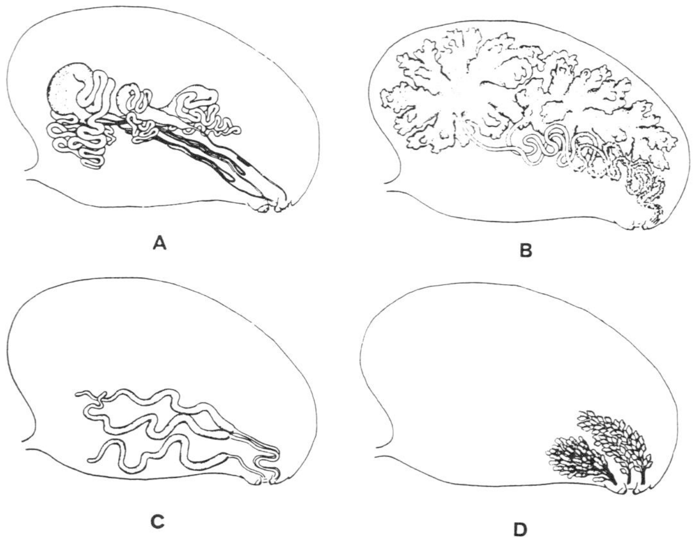

Figure 1. Diagrams showing the gross morphology and locations of the different silk gland types within the opisthosoma of an adult female *Larinioides sclopetarius*. Only the silk glands in one-half of the opisthosoma are shown. **(A)** Major ampullate (left), flagelliform (middle) and minor ampullate (right) glands. **(B)** Aggregate glands. **(C)** Cylindrical glands. **(D)** Clusters of aciniform glands opening on posterior median and posterior lateral spinnerets, and the more ventrally located cluster of pyriform glands opening on anterior lateral spinnerets. (Reproduced with permission from ref. 77. Copyright 1969 American Microscopical Society.)

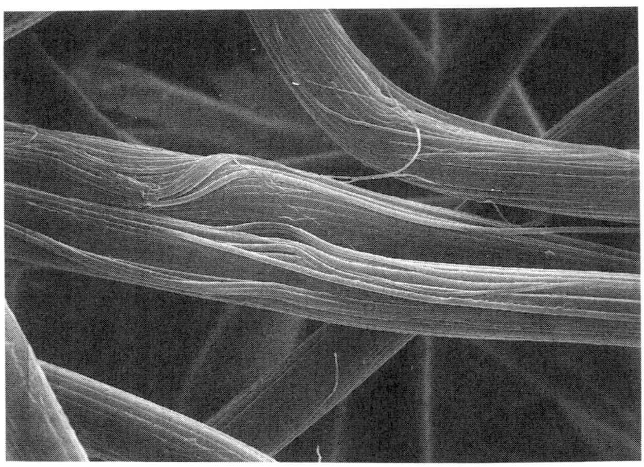

Figure 2. Eggcase of *Argiope aurantia* treated with bovine pancreatic trypsin (Sigma Chemical, T-8253) for 2 h at room temperature, rinsed three times with distilled water and prepared for SEM as described previously (*101*).

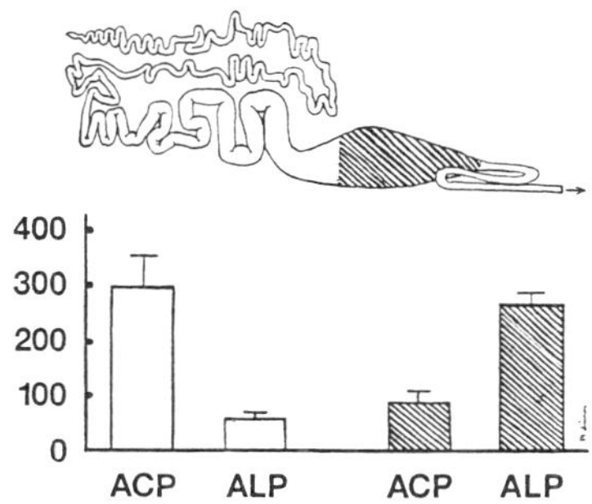

Figure 3. Acid and alkaline phosphatases (ACP and ALP, respectively) of the ampulla-specific and tail cell types of the major ampullate silk glands of adult female *Araneus cavaticus.* Glands were severed very near the junction between the two cell types, within the ampulla-specific cell type, resulting in tail homogenates that contained a very small amount of the ampulla-specific cell type while ampulla homogenates were free of the tail cell type. The junction between the two cell types occurs about three-quarters of the way into the ampulla, when approaching from the direction of the duct (on right). Both cell types were homogenized in 200 μL distilled water. Acid phosphatases were assayed by the addition of 20 μL of homogenate to 1 mL α-naphthyl phosphate (4mM, pH 5) (Sigma Chemical). Alkaline phosphatases were assayed by the addition of 20 μL homogenate to 3 mL *p*-nitrophenyl phosphate (4 mg/mL, pH 9.5) (Sigma Chemical). In both cases the changes in OD at 405 nm were monitored for 10 min at 30°C with a Milton-Roy 601 spectrophotometer. Results are given in U/L of homogenate. Error bars indicate 1 SEM. The ampulla-specific and tail cell types of 10 spiders were compared.

respectively, per minute under the conditions of the assay (see caption to Figure 3).] But given the observation that the protein concentration in the homogenate prepared from the tail cell type is approximately six times that of the homogenate prepared from the ampulla-specific cell type, were these data expressed in U/mg protein, the alkaline phosphatase of the ampulla-specific cell type would look even more prominent, whereas the acid phosphatase would appear more uniformly distributed. However, expressing activity in this way is not entirely satisfactory either since the ampulla-specific cell type is obtained free of any luminal contents while the tail cell type is unavoidably homogenized along with a significant quantity of luminal material. Despite these reservations there is no doubt that alkaline phosphatase activity, as measured in this assay system, is substantially higher in the ampulla-specific cell type than in the tail cell type. This is disconcerting since it is opposite to the results obtained by Bradfield (*46*) using *Tegenaria domestica* (family Agelenidae). By histochemical means, Bradfield observed alkaline phosphatase activity in the tail cell type of the ampullate glands (in the apical portion of the cells), but not in the ampulla-specific cell type. Activity was also seen in the secretory epithelium of the other three types of silk glands present in agelenids (see *23,24*): pyriform, aciniform and cylindrical glands. Activity in these glands was again restricted to the apical portion of the epithelial cells, but was present in all epithelial cells composing the body of the gland. Kovoor (*47*) has reported that the pyriform and aciniform glands of *Tegenaria atrica* are, like the ampullate glands, composed of two, segregated cell types. Assuming the same is true in *T. domestica*, these observations imply that both cell types in the pyriform and aciniform glands of *T. domestica* have substantial alkaline phosphatase activity, whereas only the tail cell type in the ampullate glands does.

Though the functions of phosphatases in the silk glands are unknown, the high alkaline phosphatase activity observed in the ampulla-specific cell type of *A. cavaticus* major ampullate glands helps to remind us that the ampulla is by no means a passive receptacle.

Acid phosphatase activity in the tail region of major ampullate glands of *N. clavipes* is reportedly associated, at least in part, with the Golgi complex of the epithelial cells (*48*). That these authors recognize the presence of a Golgi apparatus in the tail epithelium is in disagreement with the findings of Bell and Peakall (*49*) and Moon et al. (*29*) who were unable to locate these organelles in the tail (though Bell and Peakall did observe Golgi apparatus in the ampulla cell type). This discrepancy has been attributed to the atypical appearance of the Golgi apparatus in this epithelium and to its diminutive form when the epithelium is in a relatively inactive state (*48*). In an examination of the bipartite pyriform glands of *A. diadematus*, Kovoor and Zylberberg (*38*) noted an apparent difference between cell types in the importance of Golgi complexes to the secretion process. Thus, while secretory material was detected in the well developed Golgi complexes of the proximal cell type (that nearest the duct), it was concluded that the small, infrequent Golgi apparatus in the distal cell type did not seem to be important in the secretion of granules.

Synthesis of Fibroin (and Other Secretory Proteins)

Major Ampullate Glands. The synthesis of protein by the major ampullate silk glands can be stimulated in two ways: 1) via a cholinergic mechanism and 2) by depleting the store of silk protein in the ampulla by mechanically pulling fiber from the spigot.

Following the observation of Witt (*50*) that physostigmine (eserine) fed to *Araneus diadematus* results in increased web weights, Peakall (*51-53*) undertook an examination of the role of acetylcholine in major ampullate silk gland protein synthesis. He observed that acetylcholine stimulates the incorporation of radioactive alanine into glandular protein (*53*). This observation was confirmed by Candelas and Cintron (*54*). He also observed that the acetylcholine effect can be duplicated by physostigmine and paraoxon, both acetylcholinesterase inhibitors, and carbachol, an acetylcholine analog (*51,52*). Mechanical pulling of major ampullate silk from the spider was also found to increase the rate of silk synthesis (*51,52*). Again, this was confirmed by Candelas and Cintron (*54*). Atropine, an alkaloid that blocks muscarinic acetylcholine receptors, was able to negate the stimulatory effect of carbachol but not that resulting from mechanical silk depletion, thereby suggesting two separate mechanisms of regulation (*51,52*). When mechanical pulling was used to create unequal deficits in the two functioning major ampullate glands of adult *Araneus cavaticus* incorporation of radioactive glycine was greater in the gland with the larger deficit, indicating at least a degree of independent regulation between these glands (*55*).

Several studies have been devoted to examining changes that occur in the secretory epithelium of major ampullate gland tails following stimulation by either of the above means. Basically, stimulation substantially increases the rate of exocytosis of pre-existing fibroin and the rates of synthesis and exocytosis of new fibroin (*53,56*). In *Larinioides sclopetarius* within about 10 min after stimulation it becomes apparent that fibroin secretory granules, which were present in the tail's epithelial cells prior to stimulation, are being exocytosed into the lumen (*49,53,56*). Thus, these granules gradually congregate in the apical portion of the cell and most are reportedly secreted within 20 min after stimulation (*49*). The synthesis of new fibroin reaches a peak about 90 min after stimulation in *Nephila clavipes* (*57*) and is preceeded by three rounds of RNA synthesis. Included in the first round of RNA synthesis, peaking about 15 min after stimulation, are the small nuclear RNA U1 and a silk gland-specific alanine tRNA isoacceptor (*58-61*). Fibroin mRNA is synthesized during the second round, which peaks about 30 min after stimulation (*57*). (In studies requiring spider fibroin mRNA, mechanical silk depletion seems to be the method of choice for maximizing yields of this mRNA (*41,62*).) The third round of RNA synthesis, peaking about 45 min after stimulation, results in a tRNA population enriched in those tRNAs specific for amino acids prevalent in fibroin (*58-60*). It is, however, curious that proline tRNA increases substantially at this time (see Figure 2 in *60*), given that major ampullate luminal contents or silk from *N. clavipes*, unlike major ampullate products from several other examined araneoids (*35,43,63*; data published prior to 1987 compiled in Table 1 of *35*), contain only about 4 mole % or less of proline (*35,63,64*; compilation in *35*). The importance to fibroin synthesis of the composition of the tRNA population has been investigated using a cell-free reticulocyte lysate translation system. It was found that the in vitro synthesis of full-length polypeptide from fibroin mRNA could be achieved by supplementing the system with tRNA from stimulated major ampullate glands, but not by adding tRNA from nonglandular tissues or unstimulated major ampullate glands (*58,62*).

In addition to the apical migration and exocytosis of secretory granules, the effects of stimulation are manifested in a number of other histological and ultrastructural changes in major ampullate gland tail epithelial cells. These reportedly include distortion of nuclei during secretion of the pre-existing secretory granules and subsequent enlargement of nuclei and nucleoli (*53,56,65*), transient, synchronous movements of ribosomes within the cells (*56,65*), increased distension of rough ER cisternae (*48*) and re-configuration of portions of rough ER into whorls with protein (apparently fibroin) droplets at their center (*49,56*), proliferation and enlargement of Golgi elements (*48*), and alterations in the microvilli-lined luminal face of the cells (*48*). Experiments using actinomycin D and puromycin have shown that these protein synthesis inhibitors are able to block the synthesis of new fibroin in stimulated glands, but they do not affect the exocytosis of pre-existing fibroin (*53,56*).

Analysis by SDS-PAGE of the proteins in *N. clavipes* major ampullate gland secretory epithelium has revealed the presence of a series of polypeptides of lower molecular weight than fibroin (*57*) which are considerably more abundant in glands that have been stimulated (*66*). These polypeptides readily incorporate radioactive glycine and alanine (*62,66*), abundant in *N. clavipes* fibroin (*35,63,64*; compilation in *35*), while incorporation of radioactive histidine, a minor constituent, is slight. Moreover, such polypeptides are also formed during in vitro cell-free translation of fibroin mRNA (*62*). Thus, the interpretation has been that these polypeptides are incomplete fibroin molecules and that they are manifestations of pauses which occur at specific sites during the translation of fibroin mRNA.

Parallels to Fibroin Synthesis in the Silkworm, *Bombyx mori*. Though our topic in this section is secretory protein synthesis in spiders, we would be remiss not to mention that parallels to some of the findings discussed above have been observed in *B. mori*, where, for economic reasons, silk gland research is well ahead of that in spiders. Thus, changes in tRNA complements, including the accumulation of a silk gland-specific isoacceptor of alanine tRNA (*67,68*), also accompany fibroin synthesis in the silk glands of fifth larval instars of *B. mori* (reviewed e.g. in *69,70*). These changes reflect the fibroin's large requirements for glycine, alanine and serine and, more specifically, there is a correlation between the relative concentrations of particular tRNA isoacceptors and the frequencies of particular codons in fibroin mRNA (reviewed in *70*). Moreover, it has been observed that the addition of tRNA from *B. mori* posterior silk glands is required for efficient, full-length in vitro synthesis of fibroin in a cell-free reticulocyte system and that tRNA from other sources is not an adequate substitute (*71,72*). The interpretation of discontinuous translation of fibroin mRNA in the major ampullate glands of *N. clavipes* also has precedent in *B. mori* silk glands (*71,72*) (as well as in various systems not involved in fibroin synthesis). And, not surprisingly, distension of rough ER cisternae and development of Golgi apparatus are also linked with fibroin synthesis in *B. mori* (*73-75*). See the reports by Candelas and colleagues (*48,57,58,60,62,66*) for more detailed discussion of the similarities which exist between spider major ampullate glands and *B. mori* silk glands with regard to fibroin synthesis.

Aggregate and Cylindrical Glands. Very little work has been done on the regulation of secretory protein synthesis in spider silk glands other than major ampullate glands. Evidence has been presented indicating that physostigmine has a stimulatory effect on protein synthesis in the aggregate glands of *L. sclopetarius* and *A. diadematus*, as it does in the major ampullate glands (*76*). However, Peakall's identifications of silk glands in certain published figures, including Figure 2a of (*76*), have been called into question (*77*). Figure 2a was presented to substantiate the claim of physostigmine-induced stimulation of aggregate glands. In the same study, Peakall found no such stimulation evident in the cylindrical glands. Cylindrical glands are used only in the construction of the eggcase by mature adult females. Their development parallels that of the ovaries (e.g. *23,24,26,78*) and they exist as thin, inconspicuous structures until the adult stase. It would not be unexpected that regulation in these glands differs from that in the other types of silk glands, which are used regularly throughout much of the spider's life. Candelas et al. (*79*) have since examined fibroin synthesis in the cylindrical glands of *N. clavipes*. They found that mechanical stimulation of these glands (by which we presume it is meant that silk was mechanically drawn from the cylindrical gland spigots) resulted in increased fibroin synthesis as compared with unpulled control glands, but the response was considered slight compared with that which can be induced in major ampullate glands by the same means. They also obtained evidence indicating that pauses occur during the translation of cylindrical gland fibroin mRNA, comparable to the translational pauses reported in major ampullate glands and the silk glands of *B. mori* (see above).

Nutrition

Whereas there is little exact data on the nutritional requirements of spiders, enough is now known about the chemistry of the orb web to allow us to deduce some of the requirements for web construction. Amino acid composition data obtained from the various silk gland types has been published (*80,81*), and if one assumes that the essential amino acids for the spider are the same as those which are essential for at least most insect species (*82*), then these data can be used to estimate the percentage of the web components consisting of essential amino acids. From this analysis an interesting fact emerges. Whereas the major ampullate silk consists mostly of non-essential amino acids, about 39% of the aggregate gland protein consists of essential amino acids (Table I).

TABLE I. The essential amino acid contents of selected silk glands of *Araneus diadematus* [calculated from data of Andersen (*80*)] and *Argiope aurantia* [calculated from data of Tillinghast (*81*)]

		Essential Amino Acids (mole %)[a]	
Gland	Web Component	*A. diadematus*	*A. aurantia*
Major Ampullate	Radii, framelines, hub spiral, etc.	5.5	7.7
Flagelliform	Core fibers of adhesive spiral	15.8	19.9
Aggregate	Adhesive of adhesive spiral	40.6	37.5

[a] The following amino acids were assumed to be essential: valine, methionine, threonine, leucine, isoleucine, phenylalanine, lysine, arginine, histidine.

The aggregate gland protein appears to consist predominantly of a high molecular weight, phosphorylated glycoprotein (*83*). Amino acid analyses of partially purified preparations of this glycoprotein from *Argiope aurantia* orb webs indicate that it contains approximately 30 mole % essential amino acids, with about a third of this value arising from threonine. However, it should be made clear that this glycoprotein has, in many respects, proven difficult to work with. Most importantly, we do not know if we are dealing with a single aggregate gland-derived protein. In this regard, we can only note that individual secretion granules in the aggregate glands of *Nephila clavata* appear heterogeneous when viewed by TEM (*18*). The glycoprotein is distributed as nodules along the core fibers of the adhesive spiral of the orb web and appears to be the "glue" that secures prey to the web (*84,85*). Both the adhesive nodules and the core fibers of the adhesive spiral are covered with a concentrated solution of low molecular weight organic and inorganic compounds (LMW). These include 4-aminobutyramide, isethionic acid, choline, *N*-acetyltaurine, *N*-acetylputrescine, glycine, betaine, KH_2PO_4 and KNO_3 (*86-90*). Like the nodules, this viscid, aqueous solution is a product of the aggregate glands (*4,5*). While the roles of the various LMW remain to be established, choline, at least, appears to perform the critical function of adsorbing atmospheric moisture essential to the adhesiveness of the glycoprotein (*89,90*).

The LMW account for up to 45-65% of the desiccated weight of the araneid orb web depending on the species (*90*,Tillinghast, E.K. and Townley, M.A., unpublished data) and together with the adhesive glycoprotein may account for up to about 60-80% of the desiccated weight of the web. Not only does the adhesive glycoprotein contain a substantial content of essential amino acids, but isotope incorporation experiments indicate that choline is also essential (Tillinghast, E.K., Tugmon, C.R., and Townley, M.A., unpublished data). Apparently spiders, like insects (*82*), require a dietary source of choline. Thus, the evolutionary acquisition of aggregate glands seems to have placed nutritional demands on araneid orb weavers that may be more severe than those placed on spiders which lack these glands. The routine recycling of orb webs by ingestion (*91-94*) may be even more crucial than previously considered.

It will remain for future research to reveal any potential reserves of essential compounds useful to orb weavers during starvation. Whether spiders have reserve proteins now known to exist in insects (*95*) remains to be established. However, the spider gut contains a rich reserve of amino acids and the stercoral pocket and gut contain high levels of potassium phosphate (*96*) which might be drawn upon in times of nutritional stress. Even so, in the absence of feeding, the continued recycling of the web is probably essential to the conservation of that important cellular constituent, choline.

Additional aspects of spider nutrition are discussed in several reviews, such as those by Nakamura (*97*), Nentwig (*98*), Riechert and Harp (*99*) and Vollrath (*100*).

Molting

Molting poses special problems to some spiders. Like other arthropods they must molt. And like other ectodermal derivatives the silk glands must be remodeled. Yet at least some spiders, including araneids, require silk even during the very process of silk gland remodeling. Although they cease constructing orb webs at this time, araneids like *Araneus cavaticus* use ampullate gland silk to construct a retreat and to lay down draglines when circumstances make it necessary. Thus, the question is, how are araneid spiders able to produce ampullate gland silk while the ampullate glands are being remodeled? Before describing the mechanism that has evolved to solve this problem, some background information would be helpful.

As juveniles, araneid spiders have two obvious pairs of major ampullate silk glands and two pairs of minor ampullate silk glands, with one pair of each type larger than the other. Following the last molt the smaller pair of major ampullate glands and the smaller pair of minor ampullate glands degenerate, so that adults have only one obvious pair of each type of ampullate gland (*11*). When juveniles are closely examined, however, in addition to the four conspicuous pairs of ampullate glands, two pairs of small, vestigial-looking structures are observed, one pair attached to the major ampullate glands, the other to the minor ampullate glands (*101*). Such structures were first described by Johansson (*102*), who refered to them as accessory [ampullate] glands. Likewise, when adults are closely examined, one finds four pairs of what appear to be 'accessory' ampullate glands, two pairs attached to the single remaining pair of major ampullate glands (Figure 4) and two pairs attached to the single pair of minor ampullate glands. Thus, in passing from juvenile to adult, the degeneration of one pair of major ampullate glands and one pair of minor ampullate glands and the acquisition of two additional pairs of 'accessory' ampullate glands takes place in araneid spiders. The explanation for this reversal in complements is simply that the glands that degenerate are not totally reabsorbed; rather, they atrophy *into* 'accessory' ampullate glands (*101*).

These observations raise several questions; among them, why do juveniles apparently require two pairs of major ampullate glands and two pairs of minor ampullate glands while adults only require one pair of each? This question brings us back to the question set out above because they both have the same answer. Those ampullate glands that function only in juveniles act as 'backup' ampullate glands because they are only used during proecdysis, that is, for several days prior to ecdysis during which time the larger major and minor ampullate glands are being remodeled (*101,103*). After the spider has undergone its final molt (i.e. made the molt to adulthood), it no longer has need of these 'backup' ampullate glands because the larger major and minor ampullate glands will never be remodeled again, and they are allowed to degenerate into

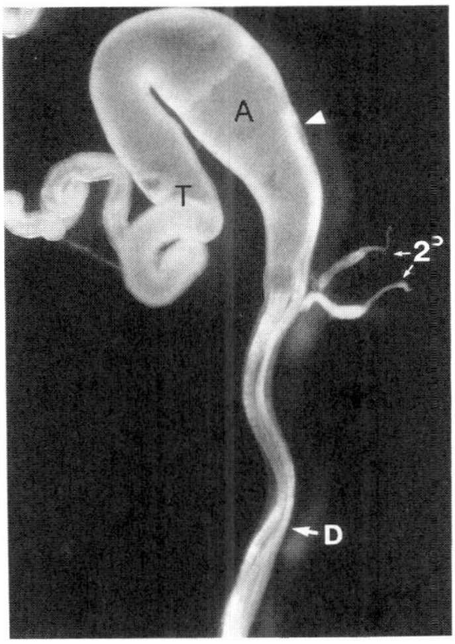

Figure 4. A primary major ampullate silk gland from an adult female *Araneus cavaticus.* Only a small portion of the tail (T) can be seen leading off to the left. The two small, secondary major ampullate glands (2°) are attached near the ampulla (A)/duct (D) junction. Because neither of these secondary ampullate glands is functional in the adult, they would both be considered 'blocked' (see "Molting" section for explanation of 'primary/open secondary/blocked secondary' terminology). Though difficult to discern in this photograph the duct follows a zigzagging course (see *10,14,77*) and a loop of the duct (arrowhead) is appressed to the right side of the ampulla. Note the conjunctive sheath, most visible where the duct segments bend slightly, which surrounds the three segments of the duct and the proximal portion of the ampulla (*14*).

'accessory' ampullate glands. They remain in this form for the remainder of the spider's life.

Another question naturally arises. If one pair of 'accessory' major ampullate glands and one pair of 'accessory' minor ampullate glands in the adult represent atrophied 'backup' ampullate glands from the last juvenile stadium, what is the origin of the other two pairs of 'accessory' ampullate glands in adults, those which are present even in juveniles? Just as the two pairs of 'backup' ampullate glands in the last juvenile instar and two of the four pairs of 'accessory' ampullate glands in the adult represent the same entities, just developed to different degrees, so too the 'accessory' ampullate glands in juveniles represent the less developed form of entities which, in their fully developed guise, are functional 'backup' ampullate glands. That is to say, the distinction between 'backup' and 'accessory' ampullate glands is, to some extent, artificial since a given gland is both a 'backup' and an 'accessory' ampullate gland at different times in the spider's development.

In basic outline, events can thus be summarized as follows. There are two sets of 'accessory/backup' ampullate glands, each set consisting of one pair of major and one pair of minor ampullate glands. In the proecdysial portion of a given juvenile stadium (which, for clarity, we will call the *n* stadium) only one of these sets is functional. They are the 'backup' ampullate glands in the *n* stadium. The other set is nonfunctional and constitutes the 'accessory' ampullate glands in the *n* stadium. In the following stadium (the *n+1* stadium), these two sets, in effect, switch identities. The 'backup' ampullate glands of the *n* stadium degenerate into 'accessory' ampullate glands and will not be functional in the *n+1* stadium, while the 'accessory' ampullate glands of the *n* stadium develop during the *n+1* stadium, so that by the time proecdysis is reached in this stadium they will be functional 'backup' ampullate glands. In the *n+2* stadium, these two sets of glands again switch identities, and so on through all the juvenile stadia. When adulthood is reached, the 'backup' ampullate glands of the last juvenile stadium regress as usual into 'accessory' ampullate glands, but the 'accessory' ampullate glands do not re-develop. Each 'accessory/backup' ampullate gland's ontogeny is, thus, cyclical, alternating functional and nonfunctional phases with each stadium.

In a paper that describes these events in greater detail (*103*), we have adopted what we believe is a more suitable nomenclature for the different ampullate glands. Those major and minor ampullate glands which are larger than their counterparts and which are functional in both juveniles and adults at all times except during proecdysis, we call primary major and primary minor ampullate glands. Those glands which are used only during proecdysis, we call secondary major and secondary minor ampullate glands. Secondary ampullate glands are subcategorized on the basis of whether or not they are or will be used during the stadium under discussion. Glands that are or will be so used, we refer to as open secondary major and minor ampullate glands. Blocked secondary major and minor ampullate glands will not, and indeed cannot, be used during the stadium under discussion. A given secondary ampullate gland is, therefore, blocked during one juvenile stadium, open during the next, blocked during the one after that, and so on.

The events described above also explain the existence of some of the structures known as nubbins (*101,104,105*), small cuticular protuberances observed in the vicinity of silk gland spigots. During proecdysis, in order for the open secondary major and minor ampullate glands to function, their ducts, obviously, must not be severed by the newly developing cuticle that forms under the old cuticle at this time. Thus, the new cuticle forms around these ducts and allows them to maintain their connections to the overlying old cuticle. At ecdysis, the old cuticle is removed, the open secondary ampullate gland ducts connected to the old cuticle are severed (thus, rendering these glands 'blocked') and the spider reverts to using the primary major and minor ampullate glands, the ducts of which connect to spigots on the new cuticle shortly before ecdysis occurs. Where open secondary ampullate gland ducts had penetrated the new cuticle during its formation, nubbins remain as the only external evidence of the occurrence. Figure 5 schematically depicts the connections made by ampullate gland ducts to spigots

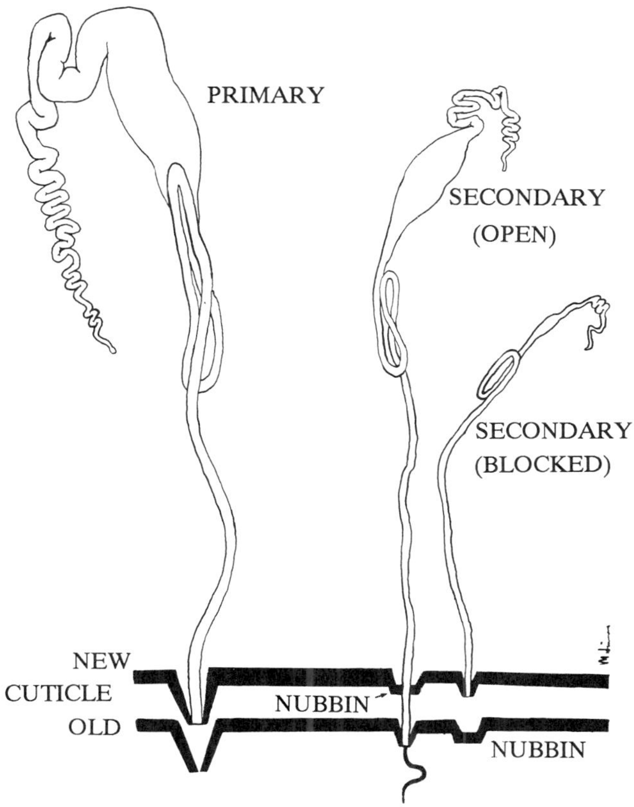

Figure 5. Schematic diagram of the ampullate silk glands serving a single spinneret and their connections to the old and new cuticles, depicted at a time just prior to ecdysis. At this time only the ducts of the open secondary ampullate glands are connected to spigots on the old cuticle and, thus, only these glands are functional. The ducts of the primary and blocked secondary ampullate glands are connected to spigots on the new cuticle. When the old cuticle is shed at ecdysis, the open secondary ampullate gland will become "blocked" and nonfunctional, the primary ampullate gland will resume functioning and the blocked secondary ampullate gland will become the open secondary ampullate gland. This latter gland will not be used until proecdysis of the stadium following that depicted in the diagram. Note that small protuberances known as nubbins remain in the cuticle at sites where open secondary ampullate gland ducts penetrated the cuticle during its formation.

on the old and new cuticles shortly before ecdysis. We should note that not all structures refered to as nubbins are formed in this way. For a more thorough discussion of nubbin (and tartipore) formation and occurrence see Townley et al. (*103*).

Acknowledgments

We wish to thank Mary Sue Potts and Dr. Arthur Borror for their generous assistance in preparing line drawings for the oral presentation on which this chapter is based and Marc Simmons for those appearing in this chapter. Those studies reported here that were products of this laboratory were supported by NIH-AREA grant R15 GM44353-01A1 and HATCH funds to the University of New Hampshire.

Literature Cited

1. Kavanagh, E. J.; Tillinghast, E. K. *J. Morph.* **1979,** *160,* 17-31.
2. Work, R. W. *Trans. Am. Microsc. Soc.* **1981,** *100,* 1-20.
3. Peters, H. M. *Acta Zool. Fennica* **1990,** *190,* 309-314.
4. Sekiguchi, K. *Annot. Zool. Japon.* **1952,** *25,* 394-399.
5. Peters, H. M. *Z. Naturforsch.* **1955,** *10b,* 395-404.
6. Apstein, C. *Arch. Naturgesch.* **1889,** *55,* 29-74.
7. Warburton, C. *Q. J. Microsc. Sci., N. S.* **1890,** *31,* 29-39.
8. Peters, H. M. *Verh. naturwiss. Ver. Hamburg (NF)* **1982,** *25,* 147-167.
9. Schimkewitsch, W. *Ann. Sci. Nat. Zool. Paléon.* **1884,** *17,* 1-94.
10. Atanasiu-Dumitresco, M. *Anal. Acad. Rom. Mem. Sect. Stiint., Ser. 3* **1941,** *16,* 773-840.
11. Sekiguchi, K. *Sci. Rep. Tokyo Kyoiku Daigaku Sect. B* **1955,** *8,* 33-40.
12. Brignoli, P. M. *A Catalogue of the Araneae Described Between 1940 and 1981;* Manchester University Press: Manchester, UK, 1983; 755 pp.
13. Platnick, N. I. *Advances in Spider Taxonomy 1981-1987;* Manchester University Press: Manchester, UK, 1989; 673 pp.
14. Kovoor, J.; Zylberberg, L. *Z. Zellforsch. mikrosk. Anat.* **1972,** *128,* 188-211.
15. Kovoor, J.; Zylberberg, L. *Ann. Sci. Nat. Zool. Biol. Anim., Sér. 12* **1974,** *16,* 5-25.
16. Kovoor, J.; Zylberberg, L. *Zoomorphologie* **1979,** *92,* 217-239.
17. Moon, M. J.; Kim, C. S.; Kim, W. K. *Korean J. Electron Microsc.* **1988,** *18,* 77-90.
18. Moon, M. J.; Kim, W. K. *Korean J. Zool.* **1989,** *32,* 211-220.
19. Moon, M. J.; Kim, W. K. *Korean J. Electron Microsc.* **1989,** *19,* 49-58.
20. Moon, M. J.; Kim, W. K. *Korean Arachnol.* **1989,** *5,* 43-55.
21. Moon, M. J.; Kim, W. K. *Korean Arachnol.* **1990,** *5,* 195-206.
22. Moon, M. J.; Kim, W. K. *Korean J. Zool.* **1990,** *33,* 354-364.
23. Kovoor, J. *Ann. Biol.* **1977,** *16,* 97-171.
24. Kovoor, J. In *Ecophysiology of Spiders;* Nentwig, W., Ed.; Springer-Verlag: Berlin, 1987; pp 160-186.
25. Kovoor, J.; Lopez, A. *Rev. Arachnol.* **1988,** *7,* 205-212.
26. Kovoor, J. *Ann. Sci. Nat. Zool. Biol. Anim., Sér. 12* **1972,** *14,* 1-40.
27. Wiśniewski, H. *Zool. Polon.* **1986,** *33,* 83-104.
28. Kovoor, J.; Peters, H. M. *Zoomorphology* **1988,** *108,* 47-59.
29. Moon, M. J.; Kim, C. S.; Kim, W. K. *Korean J. Electron Microsc.* **1988,** *18,* 91-101.
30. Hajer, J. *Acta Entomol. Bohemoslov.* **1989,** *86,* 401-413.
31. Moon, M. J.; Kim, W. K. *Korean J. Electron Microsc.* **1989,** *19,* 59-69.
32. Kovoor, J. *Acta Zool. Fennica* **1990,** *190,* 215-221.
33. Peters, H. M.; Kovoor, J. *Zoomorphology* **1991,** *111,* 1-17.

34. Work, R. W. *Trans. Am. Microsc. Soc.* **1984,** *103,* 113-121.
35. Work, R. W.; Young, C. T. *J. Arachnol.* **1987,** *15,* 65-80.
36. Palmer, J. M.; Coyle, F. A.; Harrison, F. W. *J. Morph.* **1982,** *174,* 269-274.
37. Palmer, J. M. *J. Morph.* **1985,** *186,* 195-207.
38. Kovoor, J.; Zylberberg, L. *Tissue Cell* **1980,** *12,* 547-556.
39. Peters, H. M.; Kovoor, J. *Zool. Jb. Physiol.* **1989,** *93,* 125-144.
40. Stubbs, D. G.; Tillinghast, E. K.; Townley, M. A.; Cherim, N. A. *Naturwissenschaften* **1992,** *79,* 231-234.
41. Xu, M.; Lewis, R. V. *Proc. Natl. Acad. Sci. USA* **1990,** *87,* 7120-7124.
42. Hinman, M. B.; Lewis, R. V. *J. Biol. Chem.* **1992,** *267,* 19320-19324.
43. Lewis, R. V. *Acc. Chem. Res.* **1992,** *25,* 392-398.
44. Hinman, M.; Dong, Z.; Xu, M.; Lewis, R. V. In *Structure, Cellular Synthesis and Assembly of Biopolymers;* Case, S. T., Ed.; Results and Problems in Cell Differentiation, Vol. 19; Springer-Verlag: Berlin, 1992; pp 227-254.
45. Bradfield, J. R. G. *Nature* **1946,** *157,* 876-877.
46. Bradfield, J. R. G. *Q. J. Microsc. Sci.* **1951,** *92,* 87-112.
47. Kovoor, J. *C. R. Troisième Réunion Arachnol. Exp. Fran., Les Eyzies* **1976,** 83-96.
48. Plazaola, A.; Candelas, G. C. *Tissue Cell* **1991,** *23,* 277-284.
49. Bell, A. L.; Peakall, D. B. *J. Cell Biol.* **1969,** *42,* 284-295.
50. Witt, P. N. *Proc. XVI Int. Congr. Zool.* **1963,** *2,* 7. (abst.)
51. Peakall, D. B. *Comp. Biochem. Physiol.* **1964,** *12,* 465-470.
52. Peakall, D. B. *Comp. Biochem. Physiol.* **1965,** *15,* 509-515.
53. Peakall, D. B. *Comp. Biochem. Physiol.* **1966,** *19,* 253-258.
54. Candelas, G. C.; Cintrón, J. *J. Exp. Zool.* **1981,** *216,* 1-6.
55. Tillinghast, E. K.; Townley, M. A. *J. Insect Physiol.* **1986,** *32,* 117-123.
56. Peakall, D. B. *Am. Zool.* **1969,** *9,* 71-79.
57. Candelas, G. C.; López, F. *Comp. Biochem. Physiol.* **1983,** *74B,* 637-641.
58. Candelas, G. C.; Carrasco, C. E.; Dompenciel, R. E.; Arroyo, G.; Candelas, T. M. In *Translational Regulation of Gene Expression;* Ilan, J., Ed.; Plenum Press: New York, NY, 1987; pp 209-228.
59. Arroyo, G.; Candelas, T.; Carasquillo, E.; Candelas, G. C. *J. Cell Biol.* **1990,** *111,* 106a. (abst.)
60. Candelas, G. C.; Arroyo, G.; Carrasco, C.; Dompenciel, R. *Dev. Biol.* **1990,** *140,* 215-220.
61. Carrasco, C. E.; Candelas, G. C. *J. Cell Biol.* **1990,** *111,* 106a. (abst.)
62. Candelas, G. C.; Ortiz, A.; Ortiz, N. *Biochem Cell Biol.* **1989,** *67,* 173-176.
63. Lombardi, S. J.; Kaplan, D. L. *J. Arachnol.* **1990,** *18,* 297-306.
64. Lombardi, S. J.; Kaplan, D. L. *Acta Zool. Fennica* **1990,** *190,* 243-248.
65. Peakall, D. B. In *A Spider's Web: Problems in Regulatory Biology;* Witt, P. N.; Reed, C. F.; Peakall, D. B., Eds.; Springer-Verlag: New York, NY, 1968; pp 5-28.
66. Candelas, G.; Candelas, T.; Ortiz, A.; Rodríguez, O. *Biochem. Biophys. Res. Comm.* **1983,** *116,* 1033-1038.
67. Meza, L.; Araya, A.; Leon, G.; Krauskopf, M.; Siddiqui, M. A. Q.; Garel, J. P. *FEBS Lett.* **1977,** *77,* 255-260.
68. Sprague, K. U.; Hagenbüchle, O.; Zuniga, M. C. *Cell* **1977,** *11,* 561-570.
69. Shimura, K. In *The Silkworm: An Important Laboratory Tool;* Tazima, Y., Ed.; Kodansha Ltd.: Tokyo, 1978; pp 189-211.
70. Prudhomme, J.-C.; Couble, P.; Garel, J.-P.; Daillie, J. In *Comprehensive Insect Physiology, Biochemistry and Pharmacology;* Kerkut, G. A.; Gilbert, L. I., Eds.; Pergamon Press: Oxford, UK, 1985, Vol. 10, Biochemistry; pp 571-594.
71. Lizardi, P. M.; Mahdavi, V.; Shields, D.; Candelas, G. *Proc. Natl. Acad. Sci. USA* **1979,** *76,* 6211-6215.
72. Chavancy, G.; Marbaix, G.; Huez, G.; Cleuter, Y. *Biochimie* **1981,** *63,* 611-618.

73. Tashiro, Y.; Morimoto, T.; Matsuura, S.; Nagata, S. *J. Cell Biol.* **1968,** *38,* 574-588.
74. Morimoto, T.; Matsuura, S.; Nagata, S.; Tashiro, Y. *J. Cell Biol.* **1968,** *38,* 604-614.
75. Blaes, N.; Couble, P.; Prudhomme, J.-C. *Cell Tissue Res.* **1980,** *213,* 311-324.
76. Peakall, D. B. *Nature* **1965,** *207,* 102-103.
77. Mullen, G. R. *Trans. Am. Microsc. Soc.* **1969,** *88,* 232-240.
78. Kovoor, J. *Ann. Sci. Nat. Zool. Biol. Anim., Sér. 12* **1977,** *19,* 63-87.
79. Candelas, G. C.; Ortiz, A.; Molina, C. *J. Exp. Zool.* **1986,** *237,* 281-285.
80. Andersen, S. O. *Comp. Biochem. Physiol.* **1970,** *35,* 705-711.
81. Tillinghast, E. K. *Insect Biochem.* **1984,** *14,* 115-120.
82. Dadd, R. H. In *Comprehensive Insect Physiology, Biochemistry and Pharmacology;* Kerkut, G. A.; Gilbert, L. I., Eds.; Pergamon Press: Oxford, UK, 1985, Vol. 4, Regulation: Digestion, Nutrition, Excretion; pp 313-390.
83. Tillinghast, E. K.; Townley, M. A.; Wight, T. N.; Uhlenbruck, G.; Janssen, E. In *Biomolecular Materials;* Viney, C.; Case, S. T.; Waite, J. H., Eds.; Mater. Res. Soc. Proc. 292, Pittsburgh, PA, 1993; in press.
84. Richter, G. *Naturwissenschaften* **1956,** *43,* 23.
85. Vollrath, F.; Tillinghast, E. K. *Naturwissenschaften* **1991,** *78,* 557-559.
86. Fischer, F. G.; Brander, J. *Hoppe-Seyler's Z. physiol. Chem.* **1960,** *320,* 92-102.
87. Schildknecht, H.; Kunzelmann, P.; Krauß, D.; Kuhn, C. *Naturwissenschaften* **1972,** *59,* 98-99.
88. Tillinghast, E. K.; Huxtable, R. J.; Watson, W. H., III; Townley, M. A. *Comp. Biochem. Physiol.* **1987,** *88B,* 457-460.
89. Vollrath, F.; Fairbrother, W. J.; Williams, R. J. P.; Tillinghast, E. K.; Bernstein, D. T.; Gallagher, K. S.; Townley, M. A. *Nature* **1990,** *345,* 526-528.
90. Townley, M. A.; Bernstein, D. T.; Gallagher, K. S.; Tillinghast, E. K. *J. Exp. Zool.* **1991,** *259,* 154-165.
91. Breed, A. L.; Levine, V. D.; Peakall, D. B.; Witt, P. N. *Behavior* **1964,** *23,* 43-60.
92. Peakall, D. B. *J. Exp. Zool.* **1971,** *176,* 257-264.
93. Carico, J. E. In *Spiders: Webs, Behavior, and Evolution;* Shear, W. A., Ed.; Stanford University Press: Stanford, CA, 1986; pp 306-318.
94. Townley, M. A.; Tillinghast, E. K. *J. Arachnol.* **1988,** *16,* 303-319.
95. Law, J. H.; Wells, M. A. *J. Biol. Chem.* **1989,** *264,* 16335-16338.
96. Tillinghast, E. K. *Comp. Biochem. Physiol.* **1986,** *84A,* 331-334.
97. Nakamura, K. In *Ecophysiology of Spiders;* Nentwig, W., Ed.; Springer-Verlag: Berlin, 1987; pp 287-295.
98. Nentwig, W. In *Ecophysiology of Spiders;* Nentwig, W., Ed.; Springer-Verlag: Berlin, 1987; pp 249-263.
99. Riechert, S. E.; Harp, J. M. In *Nutritional Ecology of Insects, Mites, Spiders, and Related Invertebrates;* Slansky, F., Jr.; Rodriguez, J. G., Eds.; Wiley: New York, NY, 1987; pp 645-672.
100. Vollrath, F. In *Ecophysiology of Spiders;* Nentwig, W., Ed.; Springer-Verlag: Berlin, 1987; pp 357-370.
101. Townley, M. A.; Horner, N. V.; Cherim, N. A.; Tugmon, C. R.; Tillinghast, E. K. *J. Morph.* **1991,** *208,* 175-191.
102. Johansson, B. *Acta Univ. Lund., N. S.* **1914,** *10 (5),* 1-12.
103. Townley, M. A.; Tillinghast, E. K.; Cherim, N. A. *Phil. Trans. R. Soc. Lond. B,* in press.
104. Coddington, J. A. *J. Arachnol.* **1989,** *17,* 71-95.
105. Yu, L.; Coddington, J. A. *J. Arachnol.* **1990,** *18,* 331-345.

RECEIVED June 29, 1993

Chapter 4

Molecular Map for the Silkworm

Constructing New Links between Basic and Applied Research

Marian R. Goldsmith and Jinrui Shi[1]

Department of Zoology, University of Rhode Island, Kingston, RI 02881–0816

The domesticated silkworm, *Bombyx mori*, has > 200 mapped mutations, and hundreds of inbred races which differ in economically important characters. To enable genetic mapping of such quantitative trait loci, facilitate breeding of silkworm strains for current sericultural growth areas in the tropics, and provide a physical framework for direct gene isolation, we are constructing linkage maps based on physical DNA markers. Preliminary maps using F2 crosses comprise 60 restriction fragment length polymorphisms (RFLPs) which cover 16 linkage groups and 8 unlinked markers, potentially representing 24 of 28 chromosomes. These molecular markers include 10 cloned genes correlated with conventional genetic maps, 9 retrotransposons, and 41 anonymous RFLPs isolated from a silkmoth follicle cDNA library.

The domesticated silkworm, *Bombyx mori*, has a long history as an organism of economic importance. Archeological evidence dates the oldest known silk textile at nearly 5000 years, from a find in Zhejiang province in southeastern China (*1*). The sophisticated treatment of the fiber indicates a high level of technology and argues that it was almost certainly produced long after domestication of the insect, which remains the only one which is wholly dependent on humans for its care and feeding. Despite a threat of death for revealing any secrets of the agriculture associated with raising silkworms and harvesting their silk, now known as sericulture, outside China, silk production technology was brought by immigrants to Korea in the second century, B.C., and thence to Japan where textiles were being woven with locally grown yarn by the middle of the third century, A.D. (*1*). Silkworm eggs, mulberry seeds, and silk textile technology finally reached Byzantium around 550 A.D. under the emperor Justinian (*1*)—whether smuggled out of Asia via the Silk Road by monks or merchants being a matter of legend and some controversy. There the silk industry developed slowly under strict government control, finally spreading to Europe via Spain, Italy and France from

[1]Current address: Plant Biology Division, The Samuel Roberts Noble Foundation, P.O. Box 2180, Ardmore, OK 73402

0097–6156/94/0544–0045$06.00/0

the 8th to 11th centuries (see *1*, for additional details). European sericulture, centered in Lyon, France, was devastated by a sporozoan infection, *Nosema bombycis*, diagnosed by Pasteur in 1862, too late to save the industry; this pathogen is still manageable only by destroying infected moths. Today sericulture remains rooted in Asia, but, as will be apparent, it is traveling along new routes to other areas of the world.

The silkworm's diaspora from China led to the evolution of many locally inbred "geographic" races, which differ markedly in their biological and genetic character, and today comprise an important resource for practical breeders to draw on for developing new stocks to meet demands imposed by such factors as different seasonal growing conditions, changing sericultural practices, new diseases, and the ongoing drive to improve silk yields and quality. Practical silkworm genetics has grown hand-in-hand with basic research. Over 200 mutations affecting many kinds of biological processes have been collected, most arising spontaneously during mass rearing for cocoon production; these have been analyzed in terms of their chromosomal map positions (*2*), and many have been studied in greater detail (see *3*). Indeed, today the silkworm is second only to the fruitfly, *Drosophila melanogaster*, as an insect model for genetic studies. Many other important areas of basic research have grown along with the needs of sericulture and because the silkworm is so easy to rear and maintain in the laboratory; these include insect nutrition and biochemistry, hormone physiology, development, neurobiology, pathology and immunity (see *4, 5* for reviews), making *B. mori* one of the best-studied insects known.

As a background for a broad consideration of the international silk industry, three aspects of sericulture and genetics will be presented here. First will be a brief history of the origins of scientific silkworm breeding in Japan, where sericulture has reached the highest level of technology, in part because of strong government controls and active support of all aspects of the industry. Second is an overview of the economics of the silk trade and the status of sericulture today. And third is a description of a cooperative effort recently initiated in this laboratory to generate molecular linkage maps in the silkworm which has the potential to extend our knowledge of genes involved in key biological processes, make important advances in silkworm breeding, and forge new ties between fundamental and applied research in this ancient model system.

Silkworm Breeding in Japan

The accomplishments of Kametaro Toyama, one of the founders of silkworm genetics, offer an illustration of the ease with which sericultural scientists have crossed the line between basic and applied research. His published papers include descriptions of silkworm embryology, the formation of genetic mosaics, mutations in such diverse characters as larval body markings, eggshell structure, moltinism (number of larval molts before spinning), and voltinism (number of generations per year), and one of the first reports of genetic dominance involving the gene responsible for the well-known golden yellow cocoon color in Chinese silkworm races (in *3*). Silk color is a complex character, involving genes controlling absorption of carotenoid (pink, flesh and yellow) and flavonoid (green) pigments derived from mulberry, the major food plant, by the larva's intestinal epithelium, biochemical modification of the flavonoids, and the uptake and secretion of these pigments in different sections of the silkgland (see *3* for details). Though several mutations affecting these processes have been identified in the carotenoid system, we still do not fully understand the genetics underlying green cocoon color in *B. mori* (Doira, H., Kyushu University, personal communication), nor that of the light-fast and highly prized blue-green silk of the wild Saturniid moth, *Antheraea yamamai*.

More than a theoretical scientist, Toyama spent 1902-1905 as chief sericultural expert to the government of Siam (now Thailand) where he bred new silkworm varieties by combining low productivity indigenous races with Japanese ones, increasing raw silk ratios 3-fold and overall cocoon yields by 30% (*6*). By 1906 he had published results of cross-breeding experiments between various silkworm races and began strong advocacy in Japan of using hybrids in place of the purebred races then exclusively grown for silk production. With the founding of the Sericultural Experiment Station in 1911, the Japanese government began to develop official policy on controlled silkworm breeding under Toyama's leadership, making hybrid eggs available to farmers by 1914 and eventually allowing egg production and breeding only under license. This maintained the purity of inbred stocks and insured the high quality of Japanese silk. Indeed, genetically improved strains are still officially unavailable for export, although strong pressure from many scientists recently helped relax restrictions on export of the ancient inbred geographic races, which can now be obtained for basic research. By the 1920s the Sericultural Experiment Station also began to set specific targets for breeding programs which included not only improving silk yield and health and survival of the insect, but also increasing the uniformity of the cocoon filament, its 'neatness' (smoothness) and lousiness (cross-sectional shape), and reducing the boil-off of sericin coating the fiber (*7*).

In addition to such traditional breeding goals, Japanese scientists have shown great ingenuity in achieving novel results by creative use of the geographic races. For example, in the 1930s silkworm breeders developed a high yielding strain of silkworm pupae to solve a vitamin B2 shortage (*7*). Recently, they have bred commercially useful polyphagous silkworm strains which avidly eat and grow on artificial diets lacking any mulberry constituents (*8*), as well as spinners of specialized varieties of silk, such as thick filaments suitable for stockings, thin filaments to combine with nylon in an artificial "hybrid silk", and fibers with properties fine-tuned for use as strings in traditional Japanese musical instruments (Magoshi, Y., National Institute of Sericultural and Entomological Science (NISES), Tsukuba, Japan, personal communication). The industry has also developed non-traditional uses for silk protein, for example, as an additive in cosmetics, shampoos, and hair conditioners, for artificial skin to protect burn victims, and as a matrix for enzyme immobilization (Asakura, T., Tokyo University of Agriculture and Technology, personal communication). Even the silkworm itself is being used as a living test tube for the synthesis of proteins which are harvested from its body fluids after infection by genetically engineered viruses (*9*). We can expect even greater progress in the last area once it becomes possible to carry out direct gene transfer in silkworms, which is presently under intense investigation in laboratories in Canada, France, and Japan (Iatrou, K., University of Calgary, Couble, P., Université Claude Bernard, Lyon, Tamura, T., NISES, Tsukuba, Maekawa, H., National Institute of Health (NIH), Tokyo, personal communication).

The World Silk Trade

Silk represents the lowest fraction of the world's production of textile fibers, including cotton, synthetics, cellulosic materials, and wool, in order of importance (Table I). Although silk production increased from 49,000 to 67,000 tons between 1975 and 1989, it accounted for only 0.2-0.17% of total fibers produced (*10*). This level is expected to increase modestly, to around 85,000 tons by the year 2000

Table I. World Production of Textile Fibers[a], 1975-1989

Year	*Cotton*	*Synthetics*	*Cellulosic fibers*	*Wool*	*Silk*	*Total*
1975	11,809	7,346	2,959	1,502	49	23,665
1980	13,981	10,476	3,242	1,608	55	29,372
1985	17,540	12,515	2,999	1,673	59	34,786
1989	18,800	16,000	3,200	2,000	67	40,067

[a]thousands of tons. SOURCE: UNSO/ITC Comtrade Database System, 1989. Reprinted with permission from ref. 10. Copyright 1992.

(*10*). The import market has been essentially flat in the past decade (Figure 1), with most change occurring as a switch from unprocessed and partially processed fiber to finished textiles and goods, including silk blends, knits, and other less expensive products (*10*). While Japan remains the major world importer of silk products of all categories (25% of total imports), in part because of its long tradition of making silk textiles and using silk kimonos and related articles, this situation is rapidly changing, and the United States now accounts for about half of international silk purchases in the form of textiles and finished goods ($1.71 billion in 1991), followed by Hong Kong (a major textile finishing and design center), and Western Europe (*10*).

China has dominated world production of silk for more than a decade (*11, 12*). In 1991 China's production of fresh cocoons and silk yarn accounted for approximately 90% of raw silk on the international market, followed by India, Japan, the CIS (former USSR) and Brazil (Table II). At that time total Chinese silk exports earned $2.1 billion, with over half from sales of silk textiles and finished

Table II. World Production of Raw Silk[a], 1938-1991

Producer	*1938*	*1978*	*1985*	*1986*	*1989*	*1991*
China	4,855	19,000	32,000	35,700	40,700	48,500
India	690	3,475	7,029	8,277	10,000	12,000
Japan	43,150	15,960	9,582	8,240	6,078	5,527
CIS (USSR)	1,900	3,240	3,999	4,000	4,000	4,000[b]
Brazil	35	1,250	1,458	1,780	1,900	2,034
Others	4,043	2,200	2,748	2,875	3,100	4,000[b]
Total	56,500	49,360	58,914	62,622	66,978	76,761

[a]tons; [b]estimate. SOURCE: International Silk Association. Reprinted with permission from ref. 10. Copyright 1992.

goods ($1.49 billion; *10*). Because of its key role in the international silk trade, China's gradual shifting of exports to value-added items represents an important threat to yarn importing countries, especially the preeminent silk finishers of Japan, Hong Kong and Western Europe (notably Italy and France). Though China has promised to continue releasing adequate supplies for these markets, it appears that there is a potential opening for new producers, provided they can meet the high quality demanded by the major importing countries and the prices strongly affected by Chinese domination of the market, the volatility of the fashion industry, and the international political and economic situation (*13*). Nevertheless, despite changing production and marketing patterns of silk products, the longterm prospects for the silk export/import trade appear to be reasonably stable (*13*).

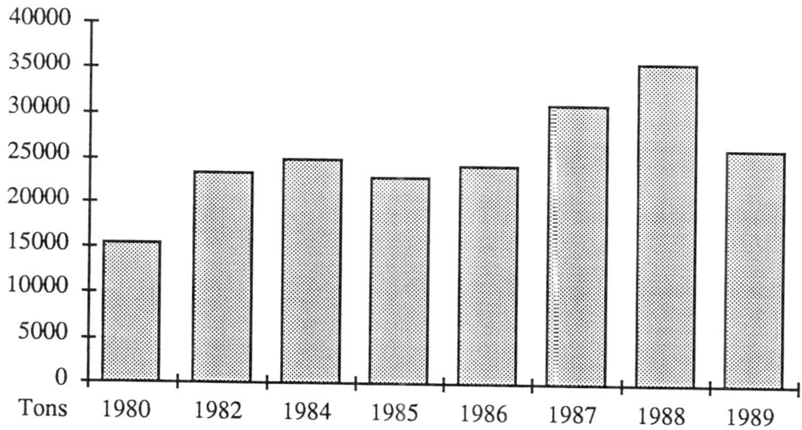

Figure 1. Total international silk imports, 1980-1989 (figures approximate). SOURCE: Commission Séricicole Internationale, La Mulatière, France.

Sericulture Today

Sericulture has always been a labor intensive agriculture requiring considerable land for the rearing of both silkworms and mulberry. For the past several years technologically developed countries such as Japan and Korea have found themselves priced out of the market by the high costs of silk production, despite enormous efforts to stave off the demise of sericulture, which has such a long history in these countries and is of great cultural significance, Even marked success in the breeding of polyphagous silkworm strains and in the formulation of nutritionally complete and cost effective artificial diets with reduced levels of the most expensive constituent, mulberry leaf powder, to enable year round rearing under controlled conditions, has not yet been sufficient to replace the use of fresh leaves during the final stage (5th instar) before cocoon spinning, when food consumption is equivalent to all previous stages and physiological problems are more likely to emerge. Silk processing also poses problems, in that high technology silk finishers, which pay the best prices, will accept only the best quality yarn, driving costs of silk production even higher. This means that, barring unexpected technical breakthroughs, the only way to keep the price of silk at an affordable level on the world market is to produce it in developing countries.

While China remains the largest silk producer and is expected to continue to be so in the foreseeable future, the major growth area for sericulture is principally in the tropics. Growth in recent years has been especially rapid in countries which have long sericultural traditions, such as India, Thailand, and Vietnam, as well as in Brazil, where Japanese immigrants introduced silk production about 70 years ago. The attractions of sericulture are multiple: widespread employment at many levels of society with opportunities for gradual technical upgrading at all levels of production, a quick turnaround on investment, with a 6 week lifecycle, year-round rearing with up to 8 crops possible, and access to hard currency in international markets (*11*). More than 30 countries in Southeast Asia, the Indian subcontinent, the Middle East, South America, Europe, and Africa are now involved in the rearing of silkworms and mulberry, and the processing of silk into yarn and textiles (*10, 14*).

Practical Silkworm Breeding

Sericulture has been developed to its highest levels in temperate countries, notably Japan, China, and Korea, where the breeding of genetically improved silkworm strains has gone hand-in-hand with optimizing rearing techniques and conditions, much as begun by Toyama at the beginning of this century. Even the best commercial hybrids are adapted for rearing at different seasons, because variations in climate, nutritional condition of the mulberry, and endemic fungal and bacterial disease can have a major impact on the growth and productivity of this relatively fragile domesticated species. Thus, as one might expect, the high yielding so-called bivoltine (two generations per year) strains are extremely weak under the severe stress of exposure to tropical heat, humidity and disease, conditions exacerbated by inadequate rearing practices and inferior mulberry leaf quality. On the other hand, relatively hardy indigenous silkworm strains are found in the traditional silk producing countries of Southern and Southeast Asia, such as India, Thailand, and Vietnam, which can withstand these adverse conditions and produce crops of some value. These multivoltine (many generations per year) strains grow rapidly and can be reared year-round, making them particularly well-suited for the tropics, where mulberry leaves are in continuous supply. Unfortunately, though hardy, they produce low yields of poor quality silk fiber which cannot compete effectively on the world market, unless turned into value-added goods such as Thai silk, which is entirely processed locally by hand and known for its special luster and distinctive dyeing and weaving qualities.

From a geneticist's point of view, these two kinds of silkworm strains represent a great wealth of genetic diversity, with desirable characters affecting silk quality and yield carried in the oriental or bivoltine races, and those affecting survival or hardiness in the multivoltines (Table III). An obvious solution to the problem of providing useful stocks for the tropics is to carry out matings or crosses between them in order to produce genetic hybrids combining the most favorable characteristics of both, much as Toyama did, or to carry out crossbreeding and selection in order to obtain true-breeding stocks with new trait combinations. The latter approach is time-consuming and laborious, requiring 9 or more generations to fix the desirable characters in the population, but the methodology is well-established, involving the selection of appropriate broods or individual offspring for further interbreeding by measuring standard economic characters (such as total cocoon weight, cocoon shell weight, silk fiber length, per cent gum, and so on) and assessing 'hardiness' (as survival, disease resistance, fertility, fecundity, etc.). Countries such as India and Thailand are actively involved in this work, and with

Table III. Inherited Silkworm Strain Characters

Affected Feature	*Trait*	*Bivoltine*	*Multivoltine*
		Yield or Quality	
Silk	cocoon shell	high	low
	fiber length	high	low
	reelability	high	low
	neatness	high	low
	lousiness	low	high
Silkworm	cocooning rate	low	high
	disease resistance	low	high
	heat tolerance	low	high
	lifecycle	long	short

great skill, luck, and excellent starting material, new strains suitable for local conditions can be produced in 1-2 years (Cunvong, B., Chul Thai Agro-Industries Co., Ltd., Petchboon, Thailand, personal communication), though the average time to introduce a new commercial line is often much longer.

Quantitative Trait Mapping

The kinds of characters under consideration here are quantitative or polygenic traits, each the cumulative result of the action of several genes controlling many biological processes. In some cases, a single major gene may be responsible for a large proportion of the observed biological effect, and under these circumstances it is possible to identify it by traditional genetic means, since one can detect it as a heritable Mendelian trait. More often, the genes involved in quantitative inheritance have small and complex additive effects which, with rare exceptions, have made them virtually impossible to study by conventional genetic means until recently.

For example, the late maturity gene (*Lm*), which has been mapped to chromosome 1 in the silkworm, is a major gene controlling relative duration of larval life. This indirectly affects silk production, since longer-lived larvae usually grow bigger, and bigger bugs produce more silk (*3*). From population studies it is known that several genes modify or modulate the effects of *Lm*, shifting expected growth rate, time to spinning, and final cocoon weight, but these have been difficult to identify. Without being able to assign map locations to economically important genes such as these and track them precisely in genetic crosses, we must rely on traditional somewhat hit-or-miss methods of strain construction which require large populations, many generations, and extensive statistical analysis to obtain useful gene combinations in a given stock.

Recently, using the tomato as a model system, Eric Lander, David Botstein and colleagues have developed new approaches for identifying and tracking these quantitative trait loci (or QTLs; *15*, *16*). The method requires construction of a molecular linkage map—that is, a map in which genetic markers are measured as physical sites in chromosomal DNA. One can localize genes having significant effects on quantitative traits to specific chromosome segments (or intervals) bounded by these molecular markers using statistical procedures derived from human linkage analysis (based on LOD scores, defined as the $\log_{10}$ of the odds ratio, the probability that, for a given set of data, a QTL exists in the interval of interest *vs*. the probability of there being no QTL; *15*). It then becomes possible to trace the inheritance of the genes underlying complex traits through various breeding experiments by tracing the molecular markers which are physically linked to them. The advantage of this approach is that the molecular markers can be identified unambiguously in DNA extracted from relatively small amounts of cellular material, without the need to express the trait genetically. Close physical linkage of specific QTLs with known molecular markers also enables use of 'marker-assisted selection' (*17*), by which one can introduce desirable QTLs directly and quickly into new genetic stocks or verify their presence in established strains, again, by tracking the associated molecular tags. This is becoming an important tool in plant breeding, and should be equally applicable to the silkworm.

QTL mapping opens up the possibility of studying the interactions of these genes, as well as their action under different environmental conditions. Not only can this lead to an improvement in sericultural practices—for example, one might find a particular gene combination that is deleterious or, conversely, beneficial, under certain circumstances—it can also provide clues for strategies to control insect pests, among which Lepidoptera (moth and butterfly) larvae are the most destructive ones known of crop plants. Finally, with a highly saturated molecular map, the markers may be close enough to the QTL (or known mutation) of interest to allow its direct isolation by now standard cloning methods, such as chromosome walking or jumping, in order to determine its structure and mechanism of action.

Constructing a Molecular Linkage Map

Our initial goals for the molecular linkage map take account of the genome size of the silkworm of 530 million base pairs (*18*), and the fact that *B, mori* has 28 chromosomes measuring a minimum genetic map distance of 1000 cM (*2*). To begin QTL mapping a minimum of 3 markers per chromosome is needed (at an average spacing of 10 cM); experience in other systems suggests that this will require mapping on the order of 300 markers (*16, 20*). Our longterm goal is to produce a map at 1 cM spacing, which will put marked sites within reach of present cloning techniques.

Construction of the map uses standard genetic techniques, in which the physical 'markers' or differences in the DNA of individual silkworms are treated as simple Mendelian traits whose inheritance patterns are characterized in the offspring of controlled genetic matings (or crosses). This involves mating two inbred parental strains which are known to be genetically divergent at the DNA level and then crossing the resulting offspring (F1 hybrids) to each other. In silkworms it is also possible to backcross F1 males to parental females, which makes analysis of the data somewhat simpler because there are only two classes of offspring, backcross parent and hybrid, but is less informative because half the genetic differences are lost (those belonging to the backcross parent, which are present in all progeny). One then scores available DNA markers in individual offspring (F2 progeny), determines the frequency with which pairs of markers are found in nonparental or recombinant arrangements, and tests for independent assortment (unlinked) using Chi square (χ^2) analysis. Two– and three– point (marker) crosses are then used to calculate the map distance between linked genes (defined as the frequency of recombinants among total offspring) and gene order; statistical routines are available for evaluating gene order with more confidence using multilocus analysis (*19*). Since silkworm females lack crossing over (chromosomal exchange), linked markers derived from the F2 female must remain in the parental arrangement. This means that often it is possible to detect linkage between closely spaced genes by inspection of the simultaneous inheritance patterns of several markers at a time.

A number of methods are available for detecting molecular markers, including the use of restriction fragment length polymorphisms (RFLPs; *20*), random arbitrary polymorphic DNA primers (RAPDs; *21, 22*), and microsatellites (*23*). In the former method, relatively large amounts of DNA are digested with sequence-specific restriction enzymes and the resulting fragments are separated by size using electrophoresis; the DNA is then transferred or 'blotted' by capillary action to a binding membrane, chemically immobilized, denatured, and reannealed (hybridized) with a labeled probe consisting of a purified, cloned sequence to detect the fragment it came from in the genome (see below). The other two approaches involve use of the polymerase chain reaction (PCR; *24*), which amplifies specific segments of DNA bounded by short, oppositely orientated sequences in the genome using a thermostable DNA polymerase. PCR requires very small amounts of DNA and little handling of samples, making it especially suitable for screening the large populations required for QTL mapping and marker-assisted selection.

We have begun to construct a map using RFLPs because their codominant inheritance makes them maximally informative genetically (characters contributed by either parent are always detectable), artifacts are relatively easy to detect, and cloned genes can serve as probes to coordinate the conventional and molecular linkage maps. The latter feature makes possible the isolation of a known mutation by positional cloning—that is, by physical methods based on knowing where the mutation is relative to a nearby molecular marker. Thirteen silkworm genes or gene complexes have been isolated whose genetic map positions are known; these include the genes encoding the fibroin heavy (*Fib H*) and light (*Fib L*) chains and sericins (*Src 1* and *2*), several hemolymph and egg storage proteins (*SP-1*, *SP-2*, *30K*, *Lsp* and *Esp*), the gene complex encoding the eggshell or chorion proteins (*Ch1-2* and *Ch3*), and two sets of homeotic genes affecting larval segment identity (*E* and *Nc*). We have used 21 cloned genes as probes to test the basic approach for constructing the map, along with nearly 100 randomly cloned fragments (or 'anonymous RFLPs') obtained from a lambda-gt11 cDNA library synthesized from early egg follicle messenger RNA.

Ten of the 21 cloned genes gave strong hybridization signals and showed polymorphism in our strains after routine digestion of silkworm DNA with 6 different restriction enzymes having 6-base pair recognition sequences which were chosen on the basis of relative cost and average fragment size generated. Among these, a retrotransposon, mag, which occurs in 10-15 copies in the silkworm genome (*25*), proved especially useful, yielding 9 polymorphic bands representing 9 distinct loci from a single probe. The random probes were amplified by PCR using 22- and 24-nucleotide primers flanking the universal cloning site and prescreened by hybridizing to test blots containing the two parental and F1 DNAs to remove clones representing ribosomal, early chorion, and egg specific protein (ESP) genes, which were present at high frequency in the library. Of 93 random clones, 10 (11%) produced hybridization patterns similar to highly repetitive DNA; 9 (10%) yielded patterns expects of multigene families; 7 (8%) produced weak signals and consequently were not used for map construction, and the remaining 67 (72%) yielded strong hybridization signals and revealed either single or low copy number hybridizing fragments.

Blots probed with different kinds of cDNA sequences are shown in Figure 2, and a typical F2 progeny test using a single copy sequence encoding a homeobox-containing segment of the silkworm homolog of the *Antennapedia* complex (pBm 3.0, a 3 kb fragment; Hara, W., NISES, Tsukuba, and Suzuki, Y., National Institute for Basic Biology (NIBB), Okazaki, personal communication) is shown in Figure 3. Each parental type carries a single hybridizing band, as expected, and the F1 hybrids carry both. To prepare the blots, DNA was isolated from individual silkworms, and, after digestion and electrophoresis, was transferred to ZetaProbe GT nylon membranes (BioRad) by capillary action in 10X SSC (1X SSC contains 0.15 M NaCl and 0.015 M sodium citrate, pH 7.0), hybridized overnight to peroxidase-labeled probes at 42°C, and subsequently washed and detected using Enhanced Chemiluminescence (ECL; Amersham) according to the manufacturer's instructions. Each sample contains approximately 10μg DNA.

Preliminary linkage maps are shown in Figure 4, based on 52 F2 progeny from a single pair mating. We have provisionally assigned 60 RFLPs to 16 linkage groups based on χ^2 contingency tests; 8 additional markers still appear to be unlinked. Ten markers correspond to known loci on standard genetic maps, as indicated, representing 8 chromosomes. The map distances noted on the figure are recombination values calculated between nearest neighbors, with provisional linkage order determined using 3 or more linked markers. Group 1 is sex-linked; linkage to chromosome 2 was verified by the presence of a dominant larval skin marker (p^3), Thus, we may have identified representative markers for 24 of the 28 silkworm chromosomes; this will have to be tested further statistically (*19*).

Future Prospects

The molecular linkage map represents a major bridge between applied and basic research in the silkworm. It offers the promise of being able to sort out the complex genetics of quantitative traits, which are of central importance to the practice of sericulture, and to enable the design of faster and more effective breeding strategies through the use of marker-assisted selection. It will also make possible the isolation and study of major genes affecting a wide range of fundamental biological processes, not only those encoding quantitative traits, but also the large collections of Mendelian mutations. Finally, the molecular linkage maps will provide a necessary framework for future introduction of foreign or altered genes into the silkworm and application of genetic engineering techniques, once gene transfer technology becomes available. Construction of the map is now a cooperative international effort. We speak for all of those involved in the project in expressing a concern that the fruits of our labors will be disseminated freely and openly via a new Silk Road that extends throughout the silkworm community.

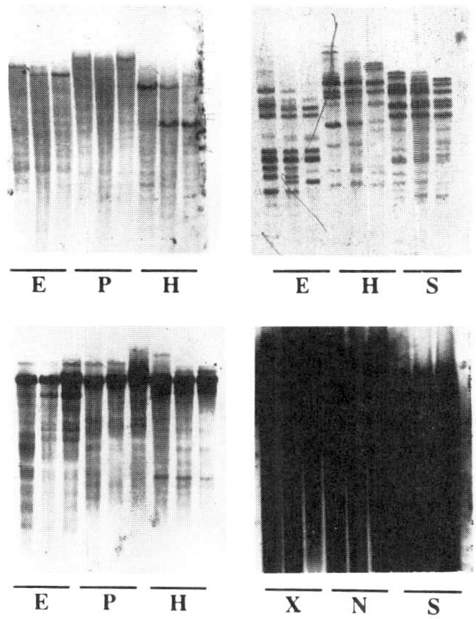

Figure 2. Representative anonymous RFLP patterns in parental and F1 silkworm DNAs. Each set of 3 samples carries DNA from p50 (left), F1 hybrid (middle), and C108 (right) strains. Clone number and sequence types are: upper left, Rcf 213, single copy; upper right, Rcf 207, multicopy (chorion ErA family); lower left, Rcf 45, ribosomal DNA; lower right, Rcf 59, highly repeated. Restriction enzymes: E: Eco RI, P: Pst I, H: Hind III; S: Sac I; N: Nde I; X: Xba I.

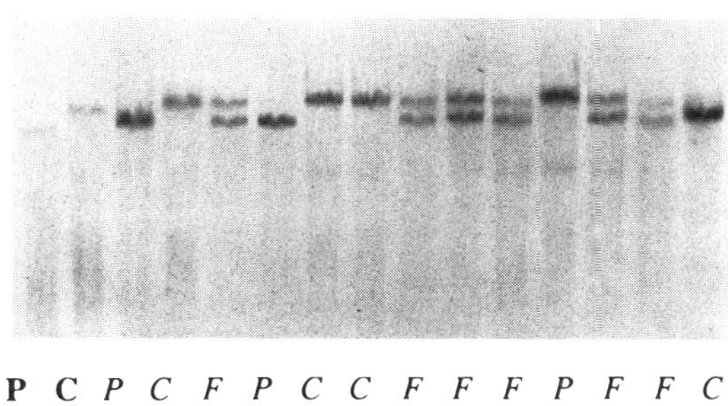

Figure 3. RFLPs for a single copy homeobox-containing gene in F2 progeny from crosses between strains p50 and C108. The first 2 samples are parentals and the remainder are F2 progeny. Samples were digested with Eco RI. P–p50 pattern; C–C108 pattern; F–F1 pattern.

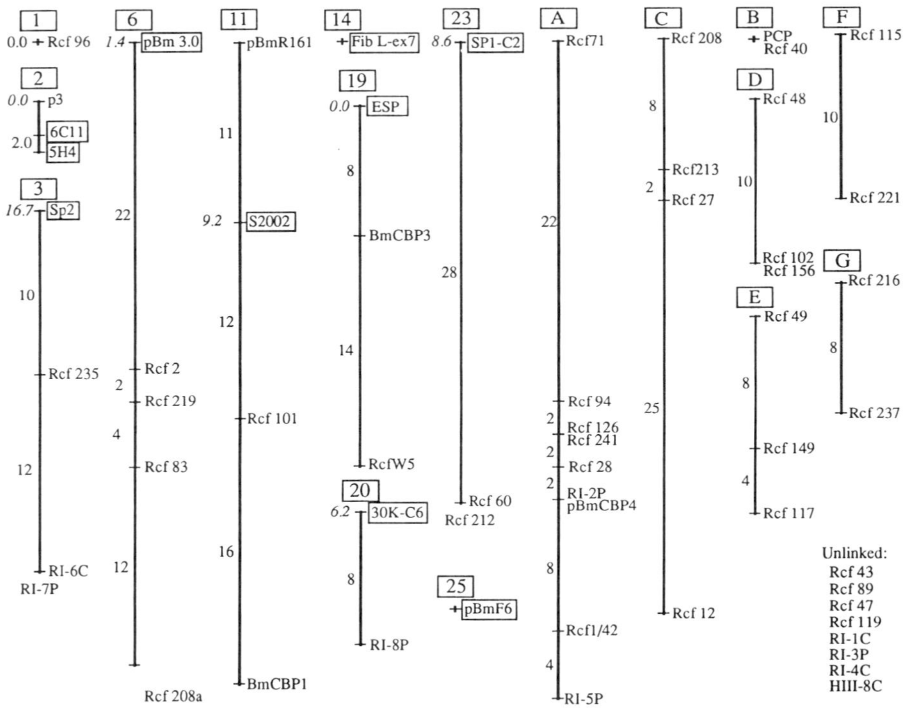

Figure 4. Preliminary linkage arrangements for *B. mori* RFLPs. Boxes indicate cloned genes identified on conventional genetic maps as follows: 6C11, 5H4, *Ch1-2*, *Ch3*; Sp2, *Pst*; pBm 3.0, *Nc*; S2002, *Src-2*; ESP, *Pes*; 30K-C6, *Lp*; Sp1-C2, *Pfl*; FibL-ex7, *Fib-L*; pBmF6, *Fib-H*. Group 1 is sex-linked. Published map locations are indicated in italics.

Acknowledgments

Construction of the molecular linkage maps is a cooperative project between this laboratory and those of T. Tamura (NISES, Tsukuba), M. Kobayashi, T. Shimada (Tokyo University), H. Maekawa (NIH, Tokyo), H. Doira, and H. Fujii (Kyushu University). We are indebted to many laboratories for generously giving us molecular tools and technical advice, and cannot thank them all here. We are especially grateful to N. Spoerel (University Connecticut Medical School, Farmington) for the follicular cDNA library, K. Iatrou and J. Drevet for PCR primers (University of Calgary), and H. Maekawa (National Institute of Health, Tokyo), T. Tamura (NISES, Tsukuba), O. Yamashita, (Nagoya University), S. Tomino (Tokyo Metropolitan University), Y. Suzuki (NIBB, Okazaki), T. Mizuno (Tohoku University), T. Eickbush (University of Rochester), A. Garel, N. Mounier, P. Couble, J.-C. Prudhomme (Université Claude Bernard, Lyon), and N. Spoerel (University Connecticut Medical School, Farmington) for identified cloned sequences. Finally, we must express deep appreciation to B. Sakaguchi, K. Koga (Kyushu University), F. C. Kafatos (Harvard University), Y. Tazima, and Y. Horie (Institute of Silkworm Genetics and Breeding, Amimachi), for their unfailing encouragement and support. This research was supported by a grant by the Carolyn and Kenneth D. Brody Foundation.

Literature Cited

1. Kuhn, D. In *Textile Technology: Spinning and Reeling*; Needham, J., Ed.; Science and Civilisation in China; Cambridge University Press: Cambridge, U. K., 1988, Vol. 5 part IX.
2. Doira, H.; *Genetical Stocks and Mutations of Bombyx mori: Important Genetic Resources*. Kyushu University, Fukuoka, Japan, 1992.
3. Tazima, Y. *The Genetics of the Silkworm*. Logos/Prentice-Hall: Englewood Cliffs, NJ, 1964.
4. Horie, Y.; Watanabe, H. *Ann Rev. Entomol.* **1980**, *25*, 49-71.
5. Goldsmith, M. R.; Wilkins, A. S., Eds.; *Molecular Model Systems in the Lepidoptera*. Cambridge University Press: New York, NY, in press.
6. Yokoyama, T. *Japan Agric. Res. Quart.* **1968**, *4*, 30-33.
7. Yokoyama, T. *Silkworm Genetics Illustrated*. Japan Society for the Promotion of Science: Tokyo, 1959.
8. Asaoka, K.; Mano, Y. *J. Seric. Sci. Jap.* **1992**, *61*, 1-5.
9. Maeda, S. *Ann. Rev. Entomol.* **1989**, *34*, 351-372.
10. *Silk review 1992, A Survey of International Trends in Production and Trade*. ITC UNCTAD/GATT: Geneva, 1992.
11. Balasubramanian, V. *Sericologia* **1988**, *28*, 497-503.
12. Greenhalgh, P. *The World Market for Silk*. Tropical Development and Research Institute: London, 1986.
13. Currie, R. *Sericologia* **1991**, *31*, 103-111.
14. *Sericulture Development in Asia*. RAPA/FAO Publication: Geneva, 1989, Vol. 5.
15. Lander, E. S.; Botstein, D. *Genetics* **1989**, *121*, 185-199.
16. Paterson, A. H.; Lander, E. S.; Hewitt, J. D.; Peterson, S.; Lincoln, S. E.; Tanksley, S. D. *Nature* **1988**, *335*, 721-726.
17. Lande, R.; Thompson, R. *Genetics* **1990**, *124*, 743-756.
18. Gage, L. P. *J. Mol. Biol.*, **1974**, *86*, 97-108.
19. Lander, E. S.; Green, P.; Abrahamson, J.; Barlow, A.; Daly, M. J.; Lincoln, S. E.; Newburg, L. *Genomics* **1987**. *1*, 174-181.
20. Helentjaris, T.; Slocum, M.; Wright, S.; Schaefer, A.; Nienhuis, J. *Theor. Appl. Genet.* **1986**, *72*, 761-69.
21. Welsh, J.; McClelland, M. *Nucl. Acids Res.* **1990**, *18*, 7213-18.

22. Williams, J. G. K.; Kubelik, A. R.; Livak, K. J.; Rafalski, J. A.; Tingey, S. V. *Nucl. Acids Res.* **1990**, *18*, 6531-35.
23. Hearne, C. M.; Ghosh, S.; Todd, J. A. *TIG* **1992**, *8*, 288-294.
24. White, T. J.; Arnheim, N.; Erlich, H. A. *TIG* **1992**, *5*, 185-189.
25. Michaille, J.-J.; Mathavan, S.; Gaillard. J.; Garel, A. *Nucl. Acids Res.* **1990**, *18*, 674.

RECEIVED June 29, 1993

Chapter 5

Importance of Unique Silk Proteins to the Ecological and Evolutionary Diversity of Araneid Spiders

Catherine L. Craig

Department of Biology, Yale University, New Haven, CT 06525

A systematic survey of the spectral properties of unpigmented silks spun by spiders showed marked differences in their reflectance properties. Phylogenetically primitive spiders in the suborder Mygalomorphae spin silks characterized by a spectral peak in the ultraviolet (UV) region of the spectrum (<400nm) while the spectral properties of silks spun by the true spiders (suborder Araneomorphae) are more diverse. In particular, comparison of silks spun by phylogenetically primitive and derived aerial web-weaving spiders show a spectral shift from those characterized by high-UV reflectance to silks that are characterized either by low UV-reflectance or silks that are spectrally flat. Correlated with this shift is a 37-fold increase in number of derived orb spinning spider species and expansion of aerial web weavers into new foraging habitats. This suggests that the molecular properties of spider silks have had a fundamental effect on the ecology of spider foraging behaviors.

One goal of evolutionary biology is to understand the processes by which adaptive novelties arise and contribute to lineage speciation. These processes, acting at the molecular, genomic and organismal levels, influence or direct evolutionary change. Silks proteins are spun by all spiders (*1*), and the consequent roles silks play in spider ecology represent one system where links among evolutionary events at the molecular, genomic and organismal levels can be explored.

Spiders produce silks throughout their life span and use them for protection, reproduction and foraging. The simplest silk producing system is found in the mygalomorph spider *Antrodiatus* (Antrodiaetidae; suborder Mygalomorphae), and is comprised of four functional spinnerets and one type of silk gland that yields two different types of protein (*2*). *Antrodiatus* digs fossorial (underground) burrows for protection and lines them with silks. Silks are also used to encase eggs. In contrast, the most complex silk producing systems are found among spiders in the superfamily Araneoidea (suborder Araneomorphae, the true spiders). The superfamily Araneoidea comprises the derived orb-web weavers and their close relatives. Some araneoid spiders produce up to eight different kinds of silks, and all have three different types of spinnerets (*3*). Even though the araneoids also spin silk retreats and use silks to protect their eggs, unlike almost all other spiders, the araneoids, and their sister taxa the deinopoids, are completely dependent on aerial silk nets to

0097–6156/94/0544–0059$06.00/0

intercept free-flying insects. Thus, it is likely that selection pressures affecting the evolution of silks and foraging behavior of non-web-spinning spiders differ from those affecting the evolution of silk-proteins spun by the aerial web weavers.

Craig, Bernard and Coddington (*4*) explored the diversity of unpigmented silks spun by mygalomorph and araneomorph spiders by comparing the silks' spectral properties. They found three main types of reflectance patterns: silks that were spectrally flat (white), silks that were characterized by a spectral peak in the ultraviolet region of the spectrum (wavelengths less than 400nm) and silks characterized by reduced reflectance in the UV (Fig. 1a; *5*). These differences in silk colors, due to the structural properties of the molecule and not secondary pigments associated with them, indicate that the silk proteins spun by spiders are diverse. Mapping the spectral data on the most recently drawn cladogram of the Araneae (*6,7*) revealed one characteristic of the silks that differed consistently across spider phylogenetic groups: the presence or absence of a spectral peak in the ultraviolet (wavelengths <400nm; Fig. 1b). All of the spiders in the phylogenetically primitive suborder, Mygalomorphae, produce retreat silks characterized by a UV peak. However, the presence or absence of UV-reflectance properties is more varied among silks spun by spiders in the suborder Araneomorphae. The most primitive of the araneomorph spiders, *Hypochilus*, spins UV-reflecting silks, as do the phylogenetically primitive, aerial web-spinning spiders, the Deinopoidea. In contrast, the derived, aerial web-spinning spiders in the families Araneidae, Theridiidae, Tetragnathidae and Theridiosomatidae (superfamily Araneoidea) spin catching silks that are refractive (*8*) and that are either spectrally flat or that are characterized by reduced UV reflectance (*5*). These catching silks are the product of the flagelliform and aggregate glands, silk glands that have evolved only in the araneoids. Nevertheless, the araneoids retain the ability to produce UV-reflecting silks, using them for other purposes. Thus, selection pressures that affect the evolution of silks spun by primitive, non-web-spinning spiders, or even primitive web-spinning spiders, are likely to be different from those affecting the evolution of silk-proteins spun by the aerial web weavers. Three ways in which silk properties have affected the foraging behavior of spiders are through 1) silk mechanical properties and the evolution of energy-absorbing nets, 2) silk reflectance properties and the evolution of web visibility and 3) silk spectral properties and the evolution of sociality.

Silk mechanical properties and the evolution of energy-absorbing nets

Spider orb-webs are dynamic, energy-absorbing nets whose ability to intercept prey is dependent on both web architecture and the material properties of web silks (*9,10*). Prey capture at spider orb-webs is a three-step process: 1) insects must first encounter or come into contact with webs (*11-14*), 2) after the insect and web make contact, the web must intercept the insect or absorb its kinetic energy without breaking (*9*) and 3) webs must retain insects long enough for the spider to reach and subdue them (*15*). Each of these events is dependent on insect flight behavior and mass as well as web architecture and the material properties of the silks. Webs that differ in kinetic properties will intercept different subsets of the prey population available to predatory spiders.

Comparison among webs spun by five, coexisting orb-weaving spiders that differed by an order of magnitude in size showed that orb-webs could be ranked along a continuum of high to low energy-absorbance. The general design of an orb contributes significantly to web energy absorption by distributing insect impact energy throughout the net and contains insect damage to the area of insect interception. High energy-absorbing webs absorb insect kinetic energy primarily by developing tension in the webs' support cables with relatively little web and fiber displacement. These webs are large, suspended under high tension and characterized by a ratio of radii to spiral turns per web that is greater than one (*10*). In contrast, webs that intercept slowly flying or light prey (insects characterized by low kinetic energies) tend to be small, to be built by small spiders, constructed under low tension

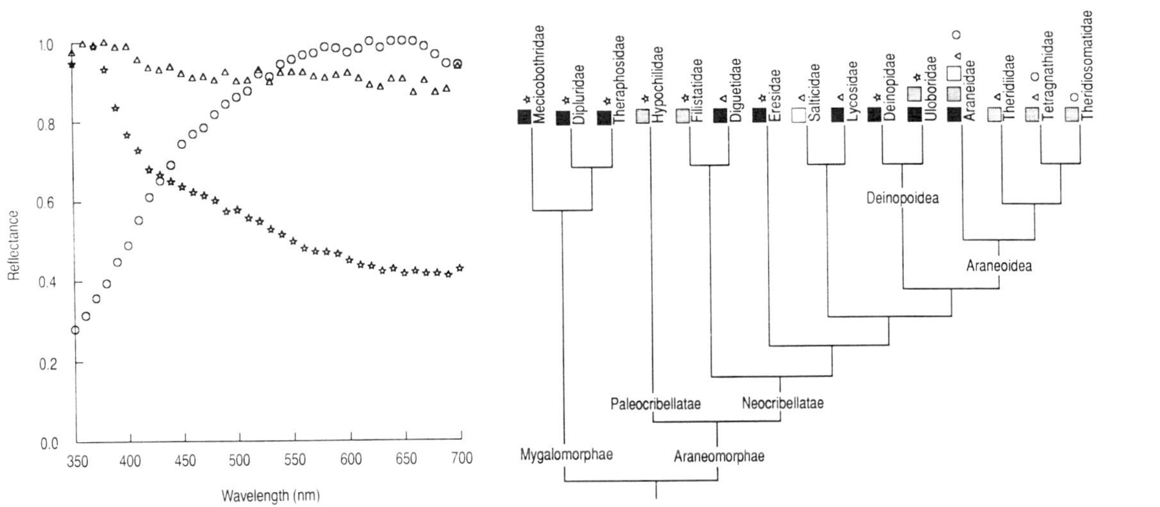

Figure 1. Left: Three types of reflectance patterns of silks spun by spiders have been observed: spectrally flat (triangles), reflecting 70–100% of light across all wavelengths; spectra with an ultraviolet (UV) peak (stars) where 30% more light is reflected across a region of about 30 nm than adjacent regions; and spectra with a gap in the UV (circles). Right: The distributions of these types of silks have been mapped onto a cladogram of the Araneae, as have the light environments in which each group of spiders forage. Spiders that forage in nocturnal light environments (filled squares) and spiders that forage in dim, diurnal light environments (hatched squares) are found among all of the phylogenetic groups studies. Web-spinning spiders that forage in bright, diurnal environments (open squares), however, are only found among the derived aerial web weavers in the superfamily Araneoidea. Cladistic analysis of the distribution of types of silks spun by spiders across the phylogenetic groups show that: both primitive and derived spiders spin silks that are spectrally flat; the mygalomorph spiders, the most primitive of the true spiders, the primitive orb web weavers, *Hypochilidae,* and *Deinopoidea,* all produce silks characterized by a spectral peak in the UV region. None of the derived orb-web weavers spin silks that selectively reflect ultraviolet light. Moreover, only the derived orb-web weavers, the Araneoidea, produce silks characterized by a gap in the ultraviolet or where reflectance is weak. (Reproduced with permission from reference 5. Copyright 1992.)

and characterized by a ratio of radii to spiral turns per web that is less than one. Low energy-absorbing webs absorb insect kinetic energy primarily by web and fiber displacement and stretch (*10*). For spiders that spin low-energy-absorbing webs, the orb design does not contribute greatly to web function.

The significance of these results lies in the apparent absence of a selective advantage to the orb architecture for low energy-absorbing webs and correlated evolutionary trend to small spiders that build webs of diverse architectural designs (*16*). Silks spun into the orb design allow spiders to intercept large and fast-flying prey whose kinetic energy is a least an order of magnitude higher than the energy-absorbing properties of their silks alone. However, the high strength and elasticity of araneoid silks alone allow the derived spiders to spin nets that intercept and absorb the kinetic energy of light or slowly flying prey, regardless of how the silks are arranged (*10*). Thus, at least with respect to web energy absorption, the evolution of the araneoid silks seems to have "released" (*17*) the derived araneoids from the orb-web constraint, thus allowing them to spin webs of diverse architectural design. One result of this apparent release is that derived spiders are able to forage in sites that can not support an orb-web but from which silks spun into irregular structures can be suspended. The evolution of the diverse architectural design of nets spun by the derived species is probably a consequence of the evolution of the unique silks spun by the araneoid spiders (*16*).

Silk reflectance properties and the evolution of web visibility

Affecting all interactions between insects and web-spinning spiders is the ability of insects to see and avoid spider webs. Spiders forage in a variety of light environments, including nocturnal, diurnal, forest shade and bright, non-forest habitats, and in all of these environments insects can see webs and avoid them (*8,18*). Web visibility is a function of the contrast between the web and its background. Silk colors that match background vegetation can effectively camouflage the web from approaching prey. However, webs spun from silks that contrast with background vegetation can easily be seen. Because insects use both color and pattern as cues to resource locations, spiders that spin high contrast silks into patterns and colors that are similar to those of resources for which insects are searching may be able to attract prey to the web area.

Insect attraction and wavelength-dependent behaviors Many insect behaviors are wavelength-dependent, meaning that a narrowly defined band of light can elicit a specific behavior. Because wavelength-dependent behaviors represent a set of fixed responses that cannot be modified by insect experience or learning, silks that reflect these wavelengths may draw insects to them. One such insect behavior is the open space response that has been found among all groups of insects tested so far. When insects are flying from closed to open space, such as away from vegetation, they are attracted to ultraviolet light because the sky is the only natural source of UV. The phylogenetically primitive orb-web weaving spiders, the deinopoids, spin silks that reflect UV light as well as light in the blue region of the spectrum (*19*). These spiders forage in vegetation, the forest understory and at night. Insects flying away from enclosed sites towards open spaces may be drawn to UV-blue light gaps in which deinopoid webs are suspended.

The potential for UV-reflecting webs to attract and intercept insects flying away from an enclosed space was tested in a series of laboratory experiments. *Drosophila* were introduced into the baseleg of a Y-shaped choice chamber and allowed to fly to either branch leg of the chamber. Webs spun from UV-reflecting silks by *Uloborus glomosus* were fixed at both of the branch legs of the chamber and illuminated. One web, however, was illuminated with white light containing a UV-component and the second web was illuminated with white light from which the UV-component had been filtered. In ten out of ten trials the majority of the flies flew to the web that reflected UV light *(20)*. Craig and Lesch (unpub. data) explored this result further by video-taping *Drosophila* (fruit flies) as they approached similarly illuminated webs in a flight chamber. When *Drosophila* approached webs spun by *Uloborus glomosus* illuminated with white light from which all UV wavelengths had been filtered, the webs were easily avoided. Nevertheless, when flies approached webs illuminated with white light containing an ultraviolet component, *Drosophila* were intercepted. Thus, *Drosophila* were attracted to webs that reflected UV light and were unable to avoid interception.

Insect attraction to decorative silk patterns Unlike primitive orb weavers, some of the phylogenetically derived orb-weaving spiders spin web catching silks that do not reflect ultraviolet or blue light but instead are characterized by reduced reflectance in the UV or are spectrally flat. These spiders spin webs in diurnal habitats where light levels are high and webs are fully illuminated throughout the spiders' foraging period. Nevertheless, the derived spiders retain the ability to produce UV-reflecting silks and use them to decorate their webs in a variety of silk patterns. The effect of silk decorative patterns on insect attraction and interception at webs was tested in a series of field experiments.

Spiders in the genus *Argiope* spin a spiral thread that is characterized by reduced reflectance in the UV region but decorate their webs with UV-reflecting silks spun into a variety of designs. By manipulating the presence of these silks, as well as the presence of the spider (spider body surfaces are also UV-reflecting), Craig and Bernard (1990) showed that insects are attracted to decorated webs, the most numerous prey being stingless bees (*21*; Craig and Ebbert, unpublished). Photographs of webs and grass flowers taken first with a filter that excluded UV light and then with a filter that excluded visible light showed that web decorations and grass flower heads are similar in orientation and design as well as color and contrast with background vegetation (Fig. 2). Because web decorations attract bees that gather pollen from the grass flowers amongst which the spiders forage, it would seem that *Argiope* has evolved to forage specifically for insects searching for floral resources.

Silk spectral properties and the evolution of sociality among spiders

The effects of UV-reflecting silks on prey attraction may also have had an important effect on the evolution of spider social systems. The social systems of web-spinning spiders have been classified as 1) solitary where spiders spin non-overlapping webs and behave aggressively towards one-another, 2) communal-territorial where spiders spin and defend individual webs connected to one another via web support threads and 3) communal non-territorial where spiders spin a common web, feed on prey communally and show no aggression to other group members (*22,23*). There are no cases of obligate communal foraging and feeding among the araneoid spiders. Nevertheless, normally aggressive orb spinners that aggregate at sites where prey activity is high will tolerate the presence of conspecifics (*24-27*). This type of behavior is thought to foreshadow the evolution of coloniality among orb weavers (*24,26-32*). However, coloniality or group foraging among arthropods may also evolve under conditions when resources are scarce or unpredictable (*33*). In this case, animals that forage in groups are, presumably, able to capture more prey than individuals that forage alone.

With the exception of spiders in the genera *Anelosimus*(*28,32*) and *Agelena* (Agelenidae) (*34,35*), true communal foraging and feeding has evolved only among spiders found to produce UV-reflecting silks (*22,23*). The physical conditions under

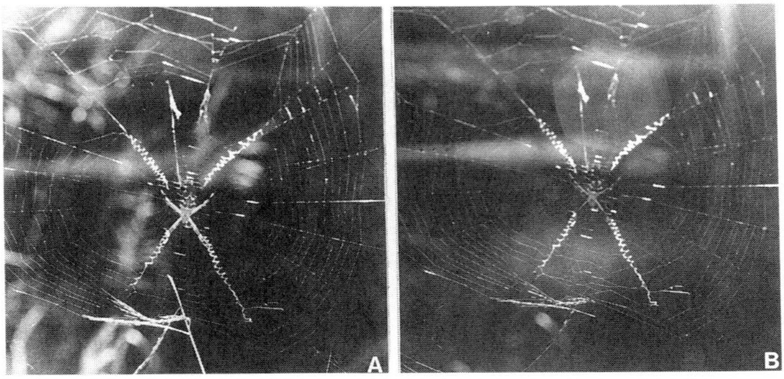

Figure 2. Spiders and web decorations photographed in visible light only or UV light only. Photograph A was taken with a long-pass filter (Corning CS 3-74) that eliminated wavelengths <410nm. Photograph B was taken in UV light with a short-pass filter (Schott UG-1) that eliminated wavelengths >400nm. Development and printing times for the pictures were identical. Photograph A shows that background vegetation that appears in visible light, is dark in the UV or completely disappears. However, the grass flower head apparent in A shows high contrast in the UV. Prey wrapping silks appears bright in both UV and visible light as do the web's decorations. (Adapted from ref. *20*)

which these spiders forage have not been quantified. Nevertheless, studies on communal-territoral (case 2) spiders living in an arid environment show that communally foraging spiders in the deinopoid superfamily capture more insects than solitary foragers (*36*). Spiders that spin UV-reflecting silks are able to lure more prey to them when foraging in a group than when foraging alone. This hypothesis was tested in a series of experiments in which the spectral properties of webs spun by solitary spiders and spiders foraging in groups were manipulated to determine the effects of UV-reflecting visual displays, non-UV-reflecting webs and group size on spider foraging performance in a spatially heterogeneous habitat (*37*).

Argiope argentata spins webs from different types of silks and forages alone as well as in aggregations of two or more individuals. The viscid silks spun by *A. argentata* are characterized by reduced reflectance in the UV (Fig. 1b) and appear blue-green to approaching insects (*38,39*). They thus show low contrast with background vegetation. *Argiope*, however, decorates its web (see above) with UV-reflecting silk designs that show high contrast with background vegetation. By manipulating the presence of the spiders and web decorations, as well as the size of the spiders' foraging groups, Craig (*37*) dissected the effects of UV-reflecting visual displays (including both decorative silks and the spiders' body surfaces) from the effects of group size on the rate that insects were intercepted.

Spiders foraging in groups of two or more individuals captured more prey than spiders foraging alone. In addition, when the interception rates of prey were compared within the groups, spiders that had spun web decorations captured more prey than spiders that had not. When spiders and web decorations were removed from the webs (the remaining portions of the web appearing blue-green to the insect eye), the rate of insect interception dropped. Furthermore, no differences in prey interception rates could be found between webs in clusters and webs spun by solitary individuals.

Silk proteins and the evolutionary diversity of aerial web-spinning spiders

It is may not be surprising that the capture silks spun by the araneoid spiders are so different from those spun by primitive web weavers as they are produced from silk glands not found in other spiders. However, these silks have had an immense influence on the evolution and ecology of the derived aerial web weaving spiders and resulted in:

- an expansion of derived spiders into new foraging environments
- an evolutionary trend to small size and subsequent evolution of webs characterized by diverse architectural designs
- a 37-fold increase in spider species number when compared to the primitive web spinners.

These correlations suggest that the evolution of the "new" silks have resulted in a complex of adaptations that opened new ways of life for aerial web-spinning spiders allowing them to meet the diverse ecological opportunities available to predators of insects.

Literature Cited

(1) Shear, W. A., Palmer, J.M., Coddington, J.A. Bonamo, J.A. *Science* **1989**, *246*, 479.
(2) Palmer, J. M., Coyle, F.A., Harrison, F.W. *Journal of Morphology* **1982**, *174*, 269.
(3) Kovoor, J. In *Ecophysiolgoy of Spiders*; W. Nentwig, Ed.; Springer-Verlag: **1987**; pp 160.
(4) Craig, C. L., Bernard, G.D., Coddington, J.A. (unpublished).
(5) Craig, C. L. *Trends Ecol. Evol.* **1992**, *7*, 270.
(6) Harvey, P. H., Pagel, M.D. *The Comparative Method in Evolutionary Biology* , Oxford University Press: New York, **1991.**
(7) Coddington, J. A.; Levi, H. W. *Ann. Rev. Ecol. Syst.* **1991**, *22*, 565.
(8) Craig, C. L., Freeman, C.F. *Behav. Ecol. Sociobiol.* **1990**, *29*, 249.
(9) Denny, M. W. *Journal of Experimental Biology* **1976**, *65*, 483.
(10) Craig, C. L. *Biol. J. Linn. Soc.* **1987**, *30*, 135.
(11) Rypstra, A. *Oecologia* **1982**, *52*, 31.
(12) Nentwig, W. *Oecologia (Berlin)* **1983**, *58*, 418.
(13) Craig, C. L. *Anim. Behav.* **1986**, *34*, 54.
(14) Chacon, P., Eberhard, W.G. *British Arachnological Society Bulletin* **1980**, *5*, 29.
(15) Nentwig, W. *Oecologia* **1982**, *53*, 412.
(16) Craig, C. L. *Am. Nat.* **1987**, *129*, 47.
(17) Hardy, A. C. In *Evolution as a Process*; J. Huxle Ford, E.B., Ed.; George Allen and Unwin: London, **1954**.
(18) Craig, C. L. *Functional Ecology* **1988**, *2*, 277.
(19) Craig, C. L.; Ebbert, K., ms.
(20) Craig, C. L., Bernard, G.D. *Ecology* **1990**, *71*, 616.
(21) Robinson, M. H.; Robinson, B. *Zool. J. Linn. Soc.* **1970**, *49*, 345.
(22) Jackson, R. R. *Insectes Socieux* **1978**, *26*, 300.
(23) Tietjen, W. J. In *Spiders: webs, behavior and evolution*; W. A. Shear, Ed.; Stanford University Press: Stanford, **1986**.

(24) Uetz, G. W., Kane, T.C., Stratton, G.E. *Science* **1982,** *217,* 547.
(25) Uetz, G. W. In *Spiders: webs behavior and evolution*; W. A. Shear, Ed.; Stanford University Press: Stanford, **1986**; pp 207.
(26) Rypstra, A. L. *Journal of Arachnology* **1985**, *13*, 71.
(27) Gillespie, R. G. *Animal Behaviour* **1987**, *35*, 675.
(28) Shear, W. A. *Bulletin of the British Arachnological Society* **1970**, *3*, 65.
(29) Lubin, Y. D. *Zoological Journal of the Linnean Society* **1974**, *54*, 321.
(30) Brach, V. *Evolution* **1977**, *31*, 154.
(31) Burgess, J. W. *Symposium of the Zoological Society of London* **1978**, *42*, 69.
(32) Buskirk, R. W. In *Social Insects*; H. H. Herman, Ed.; Academic Press: New York, **1981**.
(33) Wilson, E. O. *The Insect Societies*; Harvard University Press: Cambridge, **1971**.
(34) Darchen, R. *Revue Arachnologie* **1979**, *2*, 156.
(35) Reichert, S. E., Roeloffs, Echternacht, R.A.C. *Journal of Arachnology* **1986**, *14*, 175.
(36) Smith, D. T. *Behavioral Ecology and Sociobiology* **1982**, *11*, 149.
(37) Craig, C. L. *Functional Ecology* **1991**, *5*, 649.
(38) Menzel, R., Erber, J. *Sci. Am.* **1978**, *239*, 102.
(39) Menzel, R. In *Experimental behavioural ecology*; B. H. a. M. Lindauer, Ed.; Fortschritte der Zoologie, BD. 31. G. Fischer Verlag: **1985**; pp 55.

RECEIVED May 4, 1993

Chapter 6

Initial Characterization of *Nephila clavipes* Dragline Protein

C. M. Mello[1], K. Senecal[1], B. Yeung[2], P. Vouros[2], and David Kaplan[1]

[1]Biotechnology Division, Natick Research, Development, and Engineering Center, U.S. Army, Natick, MA 01760

[2]Department of Chemistry, Northeastern University, Boston, MA 02115

The molecular composition of the dragline fiber and the major ampullate gland protein from the golden orb-weaving spider, *Nephila clavipes* is not well defined. We report initial characterization of the major ampullate gland proteins and the dragline fiber including amino acid composition, SDS gel electrophoresis, peptide sequencing and N-terminal sequencing. Preliminary results have identified the presence of disulfide bonds within the dragline protein. Previously published cDNA sequence data of the spidroin I protein from the major ampullate gland of *N. clavipes* (*1*) corresponds with our sequence data presented below.

Silk is produced by a variety of insects and spiders. The most extensively studied silk is obtained from the larvae of *Bombyx mori*. This domestic silkworm produces one type of silk, cocoon silk, at a specific stage in its lifecycle, during the fifth larval instar just before molting. Modified salivary glands are responsible for producing the silk which is spun from the mouth. Silkworm silk is predominantly a proteinaceous fiber which contains two primary structural proteins or fibroins, a 325 kDa and a 25 kDa fibroin present in equimolar ratios, and a family of glue-like sericin proteins (*2,3*).

In contrast to the silkworm, the spider produces many different types of silk, some of which are available throughout their lifecycle. Each type of spider silk originates from a different set of glands. The major ampullate gland produces the structural silks for the orb frame, radii and dragline; the flagelliform

0097–6156/94/0544–0067$06.00/0

gland produces the viscid silk for prey capture; the aggregate gland synthesizes an adhesive silk; the minor ampullate gland is responsible for the support fibers of the orb web; the cylindrical gland produces the cocoon silk; the aciniform gland produces the silk for wrapping captured prey; and the piriform gland produces attachment silks to couple to environmental substrates (*4*). Thus, spider silks represent an evolutionarily tailored set of proteins to meet specific functions. The mechanical properties of silks and details of their secondary and tertiary molecular arrangements can help us resolve longstanding questions concerning the structure/function relationships of repetitive structural proteins.

The work presented here will focus on the major ampullate gland of the spider, *N. clavipes,* and its dragline protein fiber. The major ampullate gland contains three distinct regions, the posterior region where silk is synthesized in the lining of the gland by epithelial cells, the middle region for the storage of protein, and the anterior region which leads to the spinneret where the protein is spun into a fiber. Epithelial cells lining the posterior region of the ampullate gland of the spider, *Araneus sericatus*, have a radial microtubule system which provides for intracellular transport and secretion of fibroin in the form of secretory granules (*5*). Electron microscopy indicates that the microtubules run from the basal to luminar membranes in the cytoplasm and silk "globules" have been observed associated with these microtubules. The polypeptide chain associations, concentrations and conformations in these granules are unknown.

There is little data published on the molecular characterization of the major ampullate gland proteins. Candelas and her co-workers used radiolabelled glycine and alanine to demonstrated that the major ampullate gland of *N. clavipes* produces a single protein when stimulated by mechanical depletion of the silk gland. This protein was shown to migrate as a homogeneous band in SDS gel electrophoresis with an estimated molecular weight of 320 kDa (*6*). However. recent cDNA sequence data from the major ampullate gland of *N. clavipes* suggests that there may be two different proteins comprising the dragline fiber which are designated as spidroin I and spidroin II (*1,7*). Additional characterization of the proteins present in the major ampullate gland and in the dragline silk fiber is necessary to determine the number of different silk proteins comprising the dragline fiber. Some of these experiments are reported here and include amino acid composition, SDS gel electrophoresis, peptide sequencing of the dragline fiber and N-terminal sequencing.

Materials & Methods.

Spiders. Spiders, *N. clavipes*, were collected in Florida and Panama and maintained in the laboratory.

Recovery of Dragline Silk Fibers and Major Ampullate Glands. Collection of the dragline fiber by controlled silking of the spiders was performed according to Work and Emerson (*8*). Dissections were carried out in sterile SSC buffer (0.1M sodium chloride and 0.015M sodium citrate, pH 6.4) containing 5mM phenylmethylsulfonyl fluoride (PMSF) and the major ampullate glands were placed in liquid nitrogen immediately following excision and stored at -70°C.

Solubilization of the Dragline Fiber. Dragline fiber was solubilized in 9M LiBr and incubated at 37°C overnight with shaking. Occasionally an additional 8-12 hour incubation was necessary to completely solubilize the fiber. The fiber was not readily soluble in 8M urea, however, LiBr solubilized fibers can be placed into an 8M urea, 200mM acetate, pH 4.0 buffer by ultrafiltration without sacrificing the solubility. This buffer system was used for gel electrophoresis and the determination of protein concentration. Protein concentrations were determined with the bicinchoninic acid protein assay kit from Sigma.

Chemical Cleavage of the Dragline Fiber with N-Bromosuccinimide. Dragline protein fiber was solubilized in 85% formic acid. A 10-fold excess (by weight) of N-bromosuccinimide (NBS) was added immediately following solubilization. The reaction mixture was left on ice for 1.5 hours, followed by room temperature incubation for 1.5 hours to increase the rate of the reaction. An equal volume of dry acetone was added to quench the reaction and precipitate the peptides. Peptides were pelleted by centrifugation at 16,000 x g for 1 hour. The supernatant was removed, and the pellet was dissolved in initial mobile phase used in the HPLC separation (see below). Most of the pelleted material remained soluble in the mobile phase. Any insoluble material was removed by centrifugation and filtration.

Peptide Mapping by HPLC. A Waters 625LC system equipped with a photodiode array detector and a reverse-phase C-4, YMC column (250 x 4.6 mm) was used for the separation of dragline silk peptides. The solvent system consisted of A=water, B= acetonitrile, and C=100mM potassium phosphate, pH 3.0, where solvent C was kept at 20% to maintain a constant concentration of 20mM potassium phosphate buffer in the mobile phase. A 45 minute linear gradient from 78% A to 50% A separated the peptides. Peaks were collected from consecutive runs and pooled for amino acid analysis and N-terminal sequencing. Synthetic peptides, Gly$(Ala)_4$ and $(Ala)_6$·HCl, used as standards were purchased from Sigma.

Gel Electrophoresis. An equal volume of a 2X sample buffer (0.125M Tris-HCl, pH 6.8, 20% glycerol, 4% SDS, 0.005% bromophenol blue) was added to each sample containing 25-50μg of protein in 8M urea, 200mM acetate pH 4.0. For reducing conditions, the 2X sample buffer also contained 4% 2-mercaptoethanol. All samples were incubated in sample buffer for at least 1 hour at room temperature followed by exposure to a boiling water bath for 5 minutes. A Novex precast discontinuous SDS polyacrylamide gel, based on the classical Tris-glycine buffer system of Laemmli was used (*9*). Separations were carried out at a constant voltage (125 volts) until the bromophenol blue was 0.5-1.0 cm from the bottom of the gel. Gels were stained with 0.1% coomassie blue R-250 in 40% methanol and 10% acetic acid and destained with a 10% methanol, 7.5% acetic acid solution.

Electroblotting. SDS-polyacrylamide gels of dragline and major ampullate gland proteins were electroblotted onto polyvinylidene fluoride (PVDF) membranes for amino acid analysis and N-terminal sequencing. Following electrophoresis the gel was washed with transfer buffer (10mM 3-cyclohexylamino-propanesulfonic acid, pH 11.0 (CAPS), 5% methanol, and 0.005% SDS) for 30 minutes before blotting. Electrophoretic transfer was achieved at a constant current of 200 mA for 4 hours. The membrane was air dried and washed overnight in Milli-Q water to reduce background interference.

Amino Acid Analysis. The Waters Pico-Tag system was used for amino acid analysis. Briefly, the proteins were dried under vacuum, hydrolyzed in constant boiling HCl containing 0.5% phenol, neutralized to remove any residual acid and coupled to phenylisothiocyanate (PITC) forming phenylthiocarbamyl (PTC) amino acid derivatives. PTC-amino acids were separated on a 15 cm Pico-Tag reverse phase column using a linear gradient of sodium acetate buffer (containing 0.05% triethylamine (v/v) and 0.2μg/ml of EDTA) and 60% acetonitrile (containing 0.2μg/ml of EDTA). All data were collected and integrated by VAXLAB and the Waters Expertease 860 software.

N-terminal Sequencing. Automated Edman degradation of all proteins and peptides was performed on an Applied Biosystems, Inc. Model 477A peptide sequencer equipped with an on-line PTH amino acid analyzer at the Cornell

University, Biotechnology Analytical/Synthesis Facility by Ted Thannhauser.

Results & Discussions.

Solubilization Procedures. Both dragline fiber and major ampullate gland protein were solubilized in 9M LiBr as described earlier. These samples were transferred to 8M urea, 200mM acetate, pH 4.0 by ultrafiltration and analyzed by SDS-PAGE. Both samples are virtually identical, migrating at approximately 275 kDa. Gel electrophoresis and amino acid composition (see Table I) of electroblotted bands from the solubilized fiber and extracted gland protein suggests that they are indistinguishable. These results also suggests that complete solubilization of all protein components within the fiber has been achieved due to the similarity of the amino acid compositions of the fiber before solubilization and the 275 kDa electroblotted band. The estimated molecular weight of 275 kDa is lower than previously published results of 320 kDa for the glandular protein (*6*). However, the molecular weight markers used by Candelas *et al*. were *B. mori* silk fibroin (estimated at 400 KDa) and a myosin subunit (220 kDa) and the exact molecular weight has not been accurately determined for the silk fibroin of *B. mori*. The use of different molecular weight markers and slightly different electrophoretic conditions may account for the variation in molecular weight determinations. In addition, it should be recognized that proteins containing significant percentages of glycine and alanine have been reported to migrate anomalously during electrophoresis, usually resulting in estimates higher than the actual molecular weight (*10*).

N-Terminal Sequencing. Major ampullate gland protein was electroblotted onto PVDF membranes and N-terminal sequencing was done. It was often difficult to sequence more than ten or twelve residues into the protein because of a continual increase in the alanine and glycine peaks from cycle to cycle which was attributed to non-specific cleavage of the protein. Figure 1 reveals all of the sequence data obtained thus far from N-terminal sequence of major ampullate gland protein. It is predicted that these sequences are not N-terminal sequences, but non-specific cleavage products because the N-terminus is blocked (personal communications with Ted Thannhauser). Non-specific cleavage may have occurred during the isolation and electrophoretic transfer of the proteins or during Edman degradation due to acid or base hydrolysis. A repetitive motif has been identified and is reflected in sequence 1 of Figure 1. This repeat appears with 100% homology fourteen times in

Table I. Amino Acid Composition of Dragline Fiber and Major Ampullate Gland Protein[a]

Amino Acid	Dragline Fiber[1]	Gold Dragline Fiber	Electroblotted Dragline Fiber	Electroblotted Major Ampullate Gland
GLX	9.00	10.55	10.30	8.52
SER	6.90	2.58	3.60	3.51
GLY	41.50	44.57	42.20	41.66
ARG	2.00	2.13	2.38	1.28
ALA	27.00	27.19	27.52	25.25
PRO	1.10	1.95	1.66	0.78
TYR	2.70	3.41	3.68	4.20
LEU	2.00	4.19	4.52	4.82

[a] Amino acid composition was determined by the Waters Pico-Tag system as described in materials & methods. Only the most abundant amino acids are represented in this table. The gold dragline fiber was directly acid hydrolyzed to ensure a complete analysis. Solubilized samples of both the dragline fiber and major ampullate gland protein were separated on an SDS polyacrylamide gel and electroblotted onto PVDF membranes for amino acid analysis. All of the samples are very similar suggesting that the solubilized major ampullate gland protein and the dragline fiber are comprised of the same protein(s). [1] Reference (*11*).

spidroin I and in most cases, immediately preceding the repetitive motif identified below with the dragline peptide sequences. Sequence 3 in Figure 1 is similar to the repetitive motif of the dragline peptide sequences presented below and sequence 4 has not been identified in either spidroin I or II.

Reduction and Initial Protein Modification Experiments. Figure 2 depicts the effects of 2-mercaptoethanol on the fiber and gland protein. When the solubilized proteins are reduced, two major bands occur with an apparent molecular weight of 195 and 220 kDa. The dragline fiber in the reduced state produces an additional band that is not present in the gland preparation at a molecular weight of 180 kDa. Amino acid analysis of the reduced bands of the major ampullate gland protein presented in Table II indicates that the two most prominent bands are very similar. Unfortunately, the dragline protein would not transfer well enough to achieve adequate amino acid composition of the reduced bands. As a result, modified electroblotting protocols and alternative methods are being pursued for the determination of the amino acid composition and contribution of cysteine/cystine residues in the primary and secondary structures of the dragline protein(s).

1. Ala-Ala-Gly-Gly-Ala-Gly-Gln-Gly-Gly

2. Ala-Ala-Ala-Ala-Gly-Ala-Gln-Gly-Gly-Tyr-Gly-Ala/Gly-Leu-Ala-Ala-Ala/Gly-Ala

3. Gly- ? -Gln-Gly-Ala-Gly/Ala-Arg

4. Thr-Thr/Gly-Gln-Ala/Gly-Ala-Ala-Ala-Gly-Ala-Gly/Ala

Figure 1. N-terminal Sequencing of Major Ampullate Gland Protein. Sequences 1 and 2 represent a repetitive region (underlined) which occurs within the spidroin I sequence. A second repeat also in the spidroin I sequence is reflected in sequence 3 and has been found in peptides generated from the dragline fiber (see Figure 5). Sequence 4 does not occur in any of the published data and therfore must represent a portion of the protein which is currently unidentified.

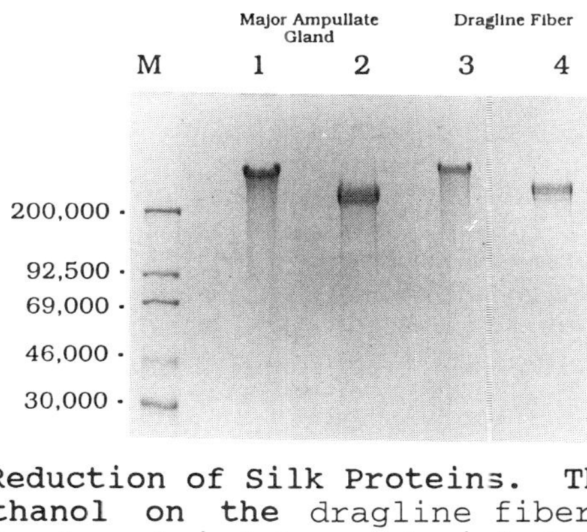

Figure 2. Reduction of Silk Proteins. The effects of 2-mercaptoethanol on the dragline fiber and major ampullate gland protein(s) are depicted above. Novex molecular weight markers are used as standards. Lanes 1 and 2 represent the major ampullate gland proteins in the non-reduced and reduced states, respectively. Lanes 3 and 4 contain the solubilized dragline fiber in the non-reduced and reduced forms. Two major bands occur with an apparent molecular weight of 195 and 220 kDa for both the fiber and gland protein(s) when reduced. The dragline fiber in the reduced state produces a band that is not present in the gland preparation at a molecular weight of 180 kDa.

Table II. Amino Acid Composition of Reduced Major Ampullate Gland Protein

Amino Acid	Major Ampullate Gland Protein	220 kDa Reduced Band	195 kDa Reduced Band
GLX	8.52	9.77	9.35
SER	3.51	2.57	2.79
GLY	41.66	45.88	44.80
ARG	1.28	1.98	2.23
ALA	25.25	28.57	28.35
PRO	0.78	0.37	0.51
TYR	4.20	3.25	3.26
LEU	4.82	4.62	4.48

Initial experiments are currently underway to quantitatively determine the content of cysteine and cystine residues in each of the individual reduced and non-reduced species. Oxidation with performic acid followed by amino acid analysis revealed approximately 0.36% cysteine in the major ampullate gland protein(s). These results indicate 7-9 possible cysteine residues and 3-5 disulfide bonds per protein chain if it is assumed that the molecular weight is 180-220 kDa. Cysteine has not been found in any of the published sequence data (*1,7*) or in our protein sequence data presented here. However, the spidroin I and spidroin II cDNAs identified by Lewis *et al.* do not represent complete sequences. Approximately 28-34% of the expected 180-220 kDa full length spidroin I protein is reflected in the cDNA sequence. A quantitative contribution of the spidroin II cDNA sequence can not be estimated because it's full length protein product has not been identified and therefore an estimate of its molecular weight is unknown. Specific protein modifications, such as, cyanylation with 2-nitro-5-thiocyano-benzoic acid to selectively cleave the protein at cysteine residues may help us to map the positions of cysteine within the dragline protein fiber.

There is evidence for a single disulfide linkage between the small and large fibroin peptide chains in silkworm cocoon silk (*12*). It has been suggested that this disulfide bond, along with hydrophobic interactions, may serve to prevent premature precipitation of the protein in the gland by interfering with the conformational transition which is thought to occur during fiber formation (*13*). Thus, these intra and interchain interactions play a critical role in maintaining solubility of the protein during synthesis, export, transport and secretion occurring before the final spinning process. These interactions are probably very important for solubility of the spider silk proteins as well. Therefore, further analysis is necessary and currently underway in our laboratory to define the potential role of disulfide linkages in these processes.

Chemical Digest. Peptides of the dragline fiber were generated by chemical digestion with N-bromosuccinimide (NBS). Figure 3 represents a 16% gel of the digested material. These peptides were separated by reverse phase HPLC and fractions were collected for amino acid analysis and N-terminal sequencing. A typical peptide map can be seen in Figure 4. Amino acid composition of a representative set of purified peptides is presented in Table III. All the peptides have very similar amino acid compositions however, the alanine and glycine content is low compared to that of the dragline fiber. Based on the amino acid composition data it appears that the alanine and glycine rich sequences are being lost on the column or during desalting procedures prior to picotag analysis. The sample (NBS peptide mixture in Table III) that is injected onto the column closely resembles the amino acid composition of dragline. Polyalanine $(Ala)_6 \cdot HCl$ and $Gly(Ala)_4$ were

Table III. Amino Acid Composition of HPLC Purified Dragline Peptides

Amino Acid	NBS Peptide Mixture	A2	A10	A23
GLX	8.60	7.92	9.65	10.71
SER	3.55	7.42	9.98	10.20
GLY	51.20	45.09	35.66	30.61
ARG	2.29	9.85	7.65	11.81
ALA	21.99	3.37	4.01	3.92
PRO	1.24	2.08	2.16	2.30
TYR	1.14	2.49	3.22	3.17
LEU	4.02	4.35	5.50	5.51

used as controls to determine if these peptides would bind irreversibly to the HPLC column. The $(Ala)_6 \cdot HCl$ was eluted from the column while the $Gly(Ala)_4$ was retained. These results indicate that there may be some irreversible binding of glycine and alanine rich sequences to the column under the gradient conditions used. Further analysis will proceed varying mobile phase conditions in an attempt to achieve recovery of all peptides present in the initial NBS-digested mixture.

Sequence data from three peptides is presented in Figure 5. All sequence data is very similar revealing a

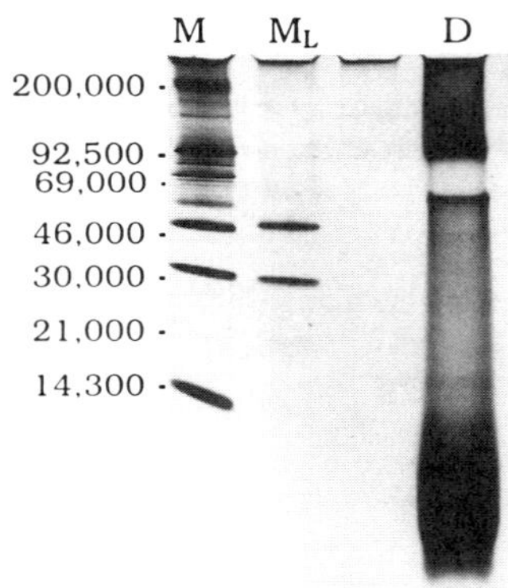

Figure 3. NBS Digest of the Dragline Fiber. Peptides generated by chemical digestion with N-bromosuccinimide are presented above in a 16% SDS polyacrylamide gel. Novex molecular weight markers were used as standards in lanes M and M_L. The digested fiber can be seen in the lane labelled D. Many peptides have been produced representing a broad molecular weight range.

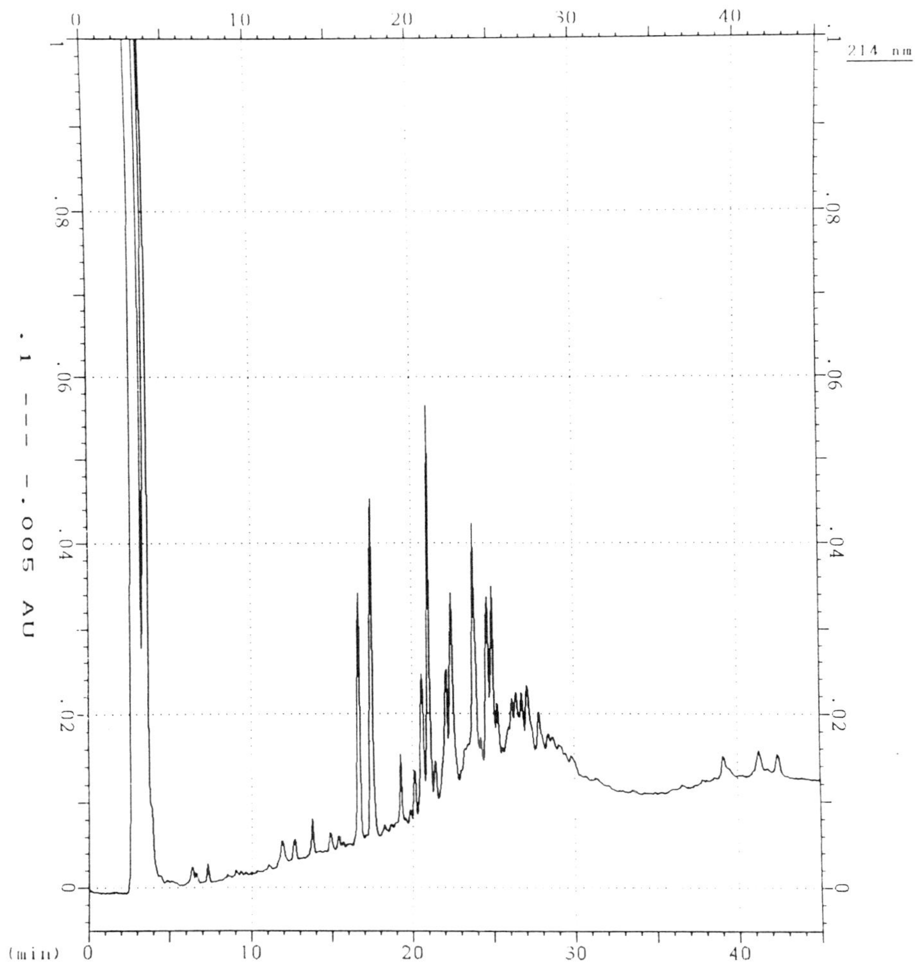

Figure 4. Peptide Map of the Dragline Fiber. Peptides generated by chemical digestion with N-bromosuccinimide were separated by reverse phase HPLC. A typical peptide map is presented above. The separations were carried out with a 45 minute linear gradient from 78% A to 50% A, where A=water, b=acetonitrile and c=100mm phosphate, pH 3.0 and solvent C was kept at 20% to maintain a constant concentration of 20mM potassium phosphate buffer.

1. ? - ? -Leu-Gly-Ser-Gln-Gly-Ala-Gly-Gln- ? -Ala/Ser-Ser/Leu- ? -Ala- ? -Ser--Ala

2. ? -Ser-Leu-Gly-Ser-Gln-Gly-Ala- ? -Asp/Gly- ? -Asp/Gly

3. ? - ? -Leu- ? -Ser-Gln-Gly-Ala-Gly-Arg- ? -Ala-Leu

Figure 5. N-Terminal Sequence of NBS-Dragline Peptides. Sequence data from three of the NBS-peptides are presented above. All sequence data is very similar and reveals a repeat (underlined) in the spidroin I protein. Analysis of these sequences also confirms that specific cleavage is occurring at the C-terminal

repeat in the spidroin I protein (*1*) with the consensus sequence Tyr-Gly-Gly-Leu-Gly-Ser-Gln-Gly-Ala-Gly-Arg-Gly-Gly. This repetitive sequence occurs with 100% homology ten times in spidroin I. Analysis of these sequences also reveals that the chemical cleavage is occurring at the C-terminal of tyrosine as expected. It is not surprising that all of these sequences are nearly identical. This consensus sequence occurs with an 80-90% homology after almost every tyrosine residue in spidroin I. The second dragline protein identified by Himman and Lewis (*7*), spidroin II, also contains many tyrosine residues yet we have not identified any sequence or amino acid composition data from any of the peptides analyzed to suggest that this protein is present in the dragline fiber. It is possible however that these sequences have not been recovered as is the case with the alanine and glycine rich sequences and continued analysis of additional peptides is being pursued.

Conclusions.

Initial characterization of the dragline fiber and the major ampullate gland from *N. clavipes* is reported. The ability to maintain solubility of the dragline protein in 8M urea, 200mM acetate has provided us with the means to investigate the primary and secondary structure of the dragline protein by some classical protein chemistry techniques. A comparison of the dragline protein to the major ampullate gland protein by SDS-PAGE has revealed that they are virtually identical, migrating as one band at 275 kDa in the non-reduced state and two bands at 195 and 220 kDa in the reduced state. A minor band occurs in the reduced sample of the dragline fiber at 180 kDa which is not present in the reduced major ampullate gland protein. Isolation and characterization of this band is currently underway in our laboratory. In addition, N-terminal sequence data from the major ampullate gland and peptide sequences from NBS digested dragline resemble repetitive sequences present in the spidroin I protein from the major ampullate gland of *Nephila clavipes*. To date, we have not obtained any amino acid composition or sequence data to confirm the presence of spidroin II in the dragline fiber.

Quantitative cysteine analysis of the major ampullate gland protein reveals 0.36% cysteine. If it is assumed that the molecular weight of the protein is 180-220 kDa, there are approximately 7-9 cysteine residues and 3-5 disulfide bonds per protein chain. The absence of cysteine in the partial cDNA sequences published to date (*1,7*) suggest that there are unique domains within the dragline protein which have not been identified. Therefore, further characterization is necessary to determine the role of cysteine/cystine residues in the dragline fiber.

Acknowledgements. Angela Choate, Kevin Remillard and Scott Stockwell deserve recognition for collecting the spiders. A special thanks is extended to Ted Thannhauser and Bob Sherwood from the Cornell University Biotechnology Analytical/Synthesis facility for amino acid analysis and N-terminal sequencing.

Literature Cited.

1. Xu, M. and Lewis, R. V. *Proc. Natl. Acad. Sci. U.S.A.* **1990**, *87*, 7120.
2. Tashiro, Y., Otsuki, E., and Shimadau, T. *Biochim. Biophys. Acta*, **1972**, *257*, 198.
3. Sasaki and Noda, *Biochim. Biophys. Acta*, **1973**, *310*, 91.
4. Gosline, J. M., Demont, M. E. and Denny, M. W. *Endeavour*, **1986**, *10*, 37.
5. Bell, A. L. and Peakall, D. B. *J. Cell Biol.* **1969**, *42*, 284.
6. Candelas, G. C. *et. al. Biochem. Biophys. Res. Commun.* **1983**, *116*, 1033.
7. Himman, M. B. and Lewis, R. V. *J. Biol. Chem.* **1992**, *267*, 19320.
8. Work, R. W. and Emerson, P. D. *J. Arachnol.* **1982**, *10*, 1.
9. Laemmli, U. K. *Nature*, **1970**, *227*, 680.
10. Beavis, R. C., Chait, B. T., Creel, H. S., Fournier, M. J., Mason, T. L. and Tirrell, D. A. *J. Amer. Chem. Soc.* **1992**, *114*, 7584.
11. Zemlin, J. C. **1968** *Technical Report 69-29-CM, AD 684333*, U.S. Army Natick Laboratories, Natick MA
12. Shimura, K., Kikuchi, A., Ohtomo, K., Katagta, Y. and Hyodo, A. *J. Biochem.* **1976**, *80*, 693.
13. Lotz, B. and Colonna-Cesari, F. *Biochimie*, **1979**, *61*, 205.

RECEIVED June 29, 1993

Chapter 7

Silk and Silk Proteins from Two Aquatic Insects

Steven T. Case[1], Jason Powers[1], Robert Hamilton[2], and Mary Jane Burton[2]

[1]Department of Biochemistry, University of Mississippi Medical Center, 2500 North State Street, Jackson, MS 39216–4505
[2]Department of Biology, Mississippi College, Clinton, MS 39058

Silk is spun by aquatic larvae belonging to several insect orders. To gain a broader perspective of silk structure, we began to collect comparative biochemical data for silks obtained from larvae of the midge (*Chironomus tentans*, Diptera) and caddis fly (*Pycnopsyche guttifer*, Trichoptera). The amino acid composition of both silks is relatively simple; five residues account for about 60% of all amino acids. However, the composition of each silk is different. The lumens of silk-producing glands from midge and caddis fly contain about a dozen proteins of various sizes. Most notably, both types contain at least one huge protein with an apparent molecular weight around 1000 kDa. These data will be discussed here in the context of compositional and structural similarities of silks from several aquatic (midge and caddis fly) and non-aquatic (silkworm) insects and spiders.

The term "silk" summons up visions of artistic webs created by spiders and commercial contributions of the silkworm. However, hidden from sight, in streams and lake beds, the aquatic larvae of numerous insects also spin silk (1). While the taxonomic distribution of silk-spinning insects is broad, their utilization of silk is similar: construction of tubes for protective housing, assistance in foraging and filter-feeding and modification of tubes for pupation.

Are all silks alike? Considering the extremes in environmental occurrences of silk (in air and underwater) and the diversity of organisms that spin it (arachnids and several insect orders), one wonders about the common structural denominators for these apparently similar biopolymers. This paper reports the first in a series of steps comparing the biochemical and physical properties of silk from aquatic larvae of two insects, the midge, *Chironomus tentans* (Diptera) and the caddis fly, *Pycnopsyche guttifer* (Trichoptera). While emphasis will be placed on the proteins that form aquatic silk, the results will be related to a rapidly growing body of

0097–6156/94/0544–0080$06.00/0

knowledge about silk and silk protein structure in the midge (2), silkworm (3) and spider (4).

Materials and Methods

Animals. *C. tentans* larvae were obtained from laboratory cultures raised as described (5). *P. guttifer* larvae were collected in local streams and maintained for several weeks in the lab.

Silk. Silk was obtained as fibers spun on glass beads. Individual midge larvae were placed in the bottom section of a 35 mm diameter petri dish covered by a 3 to 5 mm thick layer of 1000 μm diameter glass beads. A perforated lid was placed over the petri dish and up to 6 dishes were submerged in 5 cm of 0.4% NaCl in a 10 x 20 cm glass dish. Larvae were maintained for several days with aeration but without food. Individual caddis fly larvae were placed in a plastic beaker containing a 1 cm layer of glass beads and 6 cm of pond water.

Amino Acid Analysis. Glass beads with adhering silk were removed from their containers, rinsed with deionized H_2O and soluble material removed by extraction for 1 hour at room temperature in 6 M guanidine hydrochloride, 40 mM Tris-HCl, pH 8.8, and 20 mM EDTA. Some extractions included 100 mM dithiothreitol followed by the addition of 300 mM iodoacetamide for 1 hour in the dark. None of these extracts yielded protein detectable by amino acid analysis. Beads were rinsed twice with H_2O, once with 80% acetone, air dried and transferred to glass tubes for vapor phase hydrolysis in 6 N HCl for 70 minutes at 150°C. Hydrolysates were dried in a vacuum and amino acids detected using ninhydrin on a Hewlett Packard AminoQuant. Values reported here are the mean of duplicate determinations of at least two samples. Glass beads without silk treated in parallel yielded no detectable amino acids.

Silk Proteins. Silk-producing glands were fixed in 70% (v/v) ethanol for at least 1 hour and the cells surrounding the lumen removed by dissection. The gelatinous contents of the lumen were denatured, reduced, carboxymethylated, precipitated with acetone, and examined by SDS-polyacrylamide gel electrophoresis (SDS-PAGE) on concave exponential gradients of 3-20% polyacrylamide, as described (6). Proteins were visualized by staining with Coomassie blue.

Results

The Amino Acid Composition of Aquatic Silks. Midge and caddis fly larvae spun silk fibers attached to glass beads. The fibers were microscopic, but their presence was evident when the beads were gathered into tube-like structures (Figure 1). The silks were insoluble in various aqueous and organic solvents; however, they were readily hydrolyzed in acid and amino acid analysis confirmed that both silks consisted primarily of protein (Table I).

The amino acid compositions of midge and caddis fly silks were both

Table I. Amino Acid Composition of Midge (*C. tentans*) and Caddis Fly (*P. guttifer*) Larval Silks[1]

	Midge	*Caddis Fly*[2]
Amino Acid	*Mole %*	
Ala	6.1	4.7
Arg	14.4	10.4
Asx	5.5	8.0
Cys	7.3	0
Glx	9.2	4.1
Gly	9.7	24.9
His	1.1	1.5
Ile	1.5	4.8
Leu	2.6	8.3
Lys	11.1	1.5
Met	0.9	0.1
Phe	2.6	1.1
Pro	11.2	4.0
Ser	12.0	11.5
Thr	2.3	3.4
Tyr	0.5	4.9
Trp	0	0
Val	2.0	6.8

[1]Values are the mean of duplicate runs from at least two samples.

[2]Caddis fly silk was processed in parallel for comparative purposes; the values obtained are essentially identical to published values (8).

biased. Five amino acids comprised over half the residues: about 58% of midge silk is composed of Arg, Ser, Pro, Lys and Gly; about 63% of caddis fly silk is composed of Gly, Ser, Arg, Leu and Asx. These data implied that midge and caddis fly silks contain proteins composed of repeated amino acid sequences.

The Number and Size of Silk Proteins. While the morphology of midge and caddis fly silk glands is dissimilar (2,7), in both instances, silk proteins are secreted into a lumen where they exist in a concentrated but soluble state. In the midge, the lumen contains about a dozen silk proteins; four are giant (1000 to 1500 kDa), the others range from 18 to 220 kDa (2). SDS-PAGE of cellular and lumenal proteins from the caddis fly silk gland demonstrated a similar size distribution of proteins (Figure 2). The slowest migrating protein was of particular interest. When compared to the largest midge proteins, its relative electrophoretic mobility corresponded to an apparent molecular weight of about 800 kDa (Figure 3).

Discussion

Silks Consist of "Simple" Proteins. Amino acid analyses indicate that midge (Table I), caddis fly (8), silkworm (9), and spider (10,11) silks are all protein-based biopolymers. Their biased composition predicts two additional points. Firstly, each silk is likely to be composed of "simple" proteins. In the case of midge (2), silkworm (12,13) and spider (4) proteins, "simple" can be rephrased as "tandem repeated amino acid sequences". Secondly, the composition of each silk is different (Table II). Small side chain amino acids vary from 27.8% (midge) to 86.1% (silkworm), polar residues from 21.4% (silkworm) to 56.1% (midge), acidic residues from 2.3% (silkworm) to 14.7% (midge), and basic residues from 0.9% (silkworm) to 26.6% (midge). With the notable exception of basic residues in midge silk, all these values are within the ranges reported for eight different silks obtained from various glands of the spider, *Araneous diadematus* (10). While the significance of these differences is not yet clear, it will interesting to note whether or not differences in the silks' physical properties can be predicted by their amino acid compositions.

Besides its high content of basic amino acids, midge silk is unique in its Cys content. The distribution and potential importance of this residue in midge silk proteins will be discussed below and in an accompanying paper by Case and Smith.

Silk Proteins Contain Tandem Repeats. First, and foremost, the silk gland of each organism synthesizes at least one unusually large structural protein primarily composed of tandem repeated amino acid sequences. Silkworm fibroin has a molecular weight of about 350 kDa (14) and consists of unique termini flanking a central core of 59-residue repeats (12). Spider dragline silk has a 350-kDa protein (3) which presumably contains repeats of 53 and/or 34 residues (4).

The largest aquatic silk proteins are bigger. For example, the midge has an family of four very similar giant silk proteins (spIa, spIb, spIc and spId) that range from about 1000 to 1500 kDa. The remarkable size of these proteins has been estimated by biochemical methods (15,16), supported by the size of their 37 kb mRNAs (5) and confirmed by molecular cloning (17-19). Each share unique

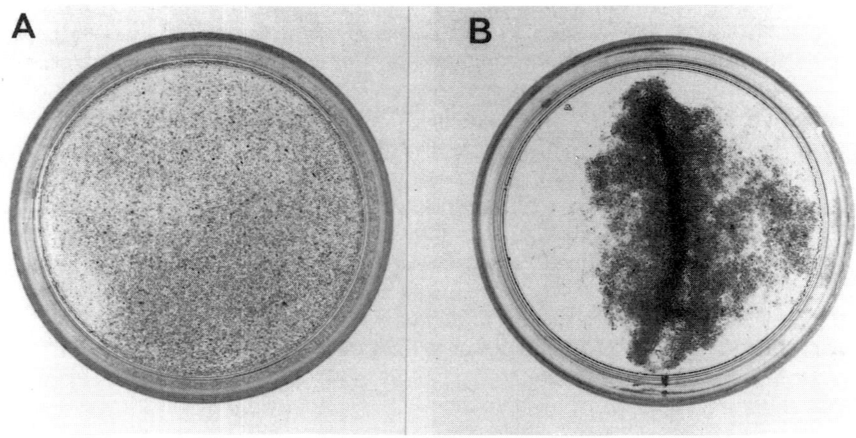

Figure 1. Collection of midge silk on glass beads. (A) The bottom section of a 35 mm-diameter petri dish with about a 3 to 5 mm-thick layer of glass beads. (B) Petri dish and glass beads two days after the addition of a larva. Microscopic examination revealed that the beads were held together by silk fibers.

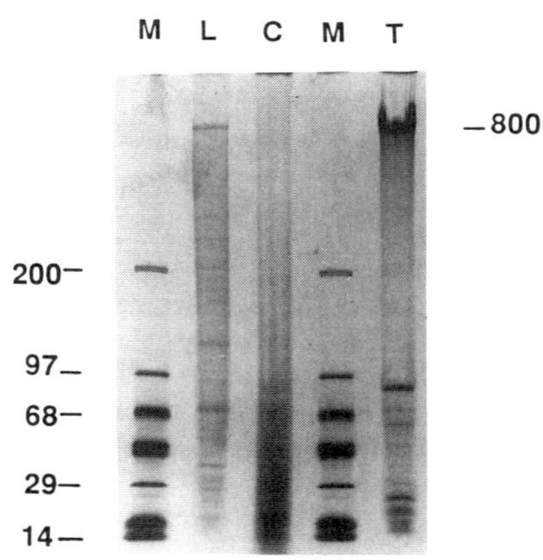

Figure 2. Electrophoretic separation of proteins from a caddis fly silk gland. Lumenal (L), cellular (C), and total (T) silk gland proteins were denatured, reduced, alkylated and separated by SDS-PAGE, as described (6). Molecular weight markers (M) are: myosin, 200 kDa; phosphorylase B, 97 kDa; bovine serum albumin, 68 kDa; ovalbumin, 43 kDa; carbonic anhydrase, 29 kDa; ß-lactoglobulin, 18 kDa; lysozyme, 14 kDa.

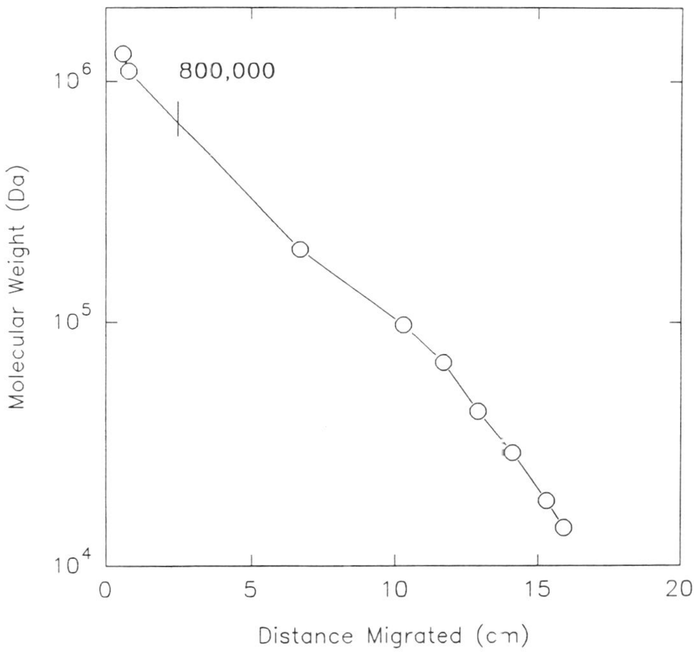

Figure 3. Estimation of the apparent molecular weight of the largest caddis fly silk protein. This plot was made from a representative gel containing several samples of caddis fly and midge silk proteins and molecular weight markers (see Figure 2).

Table II. Comparison of Selected Amino Acids in Silks from Aquatic and Non-aquatic Organisms

Common name	*Midge*[1]	*Caddis Fly*[2]	*Spider*[3]	*Silkworm*[3]
Genus and species	*C. tentans*	*P. guttifer*	*N. clavipes*	*B. mori*
Type of silk	*larval*	*larval*	*dragline*	*larval*
Residues with small side chains (Ala+Gly+Ser)	27.8%	41.1%	62.7%	86.1%
Polar residues (Asp+Arg+Glx+His+Lys+Thr+Tyr+Ser)	56.1%	45.3%	26.5%	21.4%
Acidic residues (Asx+Glx)	14.7%	12.1%	11.7%	2.3%
Basic residues (Arg+His+Lys)	26.6%	13.4%	8.6%	0.9%

[1]Data taken from Table I.

[2]Data taken from Table I; see also (8).

[3]Data taken from (3).

termini flanking 130 to 150 tandem copies of a complex core repeat (Table III), the details of which will be discussed below. The largest caddis fly silk protein is also huge (about 800 kDa, Figure 3). While primary structural data are not yet available, we suspect that this protein also contains repeats.

At least three of these organisms synthesize additional silk proteins. Silkworm sericin consists of about ten proteins that range from 20 to 300 kDa (13,14). The largest of these contains tandem repeats of 38 residues (13). The caddis fly (Figure 2) and midge (2) also have additional proteins in the 18 to 220 kDa range. Several of these midge proteins also contain tandem repeats (Table III). While the significance is unknown, a pattern is emerging among silk proteins whereby repeat lengths generally decrease as the size of the protein decreases.

Evolution of Repeats in Midge Silk Proteins. The prominent structural features of midge silk proteins are tandem repeated sequences of hydrophilic amino acids and two conserved structural motifs: a regularly occurring pattern of Cys residues and the tripeptide, + Pro - [basic residue (Lys or Arg)-Pro-acid residue (Glu, phosphoSer or phosphoThr)].

Sequence comparisons suggest an evolutionary relationship whereby genes encoding the smallest silk proteins were prototype units for the larger genes. This process most likely involved duplication and limited divergence of the Cys-containing motif or the + Pro - motif or the entire prototype containing both motifs (2). The extreme of this process is illustrated by the repeats in spIs.

The primary structure of spIs is similar to silkworm fibroin; unique amino- and carboxyl domains flank a central core of tandem repeats (17-19). However, the core is huge, consisting of 130 to 150 tandem copies of a complex repeat. Each core repeat contains 60-90 amino acids and has two domains: a constant (C) region and a subrepeat (SR) region. The C region contains 30 residues including 4 Cys, 1 Met and 1 Phe which are invariant. SR regions contain the + Pro - motif and range from 30 to 60 residues in length. Within a protein, SR regions may be regular or variable in length due to the size and number of shorter repeated amino acid sequences therein.

Higher-Order Structure of Core Repeats. To measure the higher order structures of spIs, peptides were synthesized that correspond to model C and SR regions. These peptides were subjected to circular dichroism and Fourier transform infrared spectroscopy (20). The results indicated that the C domain of spIa is primarily α-helical with one or two ß-turns. In marked contrast, an spIa SR domain appears to be a more extended, poly(Gly)II-type helix interrupted by two or more ß-turns. Thus midge spIs contain a novel pattern of alternating helical structures which probably have contrasting physical properties such as elasticity, flexibility and tensile strength.

A Role for Core Repeats in the Assembly of spI Fibers. The structures of C and SR peptide have contributed to our understanding of the assembly of spIs into fibers. It has been proposed (21-23) that core repeats facilitate fiber formation by providing multiple in-phase alignments of juxtaposed SR regions on neighboring spI molecules so that intermolecular electrostatic interactions can occur. Formation of

Table III. Summary of the Size and Primary Structure of Midge Silk Proteins[1]

Secretory Protein	*Apparent Molecular Weight (kDa)*	*Major Structural Element*
spIa	1300-1400	82-residue tandem repeats; $[C+SR]_n$
spIb	1000-1200	50-90 residue tandem repeats; $[C+SR]_n$
spIc	1000-1100	79-residue tandem repeats; $[C+SR]_n$
spId	1100-1500	74-residue tandem repeats; $[C+SR]_n$
sp195	195	25-residue simple tandem repeats
sp185	185	Cys-X-Cys-X-Cys every 22-26 residues
sp140	140	14-residue simple tandem repeats
sp115	115	14-residue simple tandem repeats
sp40	40	244-residue prototype unit
sp38	38	253-residue prototype unit

[1]Condensed (2) and updated for size estimates for spIs (18).

intermolecular disulfide bonds between neighboring C regions would provide covalent linkage of adjacent spIs. If a C region is essentially α-helical, then each of its cognate Cys side chains would project in a different direction. As a consequence, C regions may interact as parallel bundles of helices entwining numerous spI molecules into a seemingly random, three-dimensional network of fibers (2). In fact, spIs can be reassembled *in vitro* and display structures consistent with this notion (24).

Do All Silks Contain ß-Sheets? X-ray diffraction data indicate that ß-sheets are present in non-aquatic (spider and silkworm) and aquatic (caddis fly) silks, though the structure of the latter is remarkably different (8, 25). What about midge silk?

Diffraction data are lacking for midge silk; however, the infrared spectra of C and SR peptides lacked evidence for ß-sheet (20). Moreover, identical results have been obtained with soluble midge silk (S.E. Wellman and S.T. Case, University of Mississippi Medical Center, S.J. Hamodrakas, University of Athens, unpublished data). It would seem worthwhile to obtain X-ray diffraction data to ascertain if midge silk fibers truly lack ß-sheets. Alternatively, it is possible that ß-sheet formation is a consequence of dynamic structural alterations that take place during the transition of silk from an ordered liquid to solid state (4,26). This transition may involve higher-ordered assembly by the addition of other silk proteins and/or processing by mechanical shear. Future investigations will be aimed at resolving this dilemma.

Acknowledgments

We thank Mike Wallace for performing the amino acid analyses. This research was supported by U.S. Army Research Office grants DAAL03-91-G-0239 and DAAL03-92-G-0105.

Literature Cited

1. Wallace, J. B. *Annu. Rev. Entomol.* **1990**, *25*, 103-132.
2. Case, S. T.; Wieslander, L. In *Structure, Cellular Synthesis and Assembly of Bio-polymers*; Case, S.T., Ed.; Results and Problems in Cell Differentiation; Springer-Verlag: Berlin, Heidelberg, 1992, Vol. 19; pp 187-226.
3. Kaplan, D. L.; Fossey, S.; Mello, C. M.; Arcidiacono, S.; Senecal, K.; Muller, W.; Stockwell, S.; Beckwitt, R.; Viney, C.; Kerkam, K. *Mat. Res. Soc. Bull.* **1992**, *17*, 41-47.
4. Hinman, M.; Dong, Z.; Xu, M.; Lewis, R. V. In *Structure, Cellular Synthesis and Assembly of Biopolymers*; Case, S.T., Ed.; Results and Problems in Cell Differentiation; Springer-Verlag: Berlin, Heidelberg, 1992, Vol. 19; pp 227-254.
5. Case, S. T.; Daneholt, B. *J. Mol. Biol.* **1978**, *124*, 223-241.
6. Kao, W. Y.; Case, S. T. *J. Cell Biol.* **1985**, *101*, 1044-1051.
7. Engster, M. S. *J. Morphol.* **1976**, *150*, 183-212.
8. Engster, M. S. *Cell Tiss. Res.* **1976**, *169*, 77-92.

9. Lucas, F.; Rudall, K. M. In *Comprehensive Biochemistry: Extracellular and Supporting Structures* Florkin, M.; Stotz, E.H., Eds.; Elsevier: Amsterdam, 1968; pp 475-558.
10. Andersen, S. O. *Comp. Biochem. Physiol.* **1970**, *35*, 705-711.
11. Work, R. W.; Young, C. T. *J. Arachnol.* **1982**, *15*, 65-80.
12. Strydom, D. J.; Haylett, T.; Stead, R. H. *Biochem. Biophys. Res. Commun.* **1977**, *79*, 932-938.
13. Okamoto, H.; Ishikawa, E.; Suzuki, Y. *J. Biol. Chem.* **1982**, *257*, 15192-15199.
14. Sprague, K. *Biochemistry* **1975**, *14*, 925-931.
15. Rydlander, L.; Edström, J.E. *Chromosoma* **1980**, *81*, 85-99.
16. Hertner, T.; Meyer, B.; Eppenberger, H. M.; Mahr, R. *Wilhem Roux's Arch. Dev. Biol.* **1980**, *189*, 69-72.
17. Wieslander, L.; Paulsson, G. *Proc. Natl. Acad. Sci. USA* **1992**, *89*, 4578-4582.
18. Paulsson, G.; Bernholm, K.; Wieslander, L. *J. Mol. Evol.* **1992**, *35*, 205-216.
19. Paulsson, G.; Höög, C.; Bernholm, K.; Wieslander, L. *J. Mol. Biol.* **1992**, *225*, 349-361.
20. Wellman, S. E.; Hamodrakas, S. J.; Kamitsos, E. I.; Case, S. T. *Biochim. Biophys. Acta* **1992**, *1121*, 279-285.
21. Hamodrakas, S. J.; Kafatos, F. C. *J. Mol. Evol.* **1984**, *20*, 296-303.
22. Wieslander, L.; Höög, C.; Höög, J. O.; Jornvall, H.; Lendahl, U.; Daneholt, B. *J. Mol. Evol.* **1984**, *20*, 304-312.
23. Lendahl, U.; Wieslander, L. *Cell* **1984**, *36*, 1027-1034.
24. Wellman, S. E.; Case, S. T. *J. Biol. Chem.* **1989**, *264*, 10878-10883.
25. Fraser, R. D. B.; MacRae, T. P. In *Conformation in Fibrous Proteins*; Academic Press: New York, 1993; pp 292-343.
26. Viney, C. In *Structure, Cellular Sythesis and Assembly of Biopolymers*; Case, S.T., Ed.; Results and Problems in Cell Differentiation; Springer-Verlag: Berlin, Heidelberg, 1992, Vol. 19; pp 255-278.

RECEIVED May 4, 1993

Chapter 8

Synthetic and Recombinant Domains from a Midge's Giant Silk Protein

Role for Disulfide Bonds

Steven T. Case and Stanley V. Smith

Department of Biochemistry, University of Mississippi Medical Center, 2500 North State Street, Jackson, MS 39216–4505

Silk proteins from the midge (*Chironomus tentans*, Diptera) are unique, in comparison to silkworm, spider and caddis fly silks, in their containing appreciable quantities of cysteine. At least six midge silk proteins are composed of tandem repeats containing cysteines in highly conserved positions. Repeats in the spI family of proteins can be divided into subrepeat (SR) and constant (C) regions, the latter of which contains four invariant cysteines. A combination of gel electrophoresis, sulfhydryl assays, and amino acid analyses were used to demonstrate that a synthetic C region peptide and a recombinant C+SR protein can fold and form two intramolecular disulfide bonds. These results suggest an additional role for cysteines in the pathway of midge silk protein assembly.

Cysteines in proteins serve several important functions: stabilization of folding through formation of intramolecular disulfide bonds, covalent coupling of protein subunits by formation of intermolecular disulfide bonds, modification of enzyme activity by thiol-disulfide exchange and coordination of metals (1-4). The importance of these reactive and potentially interactive sites is often highlighted by their evolutionary conservation in homologous proteins. For example, subunit chains are joined by invariant cysteines in vertebrate immunoglobulins, B and T lymphocyte antigen receptors and major histocompatibility proteins (1).

The presence of cysteine makes silk from the midge, *Chironomus tentans*, unique among silks from aquatic and non-aquatic insects (such as caddis flies and silkworms, respectively) and spiders. Six of the midge's ten major silk proteins are composed of tandem repeats that include a conserved pattern of two or more cysteines (5), implying an important role for this residue.

The spI family of midge silk proteins consists of four 1000 to 1500 kDa proteins composed of 130 to 150 tandem copies of a "core repeat" (6-8). Core repeats are divisible into a variable subrepeat (SR) region and a constant (C) region.

0097–6156/94/0544–0091$06.00/0

SR regions are 30 to 60 residues long with shorter repeats containing the motif, basic residue-Pro-acidic residue. The C region contains 30 residues of which four are invariant cysteines. It has been proposed that the formation of midge silk fibers involves alignment of core repeats on adjacent segments of spI molecules (5). Intermolecular electrostatic interactions could occur between parallel SR regions; intermolecular disulfide bonds could occur between parallel C regions. Formation of these disulfide bonds would contribute to silk stability and insolubility.

In order to study the structure and molecular interaction of spI domains, C and SR peptides representative of each domain were constructed. In addition, a gene encoding a recombinant protein with C and SR domains fused as a monomeric core repeat was synthesized. This paper reports that both the synthetic C peptide (synC) and recombinant C And SR (rCAS) protein can form intramolecular disulfide bonds. These results suggest an additional role for cysteines in the pathway of silk formation in the midge.

Materials and Methods

Peptide Synthesis. The synthesis and purification of the 30-residue C domain peptide from spIa has been described (9). It is referred to here as synC and has the following amino acid sequence (single-letter code): RCGEAMRKAEAEKCARRNG-RFNAEKCRCEE.

Recombinant Protein. rCAS represents a monomeric 82-residue core repeat from spIa. Its sequence, MRKAEAEKCARRNGRFNAEKCRCEEAGKPERNEEPEK-GEKPRPEKPEKGEKPRPEKPEKGEKPRPEKPEKGEKPKPEKPEKGEKPRPER CGEAM, is encoded in a synthetic gene optimized for bacterial codon usage and expression using a T7 RNA polymerase-based system. The construction and insertion of this gene into an expression vector has already been outlined (10). rCAS was purified to apparent homogeneity by preparative isoelectric focusing and dialysis.

Reduction and Reformation of Protein Disulfides. Both synC and rCAS were denatured and disulfide bonds reduced in 6 M urea, 10 mM dithiothreitol, 10 mM Tris-HCl, pH 8, and 1 mM EDTA, followed by chromatography over a column of Sephadex G-25 equilibrated in 10 mM HCl. To study refolding, samples were diluted into 1 mM EDTA, 100 mM Tris-HCl at various pH values and incubated overnight at room temperature at peptide and protein concentrations of 250μM. Protein thiols were measured as described by Ellman (11).

Gel Electrophoresis. rCAS was examined by SDS-polyacrylamide gel electrophoresis (SDS-PAGE) as described (12). synC was examined by SDS-PAGE optimized for small proteins (13). rCAS, synC and molecular weight markers were visualized by staining with Coomassie blue.

Amino Acid Analysis. Denatured synC and rCAS were alkylated with and without prior reduction with dithiothreitol, as described above. Samples were subjected to gas phase hydrolysis in 6 N HCl at 150°C for 70 minutes. Amino acids were analyzed under non-reducing conditions on a Hewlett Packard AminoQuant.

Results

Disulfides in synC. During the course of biophysical experiments aimed at learning the structure of synC (9), aliquots were examined by SDS-PAGE. When a sample of "native" synC (stored in H_2O for several months at 4°C) was treated with reducing agents, its electrophoretic mobility appeared to slightly decrease (Figure 1A). In addition, a minor, slower migrating band disappeared that may have been a dimer. Thiol assays indicated that there was less than 1 mole of protein sulfhydryl per mole of "native" synC; however, after denaturation and reduction, synC did contain measurable sulfhydryls. Amino acid analysis of "native" synC under non-reducing conditions revealed qualitatively the presence of cystine (two cysteines joined by a disulfide bond). Carboxymethyl-cysteine was only detectable in alkylated samples that were previously reduced.

To see if disulfide bonds could reform, samples of reduced synC were incubated under various conditions (variations in ionic strength, presence of oxidizing and reducing agents) and disulfide bond formation monitored by the decrease of measurable sulfhydryls. Figure 2 illustrates the pH-dependent loss of sulfhydryl groups which is maximum at neutral to alkaline pH, the pH range where thiol groups are most reactive. The loss of detectable sulfhydryls and the predominant appearance of monomers by SDS-PAGE (Figure 1B) suggested that synC was refolding and forming two intramolecular disulfide bonds. The presence of disulfides was confirmed qualitatively by the refolding-dependent detection of cystine.

Disulfides in rCAS. To determine whether an adjacent SR domain (which lacks cysteine) would alter the ability of a C domain to form intramolecular disulfide bonds, we examined rCAS. The electrophoretic mobility of purified rCAS also decreased upon reduction (Figure 3). Results of an extensive biochemical characterization of "native" versus reduced and alkylated rCAS (including amino acid sequencing, sulfhydryl measurements, amino acid analyses and laser desorption time-of-flight mass spectrometry) were consistent with the following conclusion: purified rCAS contains two intramolecular disulfide bonds which can be reduced quantitatively to four sulfhydryl-containing cysteines.

To ascertain whether these intramolecular disulfides can reform, parallel samples of synC and rCAS were incubated overnight at pH 8. Sulfhydryls were measured before and after the incubation and aliquots examined by SDS-PAGE. Under these conditions, the sulfhydryl content of synC and rCAS declined by more than 90%. SDS-PAGE indicated that the majority of both substrates refolded as monomeric molecules, although minor quantities of putative dimers and trimers were visible (Figure 4). These results suggest that synC and rCAS both have the intrinsic ability to fold and form intramolecular disulfide bonds.

Discussion

Since synC and rCAS can form intramolecular disulfide bonds, it would appear that, in spite of the presence of the SR region, the C domain of rCAS can fold so that pairs of cysteines are sufficiently close for disulfide bond formation.

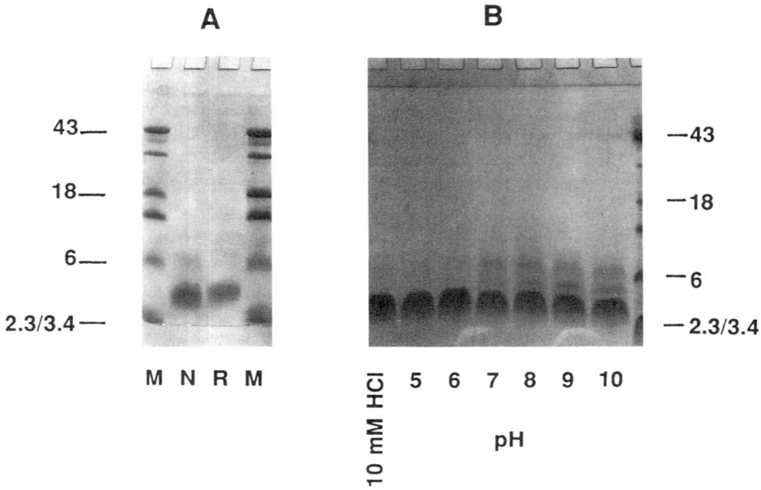

Figure 1. SDS-PAGE of synC. (A) 5 μg samples of "native" (N) and reduced (R) synC. Before electrophoresis, samples were boiled in SDS sample buffer (12) lacking β-mercaptoethanol. (B) 5 μg aliquots of samples of reduced synC that were incubated overnight at room temperature in 10 mM HCl or at the indicated pH at a peptide concentration of 250μM. Samples were run on a 16% polyacrylamide gel, as described (13). The molecular weight markers (M) on each gel are: ovalbumin, 43 kDa; carbonic anhydrase, 29 kDa; ß-lactoglobulin, 18 kDa; lysozyme, 14 kDa; bovine pancreatic trypsin inhibitor, 6 kDa; insulin, 2.3 and 3.4 kDa.

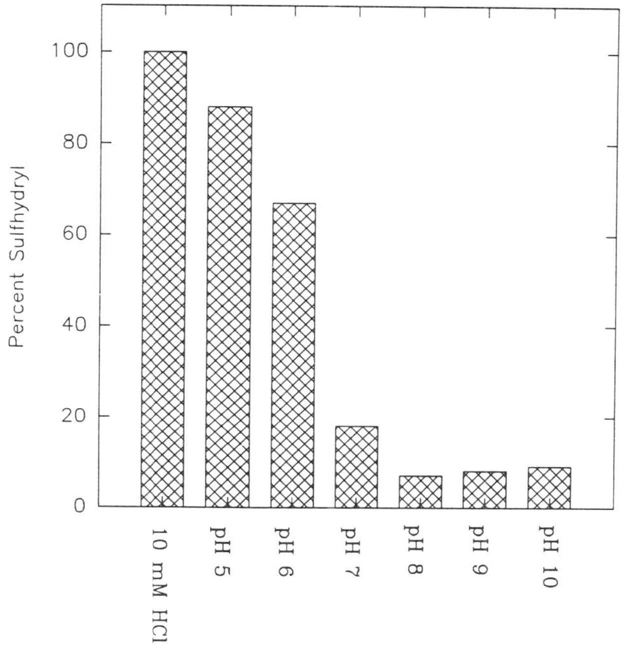

Figure 2. The relative sulfhydryl content of synC (determined as moles SH/moles of peptide) that was reduced and incubated overnight in 10 mM HCl or at the indicated pH. Sulfhydryls were measured as described by Ellman (11).

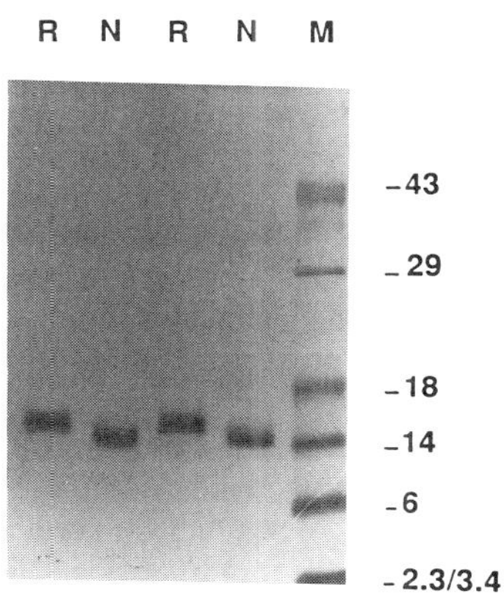

Figure 3. Differences in the electrophoretic mobility of purified "native" (N) and reduced (R) rCAS on a 15% polyacrylamide gel (12). Protein molecular weight markers (M) are the same as in Figure 1.

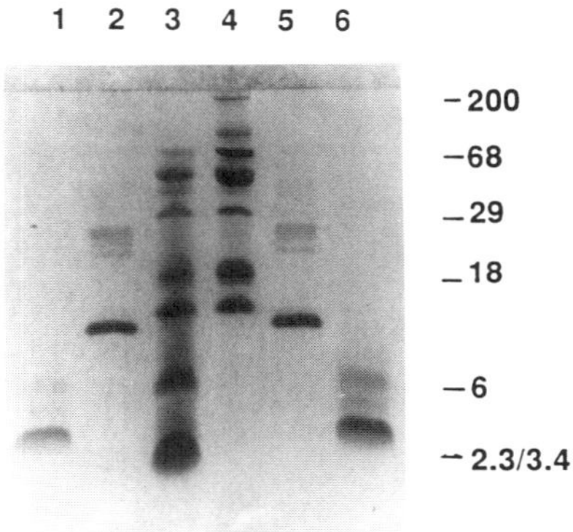

Figure 4. The electrophoretic mobility of synC (lanes 1 and 6) and rCAS (lanes 2 and 5) after reduction and refolding at peptide and protein concentrations of 250μM at pH 8, respectively. The sizes of some low (lane 3) and high (lane 4) molecular weight markers are indicated to the right. In addition to several proteins already in the low molecular weight mixture (Figure 1), high molecular weight markers include; myosin, 200 kDa; phosphorylase B, 97 kDa; and bovine serum albumin, 68 kDa.

What remains to be determined is whether or not the disulfide bonds in either "native" or refolded synC and rCAS are homogeneous (formed between unique pairs of cysteines) or heterogeneous (formed among all pairwise combinations of cysteines). We are currently sequencing tryptic fragments of rCAS to map disulfide-bonded cysteine pairs.

Whether cysteine pairs are homo- or heterogeneous, one point remains: the C domain, with or without an adjacent SR, can fold and form intramolecular disulfide bonds. By extrapolating these results to tandem repeats in spIs, one may postulate a role for the invariant cysteines in the pathway of spI assembly. Perhaps, concurrent with their synthesis, spI molecules are folded and stabilized by intramolecular disulfide bonds, one repeat at a time. This may maximize the solubility of these huge proteins by preventing aggregation prior to their secretion. In fact, the redox state of the endoplasmic reticulum does favor disulfide bond formation (14). If reduction of spI intramolecular disulfide bonds occurred after their secretion into the glandular lumen, where the protein concentration is higher than it is within the cells, their unfolding would occur in a highly concentrated protein solution favoring intermolecular interactions. Thus, if a change in redox state were encountered somewhere along the pathway of assembly and extrusion of the silk fiber, intermolecular disulfide bonds could be formed, contributing to the stability and insolubility of the ensuing silk fiber. At present, there is neither a site nor candidate molecule for accomplishing this task.

Acknowledgments

We thank Mike Wallace for performing the amino acid analyses. This work was supported by U.S. Army Research Office grant DAAL03-91-G-0239.

Literature Cited

1. Honjo, T.; Habu, S. *Annu. Rev. Biochem.* **1985**, *54*, 803-830.
2. Ziegler, D. M. *Annu. Rev. Biochem.* **1985**, *54*, 305-329.
3. Creighton, T. E.; Goldenber, D. P. *J. Mol. Biol.* **1984**, *179*, 497-526.
4. Evans, R. M.; Hollenberg, S. M. *Cell* **1988**, *52*, 1-3.
5. Case, S. T. ; Wieslander, L. In *Structure, Cellular Synthesis and Assembly of Biopolymers*; Case, S. T., Ed.; Results and Problems in Cell Differentiation; Springer-Verlag: Berlin, Heidelberg, 1992, Vol. 19; pp. 187-226.
6. Wieslander, L.; Paulsson, G. *Proc. Natl. Acad. Sci. USA* **1992**, *89*, 4578-4582.
7. Paulsson, G.; Bernholm, K.; Wieslander, L. *J. Mol. Evol.* **1992**, *35*, 205-216.
8. Paulsson, G.; Höög, C.; Bernholm, K.; Wieslander, L. *J. Mol. Biol.* **1992**, *225*, 349-361.
9. Wellman, S. E.; Hamodrakas, S. J.; Kamitsos, E. I.; Case, S. T. Biochim. *Biophys* Acta **1992**, *1121*, 279-285.
10. Smith, S. V.; Case, S. T. *Mat. Res. Soc. Symp. Proc.* **1993**, *292*, (in press).
11. Ellman, G. L. *Arch. Biochem. Biophys.* **1959**, *82*, 70-77.
12. Laemmli, U. K. *Nature* **1970**, *227*, 680-685.
13. Schagger, H.; von Jagow, G. *Anal. Biochem.* **1987**, *166*, 368-379.
14. Hwang, C.; Sinskey, A. J.; Lodish, H. F. *Science* **1992**, *257*, 1496-1502.

RECEIVED May 4, 1993

Chapter 9

In Vivo Synthesis and Structural Analysis of Alanylglycine-Rich Artificial Proteins

Eric J. Cantor[1], Howard S. Creel[2], Yoshikuni Deguchi[2], Michael J. Dougherty[1], Srinivas Kothakota[1], Mark T. Krejchi[2], Kunio Matsuki[2], Kevin P. McGrath[2], Ajay D. Parkhe[2], Edward D. T. Atkins[3], Maurille J. Fournier[1], Thomas L. Mason[1], and David A. Tirrell[2]

Departments of [1]Biochemistry and Molecular Biology and [2]Polymer Science and Engineering, University of Massachusetts, Amherst, MA 01003
[3]H. H. Wills Physics Laboratory, University of Bristol, Bristol BS8 1TL, England

A series of 22 alanylglycine-rich artificial proteins has been prepared by bacterial expression of the corresponding synthetic genes. These polymers have been isolated and purified, and subjected to structural analysis by gel electrophoresis, mass spectrometry, x-ray scattering, and vibrational and NMR spectroscopy. Strategies have been developed for the *in vivo* synthesis of artificial proteins containing non-natural amino acids.

The fibroin of commercial *Bombyx mori* silk contains large amounts of the simplest amino acids, glycine and alanine, organized into the repeating hexapeptide sequence **1**. In attempting to create simple models of *B. mori* fibroin, several groups of workers have prepared and/or examined poly(L-alanylglycine) (PLAG) and related sequential copolymers (*1-6*). Such materials form sheet-like structures in the solid state, with the chain conformation either that of the well known β-form or that of the less well defined form designated as "silk I." The silk I form of *B. mori* fibroin is obtained by quiescent drying of the silk secretion and is easily converted to the β-form by mechanical shearing forces (*7*). In this chapter we describe the synthesis and structural analysis of a family of artificial proteins containing repeating alanylglycine dyads. This family of proteins constitutes a unique set of models for *B. mori* silk fibroin.

—GlyAlaGlyAlaGlySer— **1**

0097–6156/94/0544–0098$06.00/0

Protein Design

The polymers prepared to date are listed in Table I. Sequence **2** (PLAG) is the parent polymer for the series, and has heretofore been available only via tedious chemical synthesis and only in small quantities. The chain lengths attainable by the chemical method are also modest. In contrast, our recent successful expression of PLAG in bacterial hosts (*8*), promises a ready supply of the polymer in a variety of molecular weights. Sequences **3-8** were designed to test various hypotheses regarding chain folding and crystal structure in periodic macromolecules. In each case, the repeating alanylglycine dyads were expected to adopt extended secondary structures, while the periodic sequence interruptions were conceived as potential fold points for the chain. At issue are the kinds of periodic fluctuations in molecular properties that can be exploited to control chain folding. Comparison of sequences **3** and **4**, for example, allows assessment of the importance of odd-even effects (and perhaps the unique role of proline) in defining crystallization behavior and crystal structure. Sequences **3a-d** and **4a-d** address the role of the length of the putative crystal stems and provide opportunities for exploration of potential kinetic limitations on the fold period. Sequences **5a-h** incorporate nonpolar (**5a-c**), polar (**5d-f**) or aromatic (**5g,h**) residues as sequence insertions, while **5i** and **4a** feature charged residues at those same sites. Comparative analysis of **4a** and **6** is providing insight into the consequences of local sequence inversions, and **7** and **8** are under exploration with regard to the formation of lamellar crystals with chemically differentiated fold surfaces. Finally, sequence **9** is representative of a large class of artificial proteins that we are preparing from non-natural amino acids (*9*). We describe in the following sections some general conclusions from our studies of these new materials. Detailed information can be found in references (*9-13*).

Table I. Repeating Units of Alanylglycine-Rich Artificial Proteins

Sequence No.	*Repeating Unit*
2	—AlaGly—
3[a]	—$(AlaGly)_x$ProGluGly—
4[a]	—$(AlaGly)_x$GluGly—
5[b]	—$(AlaGly)_3$ZGly—
6	—$(AlaGly)_3$GluGly$(GlyAla)_3$GluGly—
7	—$(AlaGly)_3$GluGly$(AlaGly)_3$ValGly—
8	—$(AlaGly)_3$GluGly$(AlaGly)_3$MetGly—
9[c]	—$(AlaGly)_3$SeMetGly—

[a]**a:** x=3; **b:** x=4; **c:** x=5; **d:** x=6
[b]**a:** Z=Val; **b:** Z=Met; **c**: Z=Leu; **d:** Z= Ser; **e**: Z=Asn; **f**: Z=Thr; **g**: Z=Phe; **h:** Z=Tyr; **i:** Z=Asp
[c]SeMet = selenomethionine

Expression of Artificial Proteins

A typical strategy for expression of alanylglycine-rich artificial proteins is illustrated in Figure 1 (*10*). A DNA fragment encoding the sequence of interest is prepared by chemical synthesis and cloned in a standard vector such as pUC18 (*14*). After amplification and sequence verification, the coding sequence is retrieved by digestion at *Ban*I sites flanking the coding region. *Ban*I digestion affords nonpalindromic chain ends, which polymerize in head-to-tail fashion under the influence of bacteriophage T4 ligase. The DNA multimer population is cloned at the unique *Ban*I site of the transfer vector p937.51 (*15*), and clones containing various discrete multimeric inserts are isolated. Digestion with *Bam*HI allows subcloning of each insert in the pET expression plasmids developed by Studier and coworkers (*16*) or in the pGEX vectors of Smith and Johnson (*17*). Expression in appropriate *E. coli*

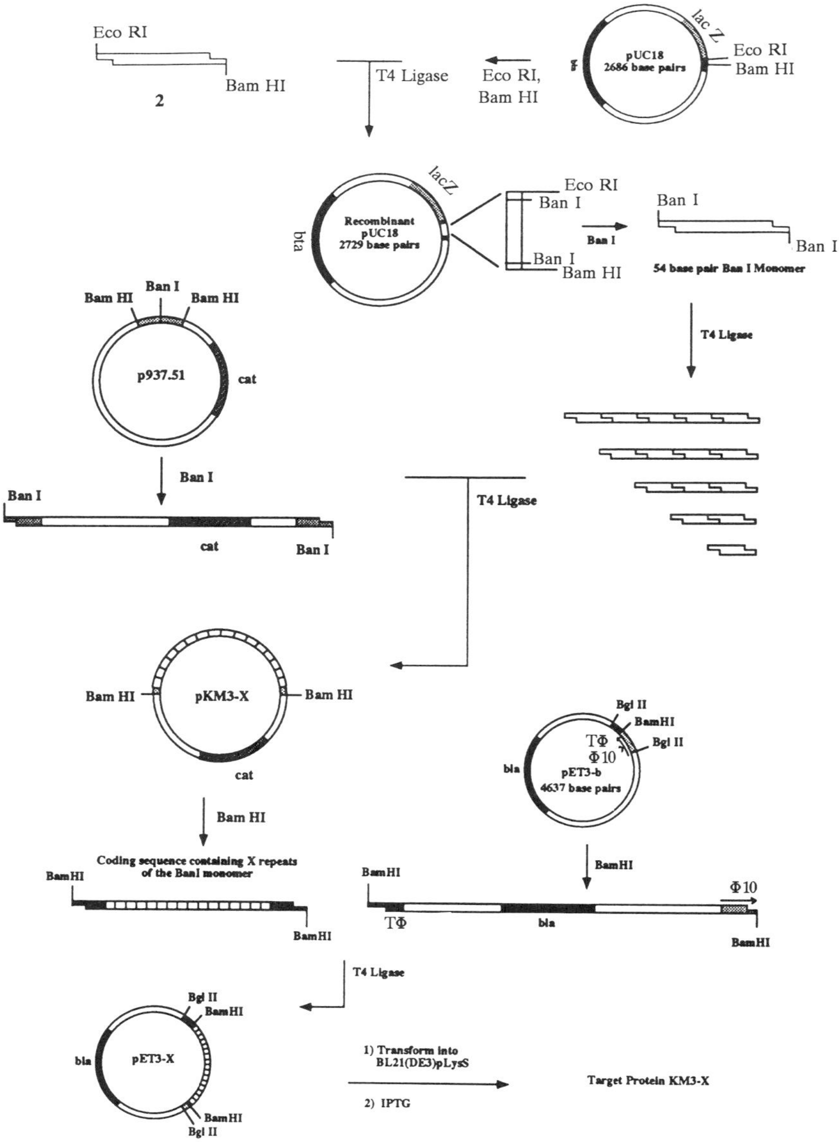

Figure 1. Strategy for cloning and expression of artificial genes encoding proteins with repeating —(AlaGly)3ProGluGly— elements. (From Ref. 10).

host strains has typically afforded ca. 50 mg of purified protein per liter of fermentation medium, with cells grown to densities of ca. 2 g/L. No systematic studies of product yield have been conducted, and we anticipate that considerably higher yields will be obtained after refinement of our fermentation methods.

Matrix-Assisted Laser Desorption Mass Spectrometry

The most powerful method for determination of the molecular structures of the polymers in Table I has been matrix-assisted laser desorption mass spectrometry (*12*). For example, the sequence of a polymer containing 54 copies of repeating unit **3a** was readily verified by the excellent agreement of the observed and calculated masses (*10*). On the other hand, certain preparations of polymers of **3b** were found to contain sequence errors after discrepancies in the calculated and observed masses were discovered (*12*). Such preparations were also found to contain significant amounts of low molar mass protein fragments, which probably arise from proteolysis, but which provide useful information regarding the protein sequence.

Electrophoretic Behavior

Many of the polymers in Table I exhibit anomalous electrophoretic migration on polyacrylamide gels containing sodium dodecyl sulfate (SDS-PAGE). Figure 2 shows a plot of mobility versus the reciprocal of the logarithm of the molar mass, for four chain-length variants of polymer **3a**. The mobilities are significantly reduced in comparison with those of standard proteins in the same molar mass range. This pattern of behavior appears to be quite general for alanylglycine-rich polypeptides. We have proposed that the reduced mobility may be a consequence of reduced binding of SDS by the negatively charged polymer of **3a** (*10*), but more recently we

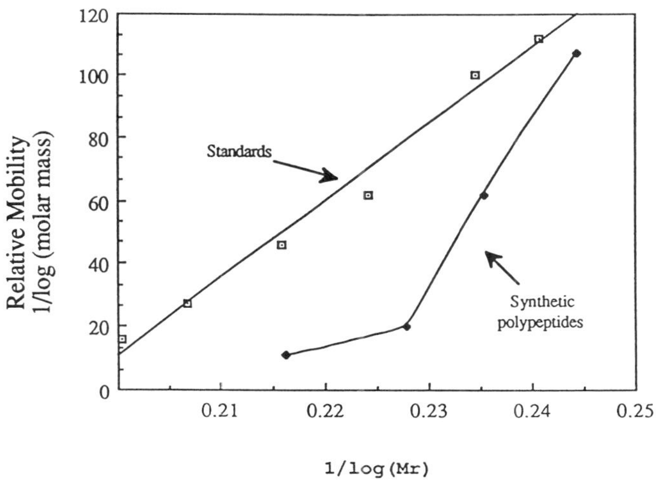

Figure 2. Plot of mobility versus the reciprocal of log (molar mass) for the series of polypeptides containing repeated (AlaGly)3ProGluGly elements. Analysis performed in 10% polyacrylamide at 15 mA constant current. Standards: α-lactalbumin, β-lactoglobulin, carbonic anhydrase, albumin (egg), bovine serum albumin, phosphorylase B (rabbit muscle). (From Ref 10).

have found anomalously low mobilities for polymers of the uncharged repeating unit **5e** (*18*). The origins of this behavior are under investigation, and seem likely to be found in an unusual structure of the polymer-SDS mixed micelle.

Solid-State Structure

The polymers shown in Table I can be obtained in any of the three solid state forms characteristic of *B. mori* fibroin, i.e., in disordered, silk I or antiparallel β-sheet structures. In general, lyophilization from aqueous solution produces disordered materials, dialysis from aqueous LiBr affords the silk I structure, and crystallization from sheared formic acid solutions gives the β-form.

Comparison of polymers of repeating units **3a** and **4a** provides insight into the coupling between sequence and structure in alanylglycine-rich polypeptides. While polymers of **3a** crystallize only reluctantly, those of **4a** form nicely oriented crystal mats that show clear evidence of the antiparallel β-sheet structure via x-ray, infrared and solid state NMR methods. The origins of this difference are under investigation, but we have suggested that it is an odd-even effect that suppresses regular chain folding and crystallization of polymers of the nonapeptide repeating unit **4a**.

Incorporation of Non-Natural Amino Acids

A major focus of our research program has been the development of strategies for the *in vivo* synthesis of artificial proteins containing non-natural amino acids. Two approaches are being explored. Each uses a host strain that is auxotrophic for the natural amino acid to be replaced by an analogue, i.e., a methionine auxotroph in the case of sequence **9**. The host is transformed with a plasmid carrying the coding sequence of interest, with the natural amino acid encoded in place of the analogue. For analogues that support growth of the host, e.g., selenomethionine, it is a simple matter of culturing the host in a medium free of the natural amino acid but containing the analogue of interest. For analogues that do not support growth, the initial stages of the fermentation are performed in a medium that does contain the natural amino acid. After induction of protein expression, a medium shift replaces the natural amino acid with the analogue, and protein synthesis is allowed to continue. This latter approach has recently been used to incorporate 5,5,5-trifluoroleucine into sequence **5c**.

Conclusions

Alanylglycine-rich artificial proteins constitute simple and instructive models of the crystalline fraction of *B. mori* silk fibroin. Such materials are readily prepared via recombinant DNA technology, and can be obtained in quantities sufficient for studies of materials fabrication and analysis. The polymers examined to date exhibit anomalous electrophoretic mobilities, and their structures are most simply verified or disproved by matrix-assisted laser desorption mass spectrometry. Intriguing sequence effects on structure and morphology have been observed and are under continuing investigation. The prospect of building non-natural amino acids into alanylglycine-rich proteins promises to add substantial breadth to the range of materials properties achievable in this class of materials.

Literature Cited

1. R.D.B. Fraser, T.P. MacRae, F.H.C. Stewart and E. Suzuki, *J. Mol. Biol.* **1965,** *11,* 706.
2. J.M. Anderson, H.H. Chen, W.B. Rippon and A.G. Walton, *J. Mol. Biol.* **1972,** *67,* 459.
3. B. Lotz and F. Colonna Cesari, *Biochimie* **1979,** *61,* 205.
4. Y. Takahashi, M. Gehoh and K. Yuzuriha, *J. Polym. Sci. Part B: Polym. Phys.* **1991,** *29,* 889.
5. B. Lotz, A. Brack and G. Spach, *J. Mol. Biol.* **1974,** *87,* 193.
6. S.A. Fossey, G. Nemethy, K.D. Gibson and H.A. Scheraga, *Biopolymers* **1991,** *31,* 1529.
7. D.L. Kaplan, S. Fossey, C. Viney and W. Muller, *Mat. Res. Soc. Symp. Proc.* **1992,** *225,* 19.
8. K. Matsuki, E.J. Cantor, H.S. Creel, M.J. Dougherty, M.J. Fournier, T.L. Mason and D.A. Tirrell, manuscript in preparation.
9. M.J. Dougherty, S. Kothakota, T.L. Mason, D.A. Tirrell, M.J. Fournier *Macromolecules* in press.
10. K.P. McGrath, M.J. Fournier, T.L. Mason and D.A. Tirrell, *J. Am. Chem. Soc.* **1992,** *114,* 727.
11. H.S. Creel, M.J. Fournier, T.L. Mason and D.A. Tirrell, *Macromolecules* **1991,** *24,* 1213.
12. R.C. Beavis, B.T. Chait, H.S. Creel, M.J. Fournier, T.L. Mason and D.A. Tirrell, *J. Am. Chem. Soc.* **1992,** *114,* 7584.
13. D.A. Tirrell, M.J. Fournier and T.L. Mason, *Mat. Res. Soc. Bull.* **1990,** *16(7),* 23.
14. C. Yanisch-Perron, J. Vieira and J. Messing, *Gene* **1985,** *33,* 103.
15. F.A. Ferrari, C. Richardson, J. Chambers, S.C. Cansey and T.J. Pollock, *U.S. Patent Appl.,* **1986,** *927,* 258.
16. F.W. Studier, A.H. Rosenberg and J.J. Dunn, *Methods Enzymol.* **1990,** *185,* 60.
17. D.B. Smith and K. S. Johnson, *Gene* **1988,** *67,* 31.
18. E.J. Cantor, M.S. Dissertation, University of Massachusetts at Amherst, 1993.

RECEIVED May 4, 1993

Chapter 10

Design, Synthesis, and Fabrication of a Novel Self-Assembling Fibrillar Protein

J. P. O'Brien[1], R. H. Hoess[2], K. H. Gardner[3], R. L. Lock[1], Z. R. Wasserman[4], Patricia C. Weber[4], and F. R. Salemme[5]

[1]DuPont Fibers, Experimental Station, Wilmington, DE 19880–0302
[2]DuPont Merck Pharmaceuticals, Experimental Station, Wilmington, DE 19880–0328
[3]DuPont Central Research and Development, Experimental Station, Wilmington, DE 19880–0356
[4]DuPont Merck Pharmaceuticals, Experimental Station, Wilmington, DE 19880–0228
[5]3-Dimensional Pharmaceuticals, Kennett Square, PA 19348–2705

Evolution has yielded a host of biopolymers and biomaterials with highly sophisticated structural organization at the molecular level. An elegant, but comparatively simple, example of such advanced structural organization is found in the fibrillar spikes protruding from the capsids of adenovirus serotypes. The shaft portion of these viral spikes is assembled from three identical protein strands and resembles a cylindrical or box beam on the molecular level. We report here the design, synthesis, and characterization of recombinant protein polymers based on multimers of 15 amino acid consensus sequences approximating the more complex natural protein. Spontaneous fibrillar aggregate formation in the genetically engineered analogs has been confirmed by various methods and the propensity of such aggregates to phase separate into liquid crystalline arrays has been observed. Moreover, it has been possible to spin filaments from liquid crystalline solutions and characterize their structure. The collective data provide what we believe is the first example of a recombinant biopolymer capable of self-organization to rigid, high-aspect-ratio assemblies suitable for anisotropic solution processing.

Polymer and materials researchers are increasingly looking to natural materials as paradigms for the design and synthesis of a new generation of structural materials. With the advent of genetic engineering methodology, it is now possible to harness polymer biosynthesis in certain biopolymer classes to control the properties and functionality of synthetic materials. Operating from a belief that future materials systems will provide value by offering efficient, specific solutions to complex material needs, this work was undertaken, in part, to determine the feasibility of synthetically replicating a sophisticated, natural polymer fiber assembly for possible commercial applications.

Figure 1 is a schematic illustration of a human serotype adenovirus. The

0097–6156/94/0544–0104$06.00/0

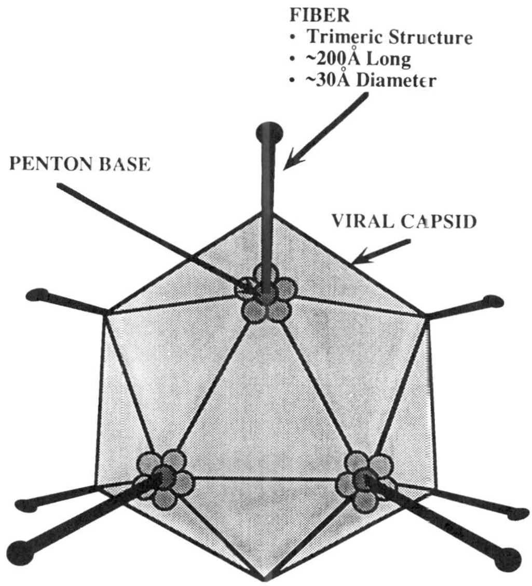

Figure 1. Schematic representation of a human adenovirus type 2 depicting placement of the capsid spikes and the approximate dimensions of the fibrillar shaft.

icosohedral capsid is featured with fibrillar spikes that emanate from 5-fold vertices (pentons) at 12 symmetrically distributed locations (*1*). The spikes, which are not covalently attached to the surface of the capsid are comprised of a shaft and a terminal globular structure, the latter of which is believed to mediate attachment of the virus to a host cell to facilitate the transfer of viral nucleic acids.

The shaft portion of the spike was of primary interest because of its apparent rigidity and the existence of prior evidence suggesting that it is comprised of a dimeric or trimeric assembly of protein molecules (*2,3*). Green et al. have suggested that the shaft proteins contain about 580 amino acid residues giving rise to a shaft length approaching 200Å. This is in agreement with previous electron diffraction studies and has been rationalized based on a model in which the proteins comprising each strand consist of ribbon-like, cross β structures containing short β strands flanked by β turns (*4*). More recently, Stouten et al. (*5*) has proposed a triple helical model for the adenovirus shaft based on features in the diffraction patterns obtained from single crystals of fiber (*6*). This model requires that individual protein strands be more chain extended than previously proposed.

Herisse et al. (*7,8)* sequenced the fibrillar region of human adenovirus type 2, leading to the identification of an uninterrupted reading frame coding for a protein polymer of mol. wt. of 62,000. Analysis of the sequence data suggested that the shaft portion of the fibrillar spike could be approximated by 15 residue consensus sequences having two short (3 residue) β strands and two alternating (4 and 5 residue) β turns.

Modeling, secondary structure analysis, and electron microscopy studies further suggested the existence of a long, narrow β sheet within the shaft region terminated by a shorter, but structurally more complex, knob region(*4*). Keying off these previous studies, we undertook synthesis of analogs of the shaft portion of adenovirus fibrillar protein via recombinant DNA methods. The constructs designed have exceptional structural regularity based on the precise repetition of monotonous 15 amino acid blocks in the polymer backbone. This approach is in contrast to previous work describing the cloning and expression of gene fragments isolated directly from adenovirus which code for the fiber portion of the gene(*9,10*). The overall approach in this work is depicted in Figure 2.

Design Considerations for Adenovirus-like Constructs

The fundamental design objective was to produce a material that could be spun into fibers with hierarchical molecular order, in which elements of structure present at the molecular level were aligned on the macroscopic scale of the fiber. A basic characteristic of natural assemblies and composites possessing hierarchical structural order is the temporal independence of structure formation at successive levels of the assembly organization. In the case of fibers, it has long been an objective to obtain fibrous materials where individual molecular polymer chains are aligned directly along the major fiber axis, an arrangement that realizes the ultimate material tensile properties. Such chain aligned order can be introduced by a variety of means, including post spin stretching or shear induced alignment from polymeric solutions. The latter process is substantially facilitated when individual polymer chains are rigid on the molecular scale, such that fibers possessing a high degree of chain alignment and associated superior mechanical properties result. Well known examples include the family of KEVLAR aramid polymers commercialized by DuPont.

In the context of the adenovirus constructs, the repetitive consensus sequence was designed to organize first as an assembly of three polypeptide chains that formed long rigid fibrils modeled on the natural adenovirus spike structural paradigm. Such fibrils might then be expected to have a high propensity for alignment when spun from concentrated solutions under appropriate solvent and concentration conditions. Since previous structural analysis had indicated that the natural structure possessed cross-beta sheet structure, an additional possibility was that post-spin processing from the prealigned fibrillar aggregates could ultimately produce chain extended polypeptide structures, perhaps possessing the tensile characteristics of high performance polyethylene, but with added thermal stability conferred by interchain hydrogen bonding between backbone amide groups. As we demonstrate below, the constructs developed do polymerize to form rigid fibrillar structures that yield lyotropic liquid crystalline solutions which can be spun into fibers.

Modeling the Adenovirus Structure

The primary objective of the protein engineering effort was to use the underlying trimeric fibril as a scaffold for anchoring chemical functional groups whose order might be preserved up to the scale of the macroscopic fiber. This objective motivated modeling studies aimed at producing structures with improved properties and additional functionality. In developing this model we followed the initial suggestion of Green et al. (*4*) that individual chains were organized as independent cross-beta sheets, but also incorporated data suggesting that the fibril was composed of 3 polypeptide chains (*2,3*). It was assumed that each protein monomer would form a cross-beta sheet with the long axis of the monomer approximately perpendicular to that of the individual beta strands, which have good hydrogen bonding between them. In the fibril model, three slightly twisted cross-beta sheets have an interchain twist characteristic of that observed in proteins with antiparallel sheets of comparable extent. These are assembled into a threefold symmetric box beam with a cylindrical cross section where the edges of adjacent sheets are overlapping (Figure 3).

The box beam, as modeled, has a gentle left-handed helical twist. Although a pitch of 140Å, arises from the interchain sheet twist, and resembles the pitch of coiled-coil helical structural proteins like tropomyosin, slight variations in local interchain geometry are possible and could result in structures whose pitch may be considerably greater. In this conception, the interior of the beam is filled with hydrophobic amino acid side chains with charged and other hydrophilic amino acid side chains located on the outside surface of the beam. It is expected that intermolecular hydrogen bonding between polar side chains at the edges of the individual sheet contributes to the stability of the proposed structural organization. The model shown in Figure 3 was assembled in several stages, using a combination of interactive computer graphics, mathematical deformations, and minimization of the potential energy. Although the model was essentially built "ab initio," one can find small segments of X-ray crystallographic structures of natural proteins which match, on a local scale, segments from the computer model, indicating that there exists some similarity between the structure of this model and that of natural proteins.

Recent analysis of the diffraction patterns obtained from paracrystals of adenovirus spike protein (*5*) have suggested an alternative model for the shaft which differs from that shown in Figure 3. Specifically, it is essentially a three-start helical structure where sheet hydrogen bonds are almost exclusively formed between segments of different chains. However, in many respects this alternative structure retains a mass distribution and local chain orientation similar to the model shown in Figure 3. Further studies of the synthetic structures will hopefully reveal more detailed features of its organization.

Synthesis of Fiber Protein Analogs

To carry out the synthesis of these novel fibrillar proteins, a suitable vector for bacterial expression was required, as well as additional strategies for generating the synthetic genes necessary to encode the target polymers. Using recombinant DNA technology, we developed a strategy to produce protein polymers of sufficently high molecular weight with defined lengths and repetitive structural motifs.

As vectors for expression of protein biopolymers, a series of pBR322 derived plasmids, designated pRHF, were constructed (Figure 4). The salient feature of these plasmids is a cassette which includes a suitable restriction site for the insertion of

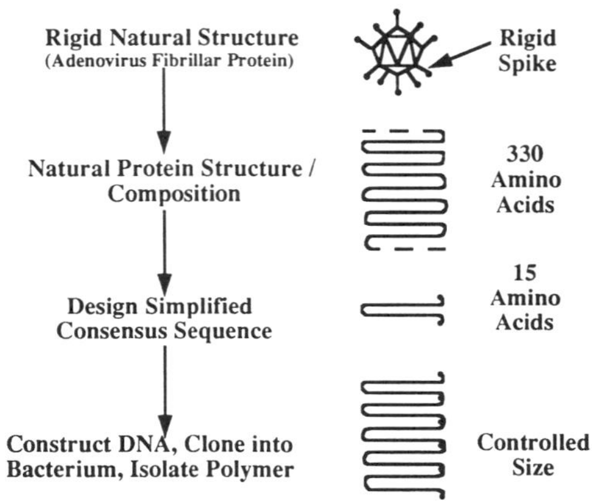

Figure 2. Overall approach to the design and synthesis of adenovirus fibrillar shaft proteins. The consensus sequences capture the basic structural elements of the more complex natural sequence.

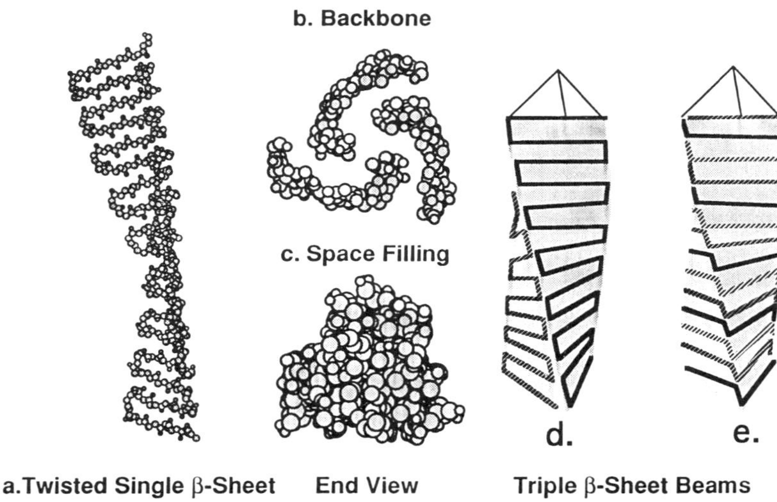

Figure 3. Computer-generated model of the target structure for the adenovirus fiber construct. a) One protein monomer, showing a backbone tracing. b) Endview of the trimeric backbone structure showing the sheet overlap. c) Endview of the trimeric space-filling structure illustrating the compact interior packing. d) A fibril schematic illustrating the chain arrangement modeled here. e) An approximate representation of the alternative structure derived in (*5*).

DNA coding for the biopolymer and the controlling sequences necessary for high levels of expression of the cloned insert. Since the synthetic biopolymer has the potential for being toxic to the host bacterium *Escherichia coli*, a conditional expression system was utilized so that construction of the gene could be completed in the absence of any expression. Once the construction of the gene was complete, the plasmid could then be transferred to a different bacterial host which would allow for expression. This precautionary measure serves to circumvent selective pressures against a gene expressing a potentially toxic gene product, and thus avoid the complication of unwanted deletions or rearrangements of the coding sequence. The sequences controlling expression of the biopolymer gene represent the promoter, originally identified from bacteriophage T7, and known to be one of the strongest prokaryotic promoters (*11*). In addition, the translational signals from the start of gene 10 (*12*) from T7 have also been incorporated to allow efficent translation of our synthetic gene. Expression of the synthetic gene is achieved by transferring the plasmid construct to an *E. coli* strain that carries a gene encoding the bacteriophage T7 RNA polymerase (*11*). The T7 RNA polymerase then recognizes its cognate promoter and transcription and translation of the biopolymer gene ensues. Since the biopolymer gene represents a highly repetitive DNA sequence, there is the likelihood that these sequences would rapidly be deleted in the bacterium as a result of homologous recombination. To avoid this possibility, the *recA* mutation that abolishes homologous recombination in *E. coli* , was introduced into the strains used for construction and expression of the biopolymer genes.

For construction of genes encoding the synthetic biopolymers, a set of DNA oligonucleotides were synthesized that encode the various Adeno sequences (Figure 5). In an effort to optimize the expression levels of these sequences, preferred codons for high level expression in *E. coli* were selected for each amino acid. The oligonucleotides coding for the Ad I and Ad III constructs were designed so that upon hybridization of complimentary strands, each double stranded oligonucleotide is left with a two base pair 5' overhang, while the Ad II oligonucleotide left a four base protruding 3' overhang. These overhanging ends allow for insertion of the oligonucleotide into the unique Acc I site of pRHF-1 (Ad I and III) or the unique Ban II site of pRHF-4 (Ad II). In both cases the overhanging ends of the oligonucleotides are nonpalindromic. Two advantages accrue from such a design. First, since we wish to construct multimers of these oligonucleotides, the nonpalindromic sequences allow the monomeric units to be ligated in only a head to tail fashion, thus ensuring that the same coding sequence will simply be repeated. Second, this also ensures that the insertion of the oligonucleotide multimer into the vector can only occur in one orientation.

The oligonucleotides for either Ad I or II were hybridized and ligated to one another to form an assortment of multimers of varying lengths. These were subsequently ligated into the appropriate vector and transformed into the *E. coli* strain DH5D*lac*. Independent ampicillin resistant transformants were grown overnight at 37°C and the plasmid DNA isolated from them. Following cleavage with EcoRI, the plasmid DNA was separated by gel electrophoresis and the size of the inserted oligonucleotide multimer estimated by comparison with the vector DNA containing no insert. Plasmids containing a substantial number of inserts were then transformed into BL21 and tested for expression. Clones pRHF-11 containing 16 repeats of the Ad I sequence, and pRHF-8 containing 26 repeats of the Ad II sequence expressed high levels of a protein of the predicted molecular weight when introduced into BL21.

For the construction of a high molecular weight version of the Ad III sequence, a somewhat different strategy was used. This approach, first reported by Kempe et al. (*13*), is outlined in Figure 6. A single insert of the oligonucleotide is made in pRHF-1. Note that in Figure 5, this oligonucleotide also incorporates two unique restriction enzyme sites, Pvu II and Hpa I. The plasmid DNA containing the

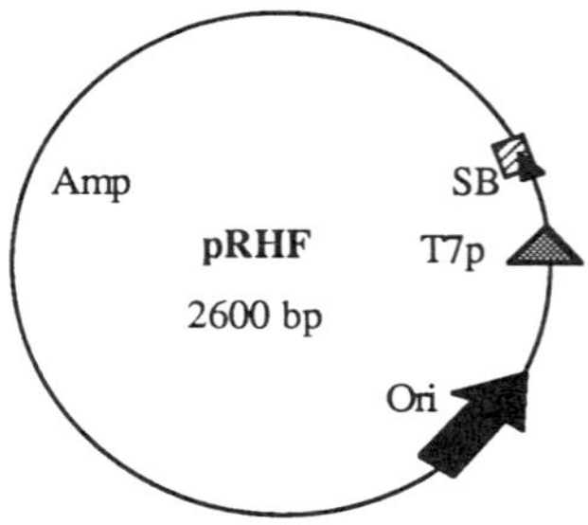

SB = Synthetic Biopolymer Insertion Site

	M	A	S	M	T	G	L	L	STOP
	5'								3'
pRHF-1	ATG	GCT	AGC	ATG	ACT	GGT	CTA	CTG	TAG
						Acc I (GGT CTA CTG)			

	M	A	S	M	T	G	L	L	G	S	G	L	L	STOP
	5'													3'
pRHF-4	ATG	GCT	AGC	ATG	ACT	GGT	CTA	CTG	GGC	TCT	GGT	CTA	CTT	TAG
								Ban II (CTG GGC TCT)						

Figure 4. Expression vector design for Ad I and Ad II.

<u>Ad I</u>

L	S	V	Q	T	S	A	P	L	T	V	S	D	G	K	
CTG	TCT	GTT	CAG	ACC	AGC	GCG	CCG	CTG	ACC	GTT	TCC	GAC	GGT	AAG	
C	AGA	CAA	GTC	TGG	TCG	CGC	GGC	GAC	TGG	CAA	AGG	CTG	CCA	TTC	GA

<u>Ad II</u>

	S	G	L	D	F	D	N	N	A	L	R	I	K	L	G	
	CT	GGT	CTG	GAC	TTT	GAC	AAC	AAC	GCT	CTG	CGT	ATC	AAG	CTG	GGC	T
CCG	AGA	CCA	GAC	CTG	AAA	CTG	TTG	TTG	CGA	GAC	GCA	TAG	TTC	GAG		

<u>Ad III</u>

L	Q	L	S	L	R	V	S	E	P	L	D	T	S	H	G	V	N	
CTG	CAG	CTG	TCT	CTG	CGT	GTT	TCT	GAA	CCG	CTG	GAC	ACC	TCT	CAC	GGT	GTT	ACC	
C	GTC	GAC	AGA	GAC	GCA	CAA	AGA	CTT	GGC	GAC	CTG	TGG	AGA	GTG	CCA	CAA	TTG	GA
	▲ *Pvu II*															▲ *Hpa I*		

Figure 5. Nucleotide sequences employed for Ad I, Ad II, and Ad III constructs.

single insert was then cleaved by a combination of either Pst I and Hpa I or Pst I and Pvu II. The appropriate DNA fragments, as outlined in Figure 6, were purified by gel electrophoresis and then religated. As a result of the religation, the insert is now doubled in size. Because the novel joint formed by ligating Pvu II and Hpa I together destroys the sequence recognized by either restriction enzyme, the plasmid is still left with a unique Pvu II and Hpa I site. The process can therefore be repeated again and again each time doubling the size of the insert. Using this method we constructed an Ad III expressing gene containing 64 repeats encoding a protein of ~100,000 mol. wt.

Large scale fermentation runs were done in order to obtain quantites sufficent for physical characterization of the Ad proteins. During overexpression in BL21, all of the Ad proteins formed inclusion bodies with the cell. By isolating the inclusion bodies, a significant factor in purification is gained. Following isolation of the inclusion bodies the pellet was first dissolved in 6 M guanidine hydrochloride followed by exchange with 6 M urea. After a desalting step, the material was passed through a QFF sepharose column and the effluent collected and dialyzed against distilled water. Samples were judged to be better than 90% pure by analyzing samples by SDS-PAGE electrophoresis.

Aggregate Formation in Synthetic Constructs

Dynamic light scattering measurements of particle diffusion coefficients can provide information about the aggregation state of a polymer in solution. Polymer solutions of Ad I (20,000 mol. wt.) were prepared by dissolving lyophilized protein in doubly distilled water which had been filtered through a 0.45 micron pore filter prior to use. Samples were then centrifuged for 20 minutes at 3000g in a refrigerated centrifuge to remove material that may not have dissolved on initial solution formation. Figure 7 shows the concentration dependence of the diffusion coefficient for Ad I solutions, together with the derived value for an apparent molecular particle diameter. The sigmoidal dependence of the viscosity with concentration to produce particles with apparent dimensions larger than 100Å is consistent with the formation of stable fibrillar aggregates at concentrations much lower than used for fiber spinning. These aggregates were observed to stably persist in solution, with slowly increasing size, for several days. Electron microscope examination of these solutions as shown in Figure 8 revealed the presence of long fibrils, behavior also suggested by measurements of the angular dependence of the solution light scattering. Ongoing image analysis of the electron micrographs of these fibrils should be very useful in providing an ultimate definition of their structural organization. It is notable that these aggregates occur in water at substantially lower concentrations than those required for lyotropic mesophase formation in hexafluoroisopropanol (HFIP) used for spinning.

Mesophase Formation and Fiber Spinning(*14*)

The purified proteins Ad I, Ad II, and Ad III were found to be readily soluble in hexafluoroisopropanol (HFIP), a known solvent for polypeptide purification (*15*) and fiber spinning (*16*).

Ad I. Solutions containing 4.7% and 9.6% by weight of Ad I (20,000 mol. wt.) in HFIP were clear and of low viscosity. They were not birefringent when placed between crossed polarizing filters in an optical microscope. A solution containing 14.6% by weight similarly exhibited the non-Newtonian rheological behavior often associated with liquid crystal solutions. On this basis, the 14.6% solution was characterized as isotropic but close to the critical concentration for anisotropy. Solutions of 15.4% and 19.5% by weight of Ad I in HFIP were translucent and more

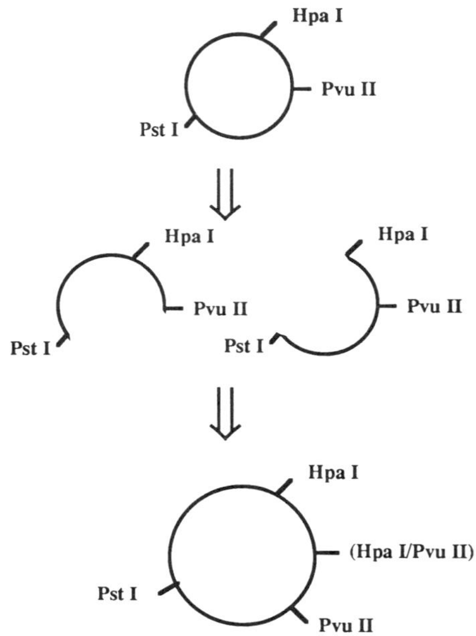

Figure 6. Expression vector strategy for Ad III.

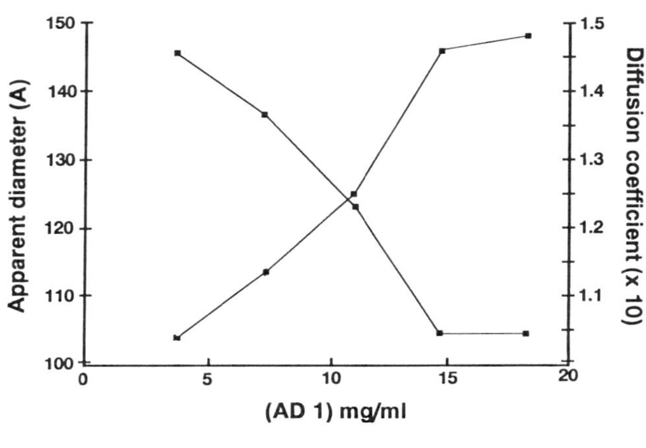

Figure 7. Concentration dependence of solution light scattering of Ad I solutions in water.

viscous with a pronounced shear thinning rheology. Samples viewed between crossed polars were found to be birefringent, showing zones which brightened and darkened as the sample was rotated in the plane of the microscope stage and were judged anisotropic.

Fibers were wet spun from a 19.5% anisotropic solution into methanol. After drying, the 80 denier fibers had a break tenacity of less than one gram/denier and an initial modulus of less than 20 grams/denier. Attempts to increase molecular orientation in the fibers by stretching before drying produced only slight improvements in mechanical properties. Figure 9 depicts the two-step process leading to mesophase formation in Ad I.

Ad II. Ad II (40,000 mol. wt) was observed to be anisotropic above 7.4% solids in HFIP. Although Ad II was soluble in HFIP and capable of forming anisotropic solutions, solubility and liquid crystal formation were improved by the addition of about one part of urea for three parts by weight of protein. In contrast to the borderline anisotropic solution at 7.4% solids, a 7.5% solution of Ad II with 2.5% by weight urea was intensely birefringent and unambiguously liquid crystalline. Urea probably promotes liquid crystal formation by disrupting intermolecular hydrogen bonding and allowing the polymer chains to fold into their minimum energy conformation. Solutions containing up to 19.1% Ad II with 5.5% by weight urea were found to be anisotropic, with rheological properties similar to the 7.5% solution described above.

Ad II fibers were spun from a 7.5% anisotropic solution in HFIP containing 2.2% weight urea into acetone. After drying, the 18 denier fibers had tensile properties similar to those mentioned above for Ad I. Increasing the molecular orientation by stretching the spun fibers 1.5x before drying and then drying under tension yielded 5 denier filaments with a break tenacity of 1.5 grams/denier, a break elongation of 16%, and an initial modulus of 45 grams/denier. Alternatively, hot stretching wet-spun, dried fibers at 200°C allowed a 2.7x stretch and yielded 6 denier fibers with tenacity 1.7 grams/denier, elongation 9%, and modulus 39 grams/denier.

Fibers were also spun from 7.4% mesophase solutions in HFIP without urea. After drying, the fibers were drawn 3x at room temperature producing 18 denier filaments with tenacity 2.6 grams/denier, elongation 30%, and modulus 38 grams/denier. These mechanical properties are comparable to those of several commercial textile fibers.

Ad III. Solutions of Ad III (100,000 mol. wt.) in HFIP were found to be isotropic, even at polymer concentrations as high as 15% by weight. Solutions of up to 15% Ad III in strong polar solvents, such as dichloroacetic acid and aqueous lithium thiocyanate, similarly showed no signs of anisotropy. We attribute this behavior to electrostatic repulsion between ionizable groups in the side chains in the Ad III construct preventing the polymer from folding as required to form the fibrillar aggregates necessary for phase separation.

Ad III fibers were spun from 15% by weight isotropic solutions in HFIP into methanol. Only modest tensile properties were achieved from the isotropic solutions similar to those obtained for Ad I.

Structural Characterization of Polymers/Fibers

Structural analysis via X-ray diffraction of Ad I, II, and III is summarized in Figure 10 which gives the powder diffraction patterns from compressed, dry powders. All three patterns are similar with reflections at 1.02nm and 0.47nm (see Table 1) and a broad scattering feature at 0.38nm. The reflection at 0.47 nm, while not observed by Devaux et al.(*6*) in adenovirus fiber crystals, is direct evidence for the presence of a β-

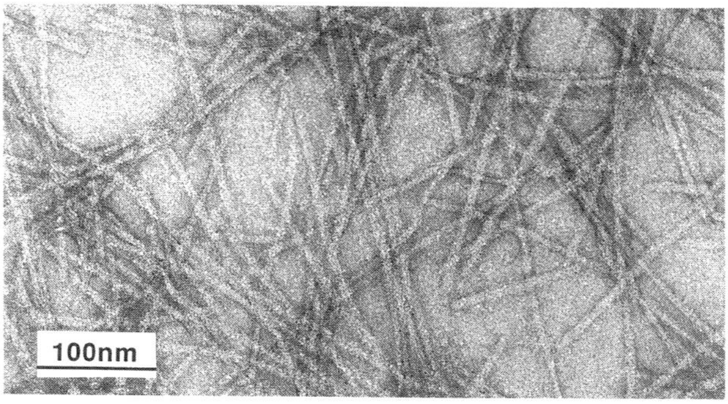

Figure 8. Ad I fibrillar aggregates. Solution was deposited on a carbon-coated copper mesh grid, stained with 1% uranyl acetate and imaged with a Hitachi 6000 electron microscope at 100 KV.

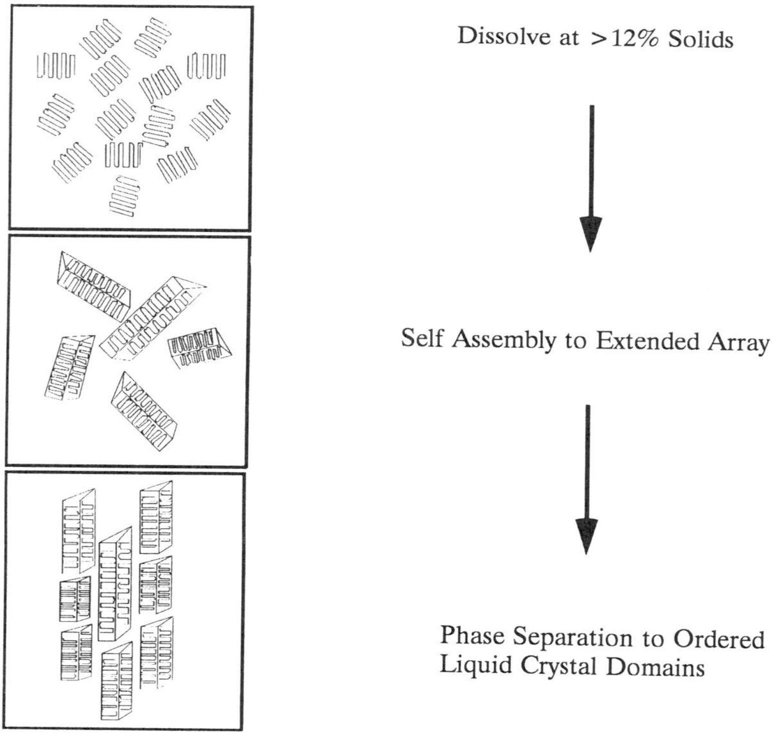

Figure 9. Mesophase formation in Ad I. Self-assembly to rodlike aggregates precedes phase separation above the critical concentration of 12% solids.

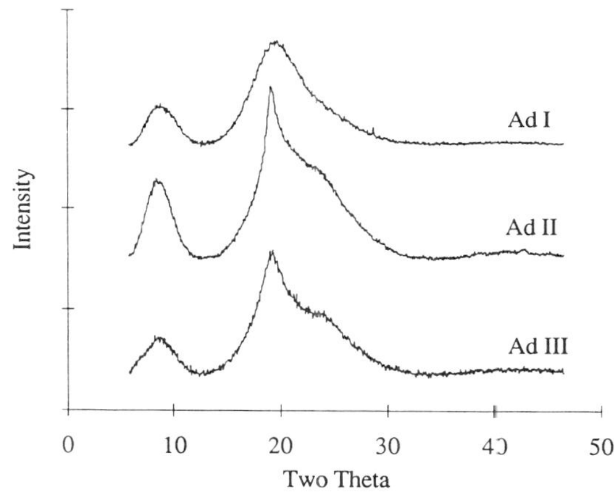

Figure 10. Powder diffraction patterns for dry adenovirus constructs.

sheet structure in these constructs. The peak is associated with the chain separation distance within the hydrogen bonded β-sheet. The absence of additional reflections in the diffraction patterns suggests that the sheets are assembled in a non-standard manner. The diffraction data also show that Ad II and Ad III have more highly developed sheets, possibly due to their higher molecular weight and/or their different composition for the fold region compared to Ad I. Diffraction patterns were obtained from fibers of the different adeno constructs described above. In general, the molecular orientation in these fibers was not sufficient to allow a determination of cross β-structure. Samples stored at high relative humidities and/or freshly prepared from water solutions show an additional reflection at 2.25nm which disappears when the samples are heat-treated and/or dried extensively. This behavior suggests that water is an integral part of the ordered structure.

Table I. X-ray diffraction spacings for Ad I, II, III

Dry			Wet	
Ad I	Ad II	Ad III	Ad I	Ad III
			2.241	2.250
1.000	1.023	1.022	1.022	1.018
0.461	0.466	0.466	0.466	0.469
0.377	0.397	0.380	0.384	2.250

The unique nature of the organization versus conventional β-silks is supported by thermal measurements. The DSC traces of these materials are characterized by a sharp melting endotherm at 219°C. In contrast, native silks melt at or above 300°C.

Conclusions

The work described in this paper represents a complete iteration through a design, synthesis, fabrication and evaluation cycle for precision protein polymer synthesis. The results have shown that it is possible through precise specification of molecular level parameters, such as polymer sequence and composition, to control both material architecture and processibility for possible downstream manufacturing operations. The facile, self-assembly of the simplified synthetic adenovirus constructs described above into sophisticated, highly ordered structures provides encouragement to the materials scientist interested in capturing the efficiency and utility of evolutionary building blocks present in natural materials. In some cases, we expect that the design and synthesis of simplified synthetic analogs to such natural structures will lead to materials which outperform their natural counterparts in applications where biologically imposed material constraints are unimportant.

Acknowledgments

The authors wish to thank Mr. Don Smith for protein purification, Dr. Richard A. Yates for fermentation development, and Dr. Catheryn L. Jackson for the electron microscopy.

Literature Cited

(1) Valentine, R. C.; Pereira, H. G. *J. Mol. Biol.* **1965,** *13*, pp. 13-22.
(2) Sundquist, B.; Pettersson, U.; Thelander, L.; Phillipson, L. *Virology* **1973,** *51*, pp. 252-256.
(3) Devaux, C.; Zulauf, M.; Boulanger, P.; Jacrot, B., *J. Mol. Biol.* **1982,** *156*, pp. 927-937.

(4) Green, N. M.; Wrigley, N. G.; Russel, W. C.; Martin, S. R.; McLachlan, A. D. *EMBO* **1983**, *2*, No. 8, pp 1357-1365.
(5) Stouten, F. W.; Sander, C., *J. Mol. Biol.* **1992,** *220,* pp. 1073-1084.
(6) Devaux, C.; Adrian, M.; Berthet-Columinas, C.; Cusack, S; Jairot, B., *J. Mol. Biol.* **1990,** *215,* pp. 567-588.
(7) Herisse, J.; Galibert, F. *Nucleic Acids Res.* **1981**, *9*, pp. 1229-1240.
(8) Herisse, J.; Rigolet, M.; deDinechin, S. D.; Galibert, M. *Nucleic Acids Res.* **1981**, *9*, pp. 4023-4042.
(9) Chatellard, C.; Chroboczek, J. *Gene* **1989**, *81*, pp. 267-274.
(10) Albiges-Rizo, C.; Chroboczek, J. *J. Mol. Biol.* **1990**, *212*, pp. 247-252.
(11) Studier, F. W.; Moffatt, B. A. *J. Mol. Biol.* **1986** *189,* pp. 113-130.
(12) Olins, P. O.; Devine, C. S.; Rangwala, S. H.; Kavka, K. S. *Gene* **1988** *73,* pp. 227-235.
(13) Kempe, T.; Kent, S. B. H.; Chow, F.; Peterson, S. M.; Sundquist, W. I.; L'Italien, J. J.; Harbrecht, D.; Plunkett, D.; DeLorbe, W. J. *Gene* **1985** *39,* pp. 239-245.
(14) Hoess, R. H.; O'Brien, J. P; Salemme, F. R. Internat. Pat. Appl., WO 92/09695, 1992.
(15) Hayashi et al., U.S. Patent No. 4,594,409.
(16) Lock, U.S. Patent No. 5,171,505.

RECEIVED May 4, 1993

Characterization, Properties, and Modeling

Chapter 11

Optical Characterization of Silk Secretions and Fibers

Christopher Viney[1,2], Anne E. Huber[1], Dwayne L. Dunaway[1], Keven Kerkam[2], and Steven T. Case[3]

[1]Molecular Bioengineering Program, Center for Bioengineering, WD-12 and [2]Department of Materials Science and Engineering, FB-10, University of Washington, Seattle, WA 98195
[3]Department of Biochemistry, University of Mississippi Medical Center, 2500 North State Street, Jackson, MS 39216–4505

Transmitted polarized light microscopy is used to identify a liquid crystalline phase both in natural silk secretions (from spiders, silkworms and aquatic insect larvae) and in reconstituted silk solutions. Factors that dictate formation of this phase are identified; drying rates are especially significant, as silk processing in nature occurs in an open system under non-equilibrium conditions. Light microscopy does not directly reveal the nature of the rod-like structures in the liquid crystalline phase; they are too small to be observed individually. On the basis of optical microstructure topologies alone, the rods could be either single molecules or molecular aggregates. However, from a combination of microstructural observations, Flory's models of phase transition behavior in lyotropic systems, and published data on the molecular conformation of fibroins in aqueous solution, we deduce that the rods are in fact aggregates of essentially globular molecules, held together by non-covalent associations that involve very few sites on any one molecule interacting with neighboring molecules. The formation of rod-like supermolecular structures enables the fibroin to retain its solubility in water, while facilitating the flow of the secretion through small spinnerets and increasing the susceptibility of the fibroin to conformational change during shear.

The silk fibers produced by spiders and the larvae of some insects exhibit many impressive engineering properties; selected examples of these properties are summarized in Table I. The mechanical properties of silk have inspired models (*2,9*) that treat the solid material as a composite of interconnected crystalline and amorphous regions. Our own work is principally concerned with the *in vivo* processing characteristics of silk secretions, i.e., with the conformational and packing changes that accompany the conversion of silk solution to a solid product exhibiting this composite microstructure. Apart from seeking to eventually achieve *in vitro* the spinning of strong, stiff, tough fibers from solutions of silk fibroin and similar molecules (regardless of their source), we note the intrinsic appeal of a room-temperature processing route that converts water-soluble polymer to insoluble high-performance fiber by physical rather than chemical means.

0097–6156/94/0544–0120$06.00/0

Table I. Selected Examples of the Sophisticated Properties of Silk

PROPERTY	COMMENTS	REF.
Mechanical		1-4
strength	up to 2 $GN.m^{-2}$ for spider drag line	
stiffness	up to 30 $GN.m^{-2}$ for spider drag line	
toughness	elongation to failure can exceed 30% for spider drag line	
Process Variables		4-6
solvent	silk fiber is spun from aqueous solution (compare to Kevlar, which is spun from solution in hot, concentrated sulfuric acid), and will not subsequently dissolve in water	
processing temperature	silk fiber is spun at ambient temperature	
processing environment	in air or under water	
fiber diameter	fibers (cribellate silk) may be as fine as 0.01μm	
Environmental		7,8
durability	resists degradation in a wide variety of environments	
biodegradability	degradation to the level of amino acids by specific proteolytic enzymes allows spiders to recycle their web and prey-swathing silk	

Interest in the Liquid Crystalline Phases of Silk

It is often desirable to carry out processing operations (fiber drawing, film forming, injection molding) on polymers in the nematic liquid crystalline state, and to subsequently retain nematic (orientational) molecular order in the solidified product (*10-14*). In the case of both liquid crystalline solutions and liquid crystalline melts, the advantages of such processing include: (1) the lower viscosity and therefore easier processability of the liquid crystalline state, compared to isotropic fluids containing the same concentration of polymer; (2) facilitated production of microstructures in which molecules are extended and globally aligned, leading to enhanced uniaxial stiffness and strength in the solid product; (3) the low thermal expansion coefficient of the solid product; (4) reduced susceptibility of the solid product to retraction (dimensional change arising from randomization of molecular orientational order) when annealed at temperatures above the glass transition but below the melting point; (5) the ability to capitalize on guest-host interactions to orient additives that are not themselves liquid crystalline. In addition, liquid crystalline melts (but not liquid crystalline solutions) exhibit relatively low solidification shrinkage compared to conventional melts, and so can be used to obtain moldings with more precise dimensional tolerances. All the properties of liquid crystalline polymers, as listed above, result from the spontaneous local alignment of extended molecules that characterizes the (fluid) liquid crystalline state. Molecular extension and alignment reduce the contributions of entanglements to the fluid viscosity, and also reduce the degree of molecular re-ordering needed during solidification.

The simplest (most disordered) class of liquid crystal is the nematic (Figure 1). It consists of rod-like structures, that are usually (but not necessarily correctly) understood to be either single molecules or rigid segments of single molecules. Locally, the rods exhibit preferred unidirectional alignment but no positional order. The axis that defines the direction of preferred local alignment is known as the *director* (Figure 1). There are also many types of liquid crystal that exhibit more complex local order (orientational order relative to more than one axis, and/or low-dimensional positional order) (*15*). However, in the present paper we are concerned with the nematic phase only.

Several factors initially led us to conjecture that natural silk secretions may form a liquid crystalline phase *en route* to being spun into fiber. One clue is provided by the high strength and stiffness of spider dragline silk, which implies a high degree of molecular alignment. When synthetic polymers are spun into fibers, significant molecular alignment requires either that the fiber be subjected to a post-spinning draw (typical draw ratios (*16,17*) lie in the range 4-10 for fibers spun from isotropic solutions and melts), or that the fiber be spun from a liquid crystalline solution or melt (*14*). Since the spider often does not subject the fiber to any post spinning draw — for example when unsupported frame silk is extruded into air or reeled out behind the spider in web construction, or when silk is extruded to make a parachute for use in ballooning to a new habitat (*5*) — a liquid crystalline route to the spun fiber is indicated implicitly. Supporting evidence is provided by the following observations, considered collectively:

- The fluid content of silk glands may be optically birefringent (anisotropic), either while still inside the gland, or as it is allowed to flow in bulk from a sectioned gland (*18-21*). In other words, the fluid silk secretion exhibits anisotropy on the macroscopic scale of the entire gland or sample. However, this observation does not by itself constitute proof that the secretion is liquid crystalline. While optical anisotropy is a characteristic property of the liquid crystalline state, it is not *uniquely* characteristic of that state. Polymer solutions and melts will similarly exhibit macroscopic anisotropy (*22*) if even a small degree of microstructural orientation is induced by extrinsic conditions such as flow or meniscus forces. Also, macroscopic optical anisotropy of a silk-containing gland may be due to anisotropy in the structure of the gland tissue through which the light has to pass too, and cannot automatically be ascribed to the gland contents if the gland tissue is not first studied on its own.
- Silk secretions are sufficiently fluid to be extrudable through a narrow orifice by organisms with limited capacity for generating excess pressure in their body cavity, even though silk secretions contain approximately 30 weight % of high molecular weight polymer (*23*). Low viscosity (and therefore enhanced processability) at high concentrations is another characteristic property of liquid crystalline fluids (*24*).
- A number of proteins that assemble *in vivo* into fibrous structures are known to form nematic liquid crystalline phases – e.g. collagen (*25-27*), and actin (*28,29*).

Experimental Section

Transmitted Polarized Light Microscopy. Simple observations made with a transmitted polarized light microscope can enable identification of liquid crystalline phases (*30,31*). A thin specimen is confined between flat glass surfaces (microscope slides or cover slips), and is viewed between crossed polars in the microscope. If the specimen is a simple nematic, a dark area (extinction) occurs everywhere that the director (or its projection onto the specimen plane) is locally parallel to the transmission direction of either polar. The observed pattern of light and dark areas (often referred to as a "texture") therefore changes as the crossed polars are rotated, and depends on the distribution of director orientations in the sample. By observing how the positions of extinction move when the crossed polars are rotated, one can determine how the preferred direction of molecular orientation varies across the sample (*30*). Thus it is *specific types of texture or extinction pattern* exhibited by a birefringent specimen, and not the mere existence of birefringence, that distinguishes specific liquid crystalline phases. A specimen is demonstrated to be liquid crystalline if (a) it is fluid, (b) it exhibits a texture typical of known liquid crystalline phases when observed between crossed polars in a microscope, and (c) removing at least one of the polars causes the specimen to become featureless at the scale of the texture (demonstrating that the texture indeed resulted from extinction, and, thus, from anisotropy in the fluid specimen).

Some points in the texture may be characterized by the confluence of two or four extinction bands, for all orientation of the crossed polars. Such points indicate the positions of *disclinations*. If a disclination is associated with *m* extinction bands, it is assigned a *strength* (*s*) equal to $\pm m/4$. The sign denotes whether rotation of the crossed polars causes the extinction bands to rotate around the disclination in the same (+) or opposite (-) sense as the polars. The topology of molecular orientation in the vicinity of disclinations (*32*) is shown schematically in Figure 2. Because disclinations of strength $\pm 1/2$ are topologically compatible with nematic order only (*15*), their presence in a microstructure is definitive proof that a liquid crystalline phase is nematic.

In the literature of liquid crystalline materials, the nematic director is implicitly associated with a consistent orientation of rod-like molecules or molecular segments over optically resolvable distances. However, nematic textures could also be observed between crossed polars if the rod-like structures were optically isotropic, and were embedded in a matrix that is also isotropic but of different refractive index than the rods. Such a microstructure will exhibit *form birefringence* if there is alignment of the rods over optically resolvable distances, because the average refractive index measured parallel to the director again is different from the average refractive index measured transverse to the director. Thus, it would not be possible to distinguish between the two cases illustrated in Figure 3 by simply looking at their textures between crossed polars; one cannot use textures alone to determine whether the rods in a fluid specimen are molecular or supermolecular (*33*). This paper argues that *supermolecular rods* (molecular aggregates) form the basis of the liquid crystalline phase of silk.

Materials. Our studies were performed on silk secretions of *Bombyx mori* silkworms, *Nephila clavipes* orb-weaving spiders, and *Chironomus tentans* larvae.

B. mori (various strains) were kindly donated by Marian Goldsmith, University of Rhode Island (Kingston, RI), or were purchased from Carolina Biological Supply Company. They were raised on mulberry leaves or on artificial diet (Carolina). Silk glands were removed from larvae in the fifth instar, i.e. during the approximately eight-day-long period that occurs between the final molt and the onset of cocoon spinning.

N. clavipes were obtained from Angela Choate, University of Florida (Gainesville, FL). They were raised on a diet of crickets (Fluker's Cricket Farm, Baton Rouge, LA) and *Drosophila* fruit flies. Studies were confined to the contents of ampullate (dragline), flagelliform (capture thread) and tubuliform (cocoon silk) glands; other glands were too small to provide useful amounts of material for the present studies.

C. tentans larvae were raised from eggs, in 0.4 wt% saline. Their diet consisted of powdered nettle leaves.

The *B. mori* and *N. clavipes* silk glands were dissected into "insect Ringer's" solution (14.0 g NaCl, 0.2 g KCl, 0.2 g $NaHCO_3$, 0.4 g $CaCl_2$, and 1000 ml distilled water (*34*)). Glands were transferred to glass microscope slides and rinsed with distilled water. Excess water was then allowed to drain. The glands were transferred to dry microscope slides, and were sectioned with a scalpel. Droplets of silk secretion from the middle division of *B. mori* glands, and from the middle region of *N. clavipes* glands, were allowed to leak onto the slides. The glands were then discarded; glass cover slips were placed over the droplets and pressed down gently, minimizing the imposed shear rate. These specimens were observed between crossed polars in a transmitted light microscope (Leitz Laborlux 12 POL).

C. tentans salivary glands were dissected from larvae and placed in a droplet of deionized water (@ 20 glands/50 µl droplet) on a siliconized coverslip on the thermostated stage (4 °C) of a dissecting microscope. The lumens were punctured with a

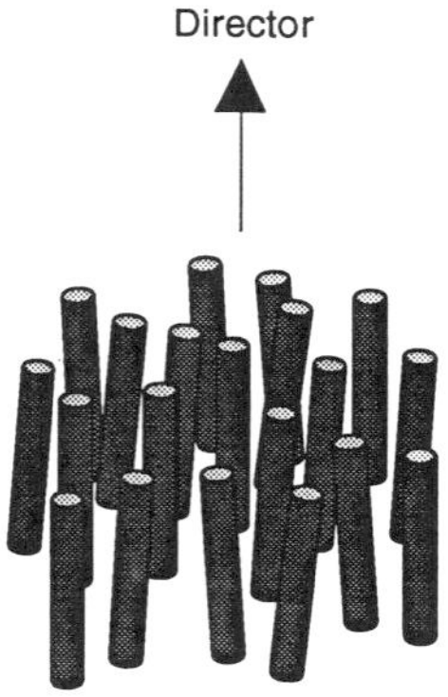

Figure 1. Illustrating local order of rod-like structures in a nematic liquid crystal, and defining the nematic director.

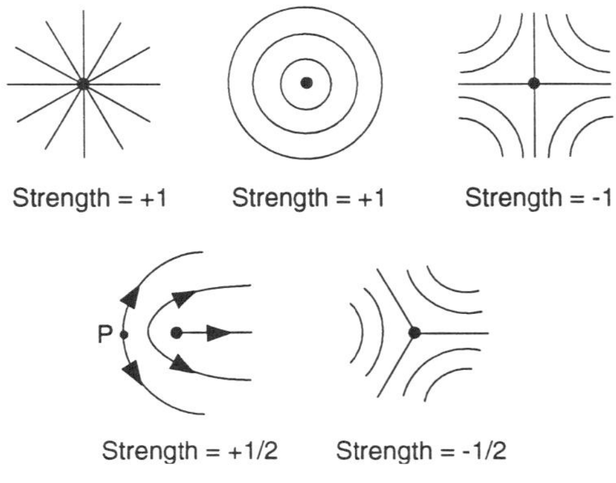

Figure 2. Rod orientation close to disclinations that occur in relatively disordered liquid crystals (*32*). Large filled circles indicate disclination cores. The arrows on the s = +1/2 disclination demonstrate that it can only occur in liquid crystalline specimens in which director orientations +n and -n are indistinguishable (a unique characteristic of nematic order). P is arbitrarily chosen as the starting position for a closed walk (either clockwise or counterclockwise) around the disclination. The first arrow that is encountered indicates the local director orientation (the opposite orientation could equally well have been chosen). The remaining arrows then consistently show the director orientation as the closed walk is followed around the disclination. On completing the circuit to P, the director orientation is seen to be exactly opposite to that defined at the start of the walk. Thus, a continuous field of rod orientations around the disclination is only possible if opposite director orientations do not represent different physical arrangements of rods.

needle; secretion was allowed to leak out for about 10 minutes. The glands were removed with dissecting needles, and the extract was transferred to a microfuge tube and centrifuged at 12,000 x *g*. Supernatants from five extracts were combined; the final sample volume after further concentration was ~50 μl. This solution was initially fluid, compared to undiluted material obtained directly from *B. mori* and *N. clavipes*. Thin samples were examined by transmitted polarized light microscopy within 48 hours of preparation. The phase changes of *Chironomus* silk are of particular interest because the natural silk fibers are produced entirely under water (*35*). The fact that solid silk structures can be produced under water very clearly demonstrates that solidification is not a process of "drying out"; rather, it is one in which the pattern of inter- and intramolecular bonding is altered.

Forced Silking. Figure 4 shows an apparatus for acquiring silk fiber at a controlled rate from spiders. It is modelled on an existing design (*36*), and consists of seven subcomponents: stand, electric motor, IR LED assembly, IR detector, power supply for the motor, stereomicroscope, and spider restraint.

The apparatus is powered by a low speed (0-200rpm) geared electric motor. A 1/4 inch Plexiglas rod with a hole drilled along its axis at one end is pressure-fitted to the shaft. A 1/16 inch hole is drilled in the other end of the rod, to allow insertion of a 1/16 inch diameter rod for use at low draw rates. Also attached to the 1/4 inch rod is a "chopper". This is a circular piece of cardboard with 12 evenly-spaced holes near the perimeter, that periodically interrupts the IR LED beam and is used for calibrating the shaft rotation rate. The IR photodetector is connected to an oscilloscope. A variable-voltage power supply is constructed from a 12V AC, 500mA transformer, connected to a full-wave diode rectifier. The output is controlled and smoothed by a capacitor and resistor attached in parallel with a potentiostat. Spiders are restrained upside down by two rubber bands that are pinned to a closed-cell styrofoam block; each band runs along one side of the spider and is used to trap four legs near the point where they join the cephalothorax.

Results and Discussion

Natural Silk Secretions Can Form a Nematic Phase. Specimens confined between a glass microscope slide and cover slip, and viewed in transmission between crossed polars, are optically isotropic. After an increase in specimen concentration, due to evaporation, textures are observed that are uniquely characteristic of the nematic liquid crystalline state (*37-40*). The nematic nature of liquid crystalline state is confirmed by the observation of disclinations that have strength ±1/2 (Figure 5). Only a small increase in concentration is required in those specimens that start out at close to their *in vivo* concentration. All the *N. clavipes* silks studied, as well as the *B. mori* silk, formed nematic textures within 10 minutes of thin specimen preparation. The *C. tentans* silk secretion only began to form nematic textures several hours after thin specimens were prepared, presumably because they initially are more dilute than the original secretions.

Thus, a wide range of silks can form a nematic phase upon partial drying. This includes spider silks that do not depend on maximized strength and stiffness. In all cases, however, the silk-producing organism can benefit from the decreased viscosity and enhanced processability associated with liquid crystalline order.

We also note that the liquid crystalline phase appears to be associated with polymer concentrations *greater* than those found *in vivo*. In living organisms, the phase change may be triggered by resorption of water (e.g. in the duct leading to the spinneret (*5*)). It is interesting to consider why the silk secretion is stored in the isotropic state, instead of in the liquid crystalline state that appears to facilitate processing. In subsequent sections of this paper, we argue that the rods in the liquid crystalline phase of silk are aggregates of individual more-or-less globular molecules. In these aggregates, molecular conformation and order will be more sensitive to shear than in independent molecules at a simi-

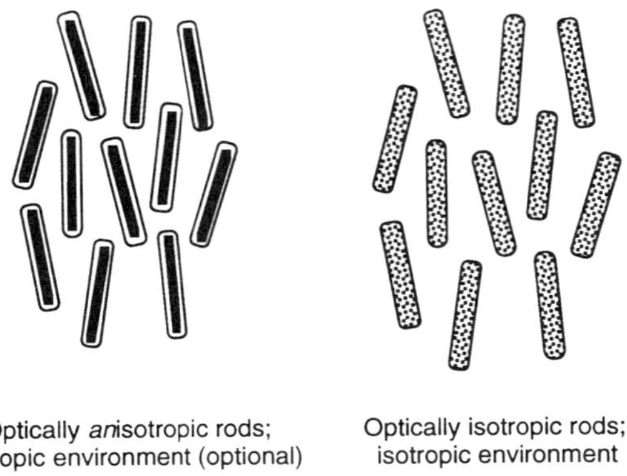

Figure 3. Left: schematic representation of optically *aniso*tropic rods (in an optional matrix that is isotropic). Right: optically *iso*tropic rods in a matrix that is also isotropic. In both cases, the material could exhibit birefringence and contribute to a texture if viewed between crossed polars, provided that rod orientation persists over optically resolvable distances; the average refractive index measured parallel to the director (N-S) differs from the average refractive index measured normal to the director (E-W). In both cases, randomization of the rod orientations would cause the material to become optically isotropic.

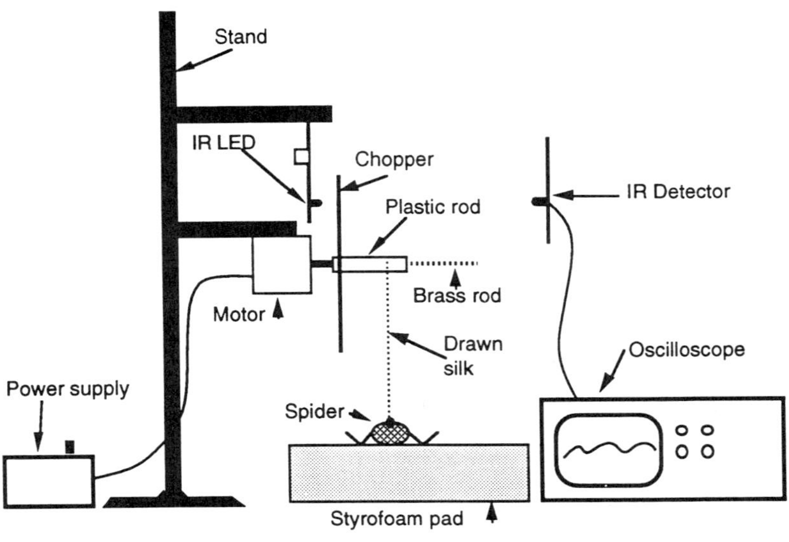

Figure 4. Apparatus used to silk spiders at a controlled rate.

lar concentration. Shear forces will be experienced by the silk gland and contents as a result of the animal's movements as it spins silk fiber. Because sheared silk undergoes a conformational transition that renders it insoluble in water, it is clearly to the animal's advantage if the silk gland contents are not excessively shear-sensitive.

Reconstituted Silk Solutions Can Form a Nematic Phase. We have observed the formation of liquid crystalline phases in reconstituted silk solutions. The broad range of solution compositions (silk : water : chaotropic salt) over which this occurs has not yet been quantified.

The following is a representative processing history that leads to material exhibiting liquid crystalline textures: *B. mori* cocoons (with pupae excised) were boiled in 0.065% NaOH for 2-3 minutes and then in tap water for another 20 minutes, to remove the sericin coating. After drying in a desiccator, silk was dissolved at a concentration of 7.5% by weight in 9.3M LiBr (*43*). Approximately 10 grams of the solution were placed in a 10-inch length of cellulose dialysis tubing (Spectra/Por; MWCO 12,000 - 14,000; previously de-sulfurized by treatment for 30 minutes in 1500 ml water + 30 g sodium bicarbonate + 0.618 g EDTA) and dialyzed against ~200 ml tap water for 2 days. The final LiBr concentration in the material retained in the dialysis tube is approximately 0.5M (approximate specimen composition is at point R in Figure 6).
The significance of this simple observation is twofold: Firstly, the liquid crystalline state of *B. mori silk* can be stabilized in the presence of a wide range of chaotropic salt concentrations (the natural secretions contain only trace amounts of salt (*3*)). Secondly, the liquid crystalline state of silk can be re-created and exploited when reconstituted silk solutions are processed.

The Liquid Crystalline State of Silk is Assembled from Supermolecular Anisotropic Structures. Several observations imply consistently that the "rods" constituting the liquid crystalline phase of silk are not individual molecules or molecular segments, but instead arise at the supermolecular level. These observations are:

- Silk proteins from different organisms have very different primary structures, none of which is recognizably nematogenic in solution; i.e., based on secondary structure prediction (*39,44,45*) and experiment, none of the silk proteins are known to have rod-like molecular conformations in solution. Hydropathy searches suggest that conformation should be insensitive to concentration in water. Nuclear magnetic resonance (NMR) and Raman spectroscopy show that the fibroin in *B. mori* silk secretion is principally random coil, a small (unquantified) fraction of material adopting an α-helical conformation (*46,47*). Most significantly, the random coil conformation in regenerated aqueous solutions of silkworm fibroin remains unchanged over a wide range of concentrations, provided that the solutions are not subjected to shear. So, even in the liquid crystalline phase, the individual molecules apparently do not contain rod-like segments. FTIR and CD spectroscopy of the principal *C. tentans* fiber-forming protein and related synthetic peptide models suggest that these molecules have a more globular than rod-like tertiary structure (*48*).
- When a thin layer of silk secretion is held between a glass microscope slide and cover slip, a concentration gradient necessarily develops across the specimen, as water is lost to evaporation from the unsealed edges. The polymer concentration at the perimeter of the specimen increases more rapidly than in the interior, so nematic textures first appear at the perimeter. A microstructure typical of biphasic systems (where the microstructure would consist of a *mixture* of isotropic and liquid crystalline regions (*49,50*)) is not resolved at the interface between the nematic periphery and isotropic interior of the specimen (*45,51*). This suggests that biphasic behavior is confined to a narrow concentration range, implying in turn that the solution has the thermodynamic characteristics of an athermal system (*52-54*). Figure 7 can therefore be used to estimate that the axial ratio of the "rods" in natural silk secretions must lie between approximately 25 and 35. Representative amino

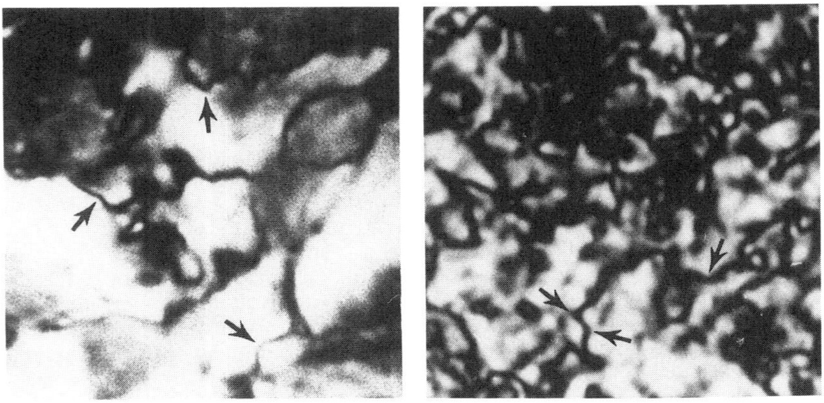

Figure 5. Left: liquid crystalline texture of fluid *N. clavipes* dragline secretion, after partial drying between a glass microscope slide and cover slip; disclinations of strength ±1/2 are indicated by arrows. Right: comparable texture of a previously well-characterized (*41,42*) synthetic nematic compound: an oligomer of *p*-hydroxybenzoic acid having a number average degree of polymerization equal to 18. The bar represents 10μm.

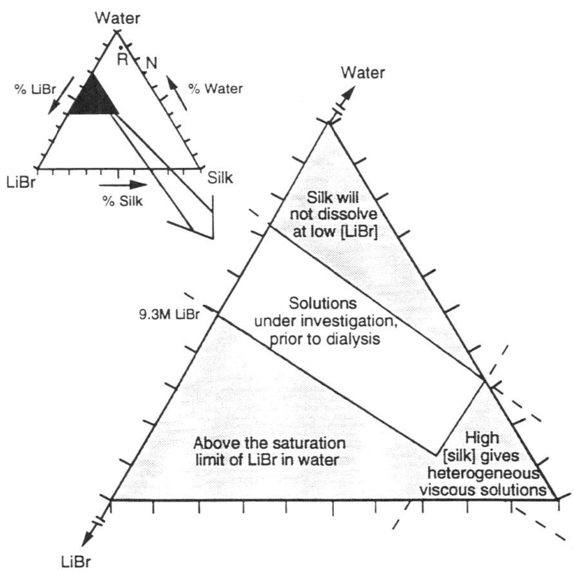

Figure 6. Ternary phase diagram for water - *B. mori* silk - LiBr. The filled-in section on the small-scale diagram is expanded in the large-scale diagram, to show the range of initial system compositions (weight %) that are accessible for study. On the small-scale diagram, N denotes the approximate composition of *B. mori* silk secretion, and R denotes the composition of the liquid crystalline reconstituted and dialyzed specimen described in the text.

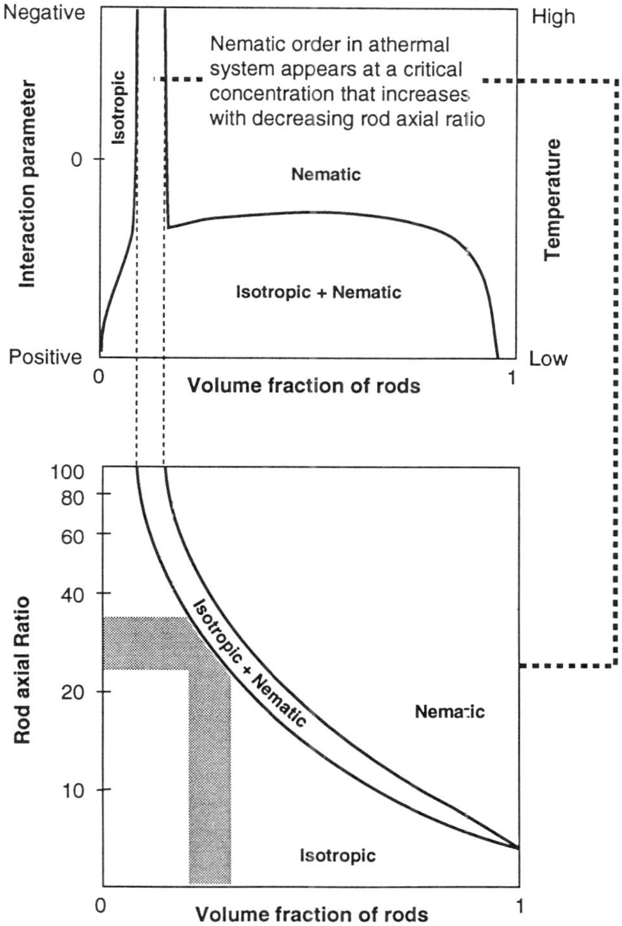

Figure 7. Top: solvent - rod phase diagram if the axial (length:diameter) ratio of the rods is 100; prediction of a simple lattice model (*52*); the system exhibits athermal behavior under conditions where the solvent - rod interactions cause the biphasic region on the diagram to form a "chimney" that encompasses a narrow range of concentrations. Bottom: the concentration range covered by the biphasic "chimney" depends on the axial ratio of the rods; prediction of a simple lattice model (*53*); the relationship to the phase diagram for rods having an axial ratio of 100 is emphasized by the two broken vertical lines extending between the diagrams. Because natural silk secretions contain approximately 30 weight % polymer (*23*), and their liquid crystalline microstructures indicate athermal behavior, the lower diagram suggests that the rods in the nematic phase have an axial ratio that lies in the range 25-35.

acid sequences for *B. mori* silk (*18,55-57*) and *N. clavipes* drag line (*58*) have been searched by standard computer algorithms in an effort to identify molecular secondary structure that could give rise to rods with such a high axial ratio in solution. These searches did not identify any significant secondary structure (*39,45*). In the case of the dominant fiber-forming protein in *C. tentans* silk (*44*) there is evidence that α-helical domains may exist over runs of approximately 10 amino acids, alternating with domains of less well-defined secondary structure. However, the α-helical domains correspond to only about 2.8 (=10/3.6) turns of the helix, or a helix length of approximately 15Å (= 2.8 times 5.4Å). Taking the diameter of the α-helix backbone as 5Å (*59*), this gives an estimate of the helix axial ratio as 3, i.e. an order of magnitude below the value implied by experiment. (In fact, an axial ratio of 3 would not even suffice to stabilize an athermal nematic liquid crystalline phase (Figure 7, lower diagram); also, the axial ratio of 3 estimated above would be even smaller if the excluded volume represented by the side chains of the amino acids were taken into account). Thus, the "rods" are unlikely to arise at the level of the secondary structure of individual molecules.

- The narrow concentration range for biphasic behavior suggests that the constituent "rods" in the nematic have little or no flexibility (*60*). However, silk fibroin molecules individually are flexible in that they can undergo a conformational transition from random coil to crystalline β-sheet upon shear.
- While silk secretions from different types of gland appear to form a nematic phase at approximately the same concentration, the scale of the nematic microstructures is different in each case (*39,45*). Differences in liquid crystalline microstructural scale can be interpreted in a number of ways (*42*), a plausible explanation in the present case being that the polydispersity of the constituent "rods" is different in each type of silk secretion. But, since the length of individual protein molecules is predetermined by its corresponding gene rather than by reaction statistics, a given protein is constrained to be monodisperse. Therefore, if polydisperse rods are present, they must be formed by aggregation of individual molecules.

Proteins that are known to form rod-like structures by aggregation of monodisperse globular molecules include actin, tubulin (an important component of the cytoskeleton) and deoxygenated hemoglobin-S (the molecule responsible for sickle-cell anemia) (*59*). Fibrous actin (or F-actin) filaments are formed by the polymerization of individual globular actin (G-actin) molecules, so that an F-actin fiber resembles two intertwined strings of beads. When G-actin is polymerized into F-actin *in vitro*, a nematic liquid crystalline phase is formed above a threshold concentration of F-actin (*29*). Also, suspensions of purified natural F-actin are liquid crystalline at concentrations of 6mg/ml or greater (*28*). The liquid crystalline phase behavior parallels that of an athermal lyotropic system, suggesting that the liquid crystalline phase forms as the result of excluded volume effects alone. In this respect, liquid crystalline F-actin and liquid crystalline silk appear to behave similarly.

It is significant that water-soluble silk fibroin molecules with a globular tertiary structure can form rod-like units (and therefore a liquid crystalline phase) by aggregation, i.e. without undergoing a change in conformation. Aggregation enables the polymer to retain its solubility (the surrounding water "sees" the same surface groups on the polymer molecules), while at the same time the solution becomes more easily processable because its viscosity is lowered. Fiber spinning at a greater-than critical rate is accompanied by a shear-induced conformational change, that enables molecules to pack and bond in a crystalline structure that is no longer water-soluble (Figure 8). This mechanism, pioneered in nature, may serve as the basis for the development of novel synthetic polymer processing routes that enable the fabrication of durable, water-insoluble fibers from molecules initially in aqueous solution.

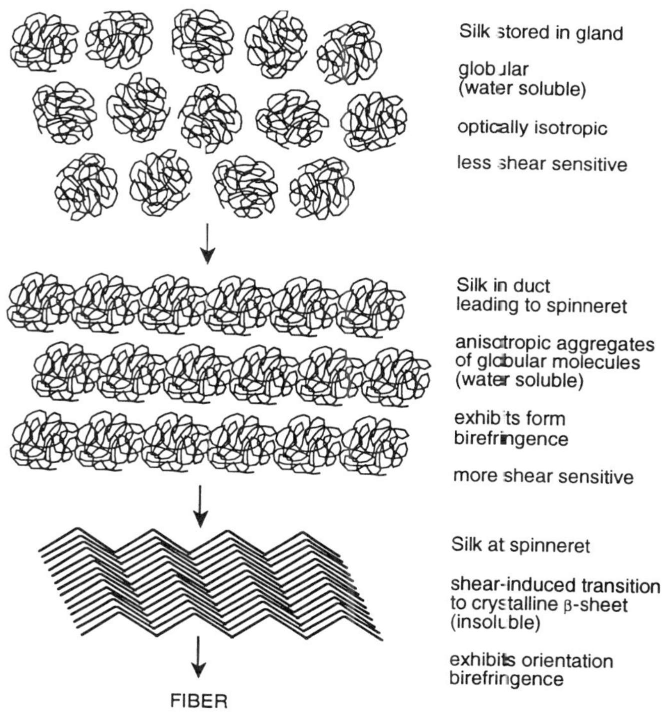

Figure 8. Schematic representation of the molecular order and properties of silk secretion at various stages in the spinning of a fiber.

Conditions that Favor the Formation of Liquid Crystalline Order in Silk Solutions. It is difficult to accurately measure and control the water content of the small samples used for microscopy. Reliable quantitative information about the concentration range of liquid crystal stability therefore is limited. Yet, even qualitative observations show that the nematic phase forms under a wide range of conditions:

- The ability to form a liquid crystalline phase is exhibited by silks of widely differing compositions, i.e it is not sensitive to the sequence of monomers; indeed, in the case of *Chironomus* silk, one is dealing with a mixture of proteins whose apparent molecular weights span two orders of magnitude (*6*).
- Silk can form liquid crystalline phases in substantially different environments – in the natural secreted state (containing only a trace of dissolved components that are not fibroin (*3*)) as well as in the presence of a significant concentration of LiBr as reported above.
- In specimens of *B. mori* and *N. clavipes* silk secretion (initial concentration ~26 vol.% or ~30 wt.% protein) that are maintained between a glass microscope slide and cover slip, a liquid crystalline phase can be recognized within minutes, i.e. after only a small change in concentration due to evaporation. A more ordered phase typically does not occur until after 2-3 days of slow drying, i.e. at a significantly higher concentration.

However, whether or not a continuously drying silk solution exhibits liquid crystalline order does appear to be *sensitive to the initial concentration, and to the rate at which water is lost* (*40*). Fluid solutions, such as water-diluted *Chironomus* silk secretions, initially form liquid crystalline phases if allowed to dry slowly between a glass slide and cover slip. If no cover slip is used, then crystalline phases are formed. Viscous solutions, such as the undiluted silk secretions from *B. mori* and *N. clavipes*, form a liquid crystalline phase whether or not a cover slip is present. A similar dependence of microstructure on initial concentration and subsequent drying rate is exhibited by other types of biological macromolecule (*51*).

Kinetics of Phase Transitions

Transformation Diagrams. *Phase* diagrams display the phases that occur under equilibrium conditions in an effectively closed system. In an open system in which equilibrium is not reached, the phases obtained depend on the driving force for transformation (how far is the system from equilibrium?) and on the extent to which the microstructure is free to rearrange in response to this driving force (what is the magnitude of the relevant diffusion coefficients?).

The kinetics of temperature-induced phase transformations in multi-component metals and ceramics are routinely displayed on Time-Temperature-Transformation (TTT) curves (*61*). These curves (Figure 9a) illustrate how microstructure changes with time in a system *of fixed composition* at different temperatures. The curves exhibit a characteristic "C" shape because:

- At small supercoolings (close to equilibrium), there is little driving force for the phase change to occur.
- At large supercoolings, there is significant driving force for the system to convert to the new equilibrium phase, but the atoms or molecules now lack the mobility to achieve this transformation quickly.
- At intermediate supercoolings, the combination of driving force and mobility enables the transformation to occur at a maximum rate.

Typically, TTT curves are drawn for 1% completion and 99% completion of the transformation, though a single curve is often adequate for illustrative purposes. The time difference between 1% completion and 99% completion is most significant for transformations that occur entirely in the solid state.

An analogous graphical device provides a useful intellectual framework for illustrating the kinetics of phase transformation in polymer solutions (Figure 9b). For transfor-

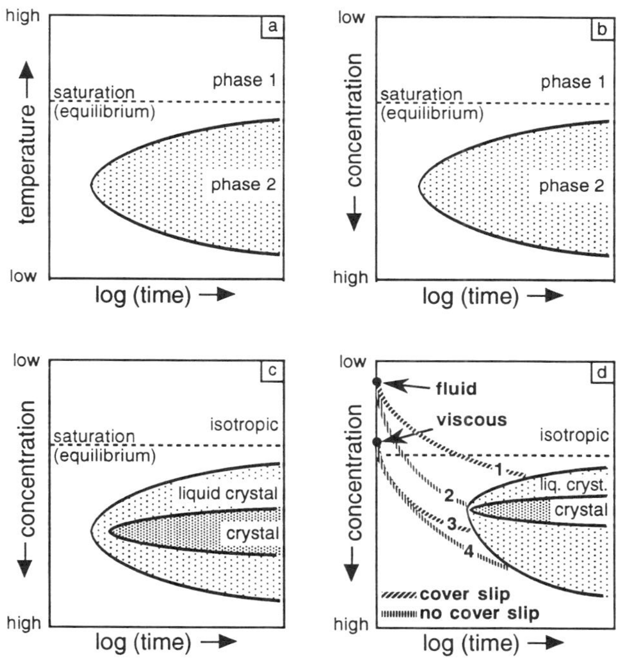

Figure 9. Transformation diagrams, schematically showing phase transition kinetics. (a) Time-Temperature-Transformation (TTT) relationship for phase changes induced by cooling from a melt. (b) Analogous Time-Concentration-Transformation (TCT) relationship for phase changes induced by an increase in concentration. (c) TCT curves for a lyotropic system that is losing solvent. (d) TCT curves schematically representing behavior of dilute and viscous solutions of silk. Adapted from ref. 40.

mations in solutions of different concentrations *at a fixed temperature*, the vertical axis is now scaled in terms of concentration. The behavior of the system at varying levels of *supersaturation* is thus shown, rather than its behavior at varying degrees of supercooling. Transformations at low supersaturations would occur slowly, due to the system being close to equilibrium, and transformations at very high supersaturations would also occur slowly, since chain interactions would limit microstructural mobility. Again, therefore, a maximum transformation rate is expected at intermediate supersaturations, where the combination of kinetic drive and molecular mobility is optimized. Given that the transformation from isotropic material to a mesophase is kinetically easier (requires less microstructural rearrangement) than the precipitation of a three-dimensionally ordered crystalline phase, the transformation curves for mesophase formation and crystallization should be nested (Figure 9c).

The format of these curves is qualitatively consistent with the observed behavior of silk solutions (Figure 9d). Dilute specimens maintained between glass slides (and thus losing water relatively slowly; curve 1) can form liquid crystalline phases, and may be able to crystallize at very long times. Dilute specimens without a cover slide will start at the same concentration as those maintained between glass surfaces, but their concentration will change more rapidly with time. Thus the "drying curve" of such a specimen (curve 2) may be able to intercept the "nose" of the crystallization curve. Initially concentrated specimens with (curve 3) or without (curve 4) a cover slide do not have sufficient microstructural mobility to crystallize immediately after transformation to the liquid crystalline state. The smooth curves (1-4) showing qualitative concentration change with time in Figure 9d are schematic; they are discontinuous at the "C" curve for transformation to the liquid crystalline state because the diffusion coefficients of solvent will be different in liquid crystalline and crystalline material.

Effect of Draw Rate on Molecular Alignment and Fiber Properties. The apparatus shown in Figure 4 was used to acquire *N. clavipes* dragline at linear velocities of 0.49 $cm.s^{-1}$, 1.06 $cm.s^{-1}$ (approximately the natural rate of fiber spinning) and 1.96 $cm.s^{-1}$. Birefringence measurements were performed with a Berek compensator (*62*), yielding values of 0.025±0.001, 0.033±0.001 and 0.041±0,001 for the three fiber draw rates, respectively (*39*). Each birefringence value is an average obtained from 5 samples of similar thickness.

The molecular order in dragline is expected to depend on silking rate, for at least two reasons. The first, is that the formation and stability of the liquid crystalline state is kinetics-dependent as demonstrated above. The second, is that the volume fraction, perfection, and global alignment of crystalline β-sheet regions will depend on the magnitude and duration of the applied load. The measurements of birefringence do not enable these factors to be assessed separately, but they demonstrate that the overall degree of molecular alignment (and therefore the magnitude of alignment-dependent properties such as the fiber stiffness) can be increased beyond the already impressive values exhibited by naturally spun material.

Acknowledgments

We have benefitted from several stimulating conversations with Drs David Kaplan and Lisa Gilliland. CV acknowledges support from the US Army Research, Development & Engineering Center (Natick MA; contract #DAAK60-91-K-005), the Donors of the American Chemical Society Petroleum Research Fund (grant #25218-AC), the National Science Foundation (grant #BCS-9202007, jointly sponsored by the Division of Biological and Critical Systems and the Division of Materials Research), the Whitaker Foundation and the Washington Technology Center. STC acknowledges support from the U.S. Army Research Office (grant #DAAL03-91-G-0239).

Literature Cited

1. Zemlin, J.C. *A study of the mechanical behavior of spider silks*, Report # 69-29-CM (AD 684333), U.S. Army Natick Laboratories: Natick, MA, 1968.
2. Gosline, J.M.; DeMont, M.E.; Denny, M.W. *Endeavour* **1986**, *10*, 37-43.
3. Kaplan, D.L.; Lombardi, S.J.; Muller, W.S.; Fossey, S.A. In *Biomaterials: Novel Materials from Biological Sources*; Byrom, D., Ed.; Stockton Press: New York, 1991; pp. 3-53.
4. Kaplan, D.L.; Fossey, S.; Mello, C.M.; Arcidiacono, S.; Senecal, K.; Muller, W.; Stockwell, S.; Beckwitt, R.; Viney, C.; Kerkam, K. *MRS Bulletin* **1992**, *17 (10)*, 41-47.
5. Foelix, R.F. *Biology of Spiders;* Harvard Univ. Press: Cambridge, MA, 1982.
6. Case, S.T.; Wieslander, L. In *Structure, Cellular Synthesis and Assembly of Biopolymers*; Case, S.T., Ed.; Springer-Verlag: Berlin, 1992; pp. 187-226.
7. Lucas, F.; Shaw, J.T.B.; Smith, S.G. In *Advances in Protein Chemistry*; Anfinsen, C.B.; Anson, M.L.; Bailey, K.; Edsall, J.T., Eds.; Academic Press: New York, 1958; Vol. 13; pp. 107-242.
8. Peakall, D.B. *J. Exp. Zool.* **1971**, *176*, 257-264.
9. Vollrath, F. *Scientific American* **1992**, *266 (3)*, 70-76.
10. Cox, M.K. *Mol. Cryst. Liq. Cryst.* **1987**, *153*, 415-422.
11. Ryan, T.G. *Mol. Cryst. Liq. Cryst.* **1988**, *157*, 577-596.
12. Collings, P.J. *Liquid Crystals: Nature's Delicate Phase of Matter*; Princeton Univ. Press: Princeton, NJ, 1990.
13. Weiss, R.A.; Ober, C.K., Eds.; *Liquid-Crystalline Polymers*; American Chemical Society: Washington, DC, 1990.
14. Donald, A.M.; Windle, A.H. *Liquid Crystalline Polymers;* Cambridge University Press: Cambridge, UK, 1992.
15. Gray, G.W.; Goodby, J.W.G. *Smectic Liquid Crystals: Textures and Structures;* Leonard Hill: Glasgow, UK, 1984.
16. Bassett, D.C. *Principles of Polymer Morphology;* Cambridge University Press: Cambridge, UK, 1981.
17. Billmeyer, F.W. *Textbook of Polymer Science;* Wiley: New York, 1984.
18. Lucas, F.; Rudall, K.M. In *Comprehensive Biochemistry: Extracellular and Supporting Structures*; Florkin, M.; Stotz, E.H., Eds.; Elsevier: Amsterdam, 1968; Vol. 26B; pp. 475-558.
19. Magoshi, J.; Magoshi, Y.; Nakamura, S. *J. Appl. Polym. Sci.: Appl. Polym. Symp.* **1985**, *41*, 187-204.
20. Magoshi, J.; Magoshi, Y.; Nakamura, S. *Polym. Comm.* **1985**, *26*, 60-61.
21. Li, G.; Yu, T. *Makromol. Chem., Rapid Comm.* **1989**, *10*, 387-389.
22. Riande, E.; Saiz, E. *Dipole Moments and Birefringence of Polymers;* Prentice Hall: Englewood Cliffs, NJ, 1992.
23. Iizuka, E. *J. Appl. Polym. Sci.: Appl. Polym. Symp.* **1985**, *41*, 173-185.
24. Dobb, M.G.; McIntyre, J.E. *Adv. Polym. Sci.* **1984**, *60/61*, 61-98.
25. Gathercole, L.J.; Barnard, K.; Atkins, E.D.T. *Int. J. Biol. Macromol.* **1989**, *11*, 335-338.
26. Bouligand, Y.; Giraud-Guille, M.-M. In *Biology of Invertebrate and Lower Vertebrate Collagens*; Bairati, A.; Garrone, R., Eds.; Plenum: New York, 1985; pp. 115-134.
27. Giraud-Guille, M.-M. *Mol. Cryst. Liq. Cryst.* **1987**, *153*, 15-30.
28. Kerst, A.; Chmielewski, C.; Livesay, C.; Buxbaum, R.E.; Heidemann, S.R. *Proc. Nat. Acad. Sci. USA,* **1990**, *87*, 4241-4245.
29. Suzuki, A.; Maeda, T.; Ito, T. *Biophys. J.* **1991**, *59*, 25-30.
30. Demus, D.; Richter, L. *Textures of Liquid Crystals;* Verlag Chemie: Weinheim, 1978.
31. Dannels, C.M.; Viney, C. *Polymer News* **1991**, *16*, 293-302.

32. Frank, F.C. *Discuss. Farad. Soc.* **1958**, *25*, 19-28.
33. Bennett, H.S. In *McClung's Handbook of Microscopical Technique*; McClung Jones, R., Ed.; Hafner Publishing Co.: New York, 1950 (repr. 1967); pp. 591-677.
34. Cokendolpher, J.C.; Brown, J.D. *Entomol. News* **1985**, *96*, 114-118.
35. Case, S.T.; Wellman, S.E.; Hamodrakas, S. In *Materials Synthesis Based On Biological Processes*; Alper, M.; Calvert, P.; Frankel, R.; Rieke, P.C.; Tirrell, D., Eds.; Materials Research Society: Pittsburgh, PA, 1991; Vol. 218; pp. 233-237.
36. Wilson, R.S. *Quart. J. Microsc. Sci.* **1962**, *104*, 557-571.
37. Kerkam, K.; Viney, C.; Kaplan, D.L.; Lombardi, S.J. *Nature* **1991**, *349*, 596-598.
38. Kerkam, K.; Kaplan, D.L.; Lombardi, S.J.; Viney, C. In *Materials Synthesis Based on Biological Processes*; Alper, M.; Calvert, P.D.; Frankel, R.; Rieke, P.C.; Tirrell, D., Eds.; Materials Research Society: Pittsburgh, 1991; Vol. 218; pp. 239-244.
39. Viney, C.; Kerkam, K.; Gilliland, L.K.; Kaplan, D.L.; Fossey, S. In *Complex Fluids*; Sirota, E.B.; Weitz, D.; Witten, T.; Israelachvili, J., Eds.; Materials Research Society: Pittsburgh, 1992; Vol. 248; pp. 89-94.
40. Viney, C.; Huber, A.; Dunaway, D.; Case, S.T.; Kaplan, D.L. In *Biomolecular Materials*; Viney, C.; Case, S.T.; Waite, J.H., Eds.; Materials Research Society: Pittsburgh, 1993; Vol. 292; pp. 211-217.
41. Economy, J.; Volksen, W.; Viney, C.; Geiss, R.; Siemens, R.; Karis, T. *Macromolecules* **1988**, *21*, 2777-2781.
42. Dannels, C.M.; Viney, C.; Twieg, R.J.; Chang, M.Y. *Mol. Cryst. Liq. Cryst.* **1991**, *198*, 341-350.
43. Yamaura, K.; Okumura, Y.; Ozaki, A.; Matsuzawa, S. *J. Appl. Polym. Sci.: Appl. Polym. Symp.* **1985**, *41*, 205-220.
44. Hamodrakas, S.J.; Kafatos, F.C. *J. Mol. Evol.* **1984**, *20*, 296-303.
45. Viney, C. In *Structure, Cellular Synthesis and Assembly of Biopolymers*; Case, S.T., Ed.; Springer Verlag: Heidelberg, 1992; pp. 255-278.
46. Asakura, T. *Makromol. Chem., Rapid Comm.* **1986**, *7*, 755-759.
47. Zheng, S.; Li, G.; Yao, W.; Yu, T. *Appl. Spect.* **1989**, *43*, 1269-1272.
48. Wellman, S.E.; Hamodrakas, S.J.; Kamitsos, E.I.; Case, S.T. *Biochim. Biophys. Acta* **1992**, *1121*, 279-285.
49. Viney, C.; Windle, A.H. *Liq. Cryst.* **1986**, *1*, 379-396.
50. Viney, C.; Yoon, D.Y.; Reck, B.; Ringsdorf, H. *Macromolecules* **1989**, *22*, 4088-4093.
51. Viney, C.; Huber, A.; Verdugo, P. In *Fundamentals of Biodegradable Materials and Packaging*; Kaplan, D.L.; Thomas, E.L.; Ching, C., Eds.; Technomic: Lancaster, PA, 1993; in press.
52. Flory, P.J. *Proc. Roy. Soc. Lond.* **1956**, *A234*, 73-89.
53. Flory, P.J.; Ronca, G. *Mol. Cryst. Liq. Cryst.* **1979**, *54*, 289-310.
54. Flory, P.J. *Adv. Polym. Sci.* **1984**, *59*, 1-36.
55. Tsujimoto, Y.; Suzuki, Y. *Cell* **1979**, *18*, 591-600.
56. Strydom, D.J.; Haylett, T.; Stead, R.H. *Biochem. Biophys. Res. Comm.* **1977**, *79*, 932-938.
57. Rawn, J.D. *Proteins, Energy, and Metabolism;* Neil Patterson Publ.: Burlington, NC, 1989.
58. Xu, M.; Lewis, R. V. *Proc. Nat. Acad. Sci. USA* **1990**, *87*, 7120-7124.
59. Stryer, L. *Biochemistry;* W.H. Freeman and Company: New York, 1988.
60. Papkov, S.P. *Adv. Polym. Sci.* **1984**, *59*, 75-102.
61. Porter, D.A.; Easterling, K.E. *Phase Transformations in Metals and Alloys*; Van Nostrand Reinhold: Wokingham (UK), 1984.
62. Viney, C. *Transmitted Polarised Light Microscopy;* McCrone Research Institute: Chicago, 1990.

RECEIVED June 29, 1993

Chapter 12

Structural Evolution of Genetically Engineered Silklike Protein Polymers

J. Philip Anderson[1], Matthew Stephen-Hassard[2], and David C. Martin[1,2]

[1]Macromolecular Science and Engineering Center and [2]Department of Materials Science and Engineering, College of Engineering, University of Michigan, Ann Arbor, MI 48109–2136

The effect of processing on the evolution of crystal structure and morphology in a silk-like protein polymer, SLPF, were examined by transmission electron microscopy. A schematic ternary phase diagram was developed for the SLPF, water and formic acid system. Polymer swelling by solvent vapor was monitored by wide angle x-ray scattering. The disruption of the SLPF crystalline structure occured in steps consistent with the Lotz and Keith crankshaft model of Silk I.

Via recombinant DNA technology, designed protein polymers can be produced to obtain specific physical, chemical and biological properties of interest (*1*). These materials incorporate functional sequences which are modeled after segments of existing natural proteins. The synthesis of well-defined macromolecules by molecular biological techniques has been used to produce a new class of materials which is rapidly expanding (*2-4*). For the optimal application and future design of these protein polymers it necessary to understand the relationship between processing, structure and function.

SLPF (silk-like polymer with fibronectin cell attachment functionality) is one such protein polymer, commercially designed and produced by Protein Polymer Technologies, Inc. for general use as a stable promoter of cellular attachment to a variety of surfaces. The amino acid sequence of SLPF is shown in Figure 1 by single character code (5). SLPF can be used to coat artificial surfaces like polystyrene culture dishes to promote both the adhesion and spreading of cells (*6-7*). To optimize performance for a variety of applications *in vitro* and *in vivo*, it is of interest to determine the influence of crystal structure and morphology on the biological and physical properties of SLPF. The utility of SLPF depends on generating autoclavible and biocompatible crystalline polymorphs and morphologies which have the ability to present cell attachment sites in an accessible manner.

The crystalline structure of natural silk has been explored by x-ray diffraction (*8-10*). In an attempt to gain more information on the fiber processing *in vivo*, the silk producing glands from *Bombyx mori* were air dried and on x-ray analysis they were found to have a noticeably different crystal structure from the fibrous silk normally extruded from the insect (*11*). The unextruded protein structure (Silk I) was found to be easily drawn into the extruded polymorph, Silk II. Lotz and Keith proposed an orthorhombic unit cell for Silk I [space group $P2_12_12_1$ (a = 0.472 nm, b = 1.44 nm,

0097–6156/94/0544–0137$06.00/0

c = 0.96 nm)] from electron microscopy, selected area electron diffraction and x-ray powder diffraction experiments on single crystals of poly(glycine-alanine) (*12*). The Lotz and Keith model for Silk I is a compact sheet form with the polypeptide backbone in a crankshaft conformation. The molecules pack in the unit cell so that the (020) distance between hydrogen-bonded sheets is on the order of 0.74 nm (Figure 2).

To provide a model consistent with the antiparallel character observed in extended forms similar to Silk I, the complete crankshaft model consists of sheets composed of the antiparallel chains, in a regular alternating pattern. As in the extended β pleated sheets, Silk I crankshaft sheets stack to fill space. The face of the resulting solid, perpendicular to the chains, is thought to be composed of regular hairpin turns, where the chains switch direction and proceed back through the crystal.

Studies of the crystal structure and morphology of SLPF films are described in more detail elsewhere (*13*). Dry SLPF powder and thin films were found to be consistent with the crankshaft model for Silk I. The fundamental component of SLPF films are whisker-shaped crystallites which have a nearly monodisperse width (12 nm +/- 2 nm) and variable lengths. These whisker crystallites associate into larger aggregates which are termed sheaves. Here we describe how crystallinity and morphology of SLPF varies with processing conditions.

This research is focused on the structural evolution of genetically engineered protein polymers during processing into films and fibers. Currently, the manner in which higher order structure develops from folding and cooperative association in macromolecular systems is a central theme of biophysics and materials science. Because of the fidelity of control which it is possible to exert with the genetic methods of polymer synthesis we expect these systems to provide access to a variety of unanswered questions concerning the nature of self-organization and phase stability of condensed macromolecules.

Experimental

SLPF (batch S4) was supplied by Protein Polymer Technologies, Inc. in powder form (ProNectin F). The theoretical sequence, predicted genetically by the synthetic nucleotide sequence which encodes SLPF, is shown in Figure 1 by the single character amino acid code (*5*) and corresponds to a molecular weight of 73 kDa. The segmented copolymer was produced by recombinantly engineered *E. coli* cells, then purified from other cellular components by a process involving the extraction of the lysate with 5M LiBr solution. The protein was precipitated using ammonium sulfate, washed, and dialyzed against deionized water. The insoluble SLPF suspension was then lyophilized to a dry powder.

Transmission electron microscopy (TEM) and wide angle x-ray scattering (WAXS) were used in conjunction with solubility and swellability experiments to obtain basic information relevant to the processing of SLPF. All experiments were performed at room temperature.

TEM. Sample preparation for TEM involved carbon coating freshly cleaved mica sheets to form a continuous amorphous film in a Denton evaporator. Formic acid (95-97%) was purchased from Aldrich Chemical Co. and used as received. Solutions containing typically one mg of SLPF per ml of formic acid were atomized and sprayed onto the carbon film. The carbon coated mica, covered with solution droplets, was left in a laboratory hood. Upon solvent evaporation, which occurred in seconds, crystallized droplets of SLPF were left on the surface of the carbon film. This film was then floated off the mica onto deionized water. Submerged 400 mesh copper TEM grids were lifted with tweezers to capture SLPF droplet-covered carbon film. The samples were examined with a Phillips EM 420 operating at 120 kV. A variety of magnifications were used, typically 3 kX to 20 kX.

{HEAD}- [(Silk-Like)$_9$Fibronectin Segment]$_{12}$-{TAIL}

=

{(fM)DPVVLQRRDWENPGVTQLNRLAAHPPFASDPMGAGS(GAGAGS)$_6$GAAVTGRGDSPASAAGY}

-[(GAGAGS)$_9$GAAVTGRGDSPASAAGY]$_{12}$-

{(GAGAGS)$_2$GAGAMDPGRYQLSAGRYHYQLVWCQK}

Figure 1. Amino acid sequence of SLPF showing the design which consists of a head, crystallizable center, and tail. Here we use the single-letter amino acid code (5). G, A, and S code for glycine, alanine, and serine, the principal amino acids in crystalline regions of *Bombyx mori* silk.

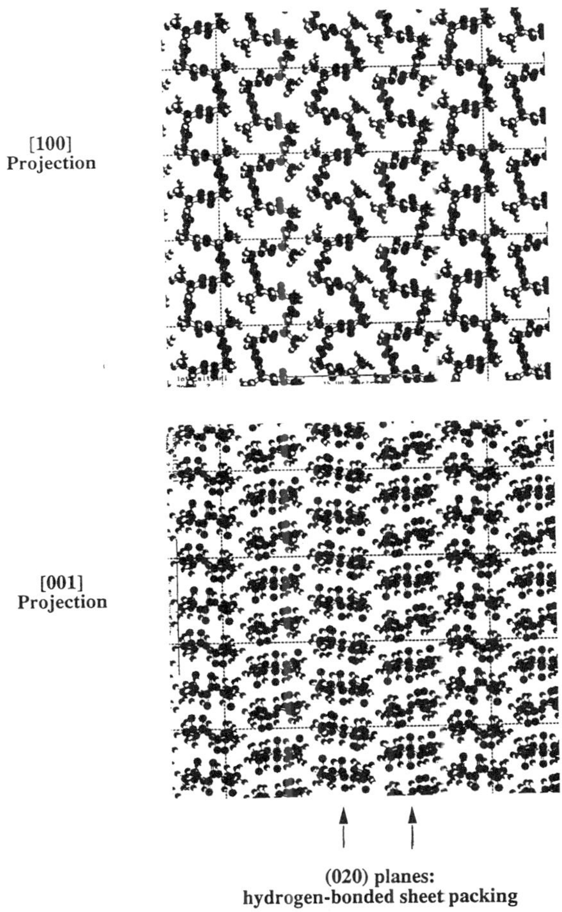

Figure 2. Schematic diagram of the Lotz and Keith crankshaft model for Silk I shown in the [100] and [001] crystallographic projections. The crankshaft conformation of the chain is evident in the [100] projection. The molecules hydrogen-bond face to face, with the result that the spacing between the hydrogen bonded sheets (i.e., the (020) spacing) is on the order of 0.7 nm.

Variations in polymer concentration and water content of the solutions lead to a variety of SLPF morphologies and degrees of crystallinity. The nature and extent of SLPF crystallinity was estimated by observation and distinction of SLPF whisker crystallites and sheaf structures. Other morphological features of the films were observed as described in more detail below.

Solubility and Swelling. SLPF on 0.5 mm etched glass WAXS sample holders were exposed to a variety of atmospheres produced by a range of water and formic acid mixtures. Dry SLPF power was held above a 50 ml formic acid/water solution, in a closed petri dish. Solution composition was varied from 0 to 100% formic acid. The duration of swelling by solvent vapor was 24 hours. Weight gain of the samples was monitored with a Mettler AE analytical balance, while structural changes were assayed by WAXS. The solubility of SLPF in mixtures of water and formic acid was examined by adding increasing amounts of polymer until precipitation occurred.

WAXS. A Rigaku Geigerflex theta-theta x-ray system was used to perform WAXS on dry SLPF powder and solvent-swollen films. The instrument was equipped with a 1.5 kW Cu Kα tube which emits x-ray radiation at a wavelength of 0.154 nm, a graphite monochromator and a NaI scintillation type detector. Detailed WAXS runs were conducted at 0.5 degrees per minute with sampling at 0.05 degree intervals. The samples are held horizontally in the instrument with both the source and detector moving during the WAXS experiment, making it possible to analyze the microstructure during transitions from swollen films and solutions to dry powder and vice versa.

Results and Discussion

Part of a ring droplet with an interior consisting of plane oriented whiskers and a periphery of radially oriented sheaves shown in Figure 3. The whisker crystallites create a patchwork structure in the droplet interior. Immediately beyond the sheaves at the edge of the droplet there was an outer ring of apparently amorphous material. The lack of contrast between the whiskers and their matrix supports the idea that uncrystallized material is surrounding these well crystallized whiskers. In ring droplets like that shown in Figure 3, there is often less SLPF in the center then at the periphery.

There is further evidence for semi-crystallinity in SLPF droplets. The photomicrograph in Figure 4 shows regions entirely devoid of SLPF, indicated by the lightest contrast. The slightly darker patches are believed to be amorphous SLPF. Crystallization consistent with dense whisker nuclei is found only in the thickest parts of the film.

Shown in Figure 5 is a highly crystalline film in which the sheaf morphology is well developed. This film is reminiscent of the spherulitic texture commonly seen in thin films of crystallizable synthetic polymers (*14*) and previously observed in natural proteins (*15*). This data indicates that synthetic and natural proteins behave in some manners consistent with conventional semi-crystalline polymers. As more information becomes available, it is reasonable to expect additional areas where proteins can serve as model systems for understanding polymer morphology and organization or vice versa.

In general, experiments have indicated that the morphology becomes less well defined as the amount of water content increases. At low water content, the whisker crystallites are large and distinct. As the amount of water increases, the amorphous fraction of the film grows as the whiskers shorten and distort.

To reveal the relationship between processing pathway and structure we have developed a ternary phase diagram scheme to describe the relative amount of protein polymer (SLPF), solvating component (formic acid) and non-solvating component (water) during film or fiber formation, as shown in Figure 6. Because of the fact that SLPF is not fully soluble in water but is nearly so in formic acid, there is two-phase

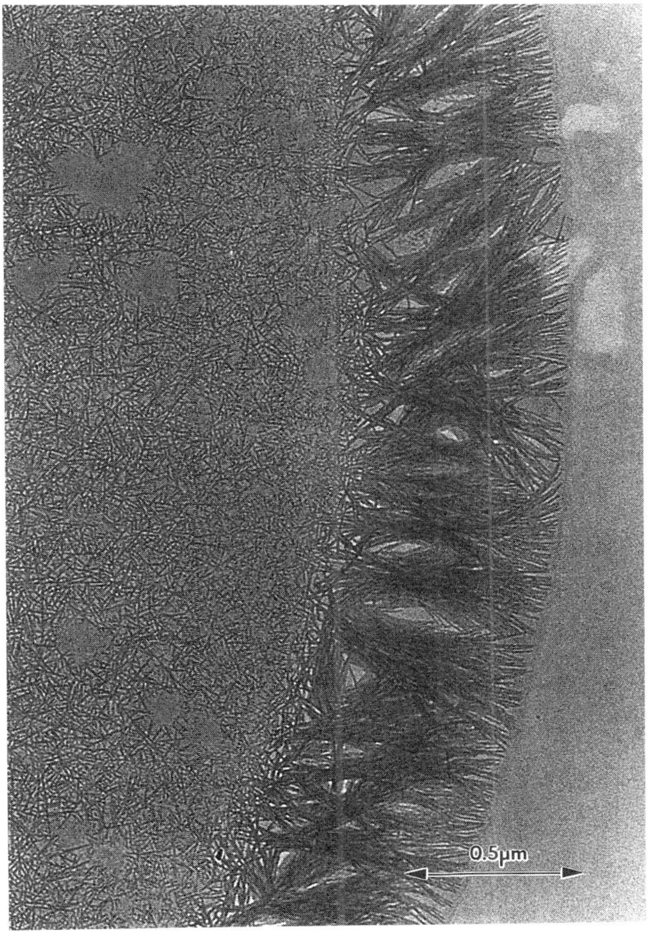

Figure 3. TEM micrograph of an SLPF droplet with patchwork whisker crystalline interior and radially oriented sheaf periphery.

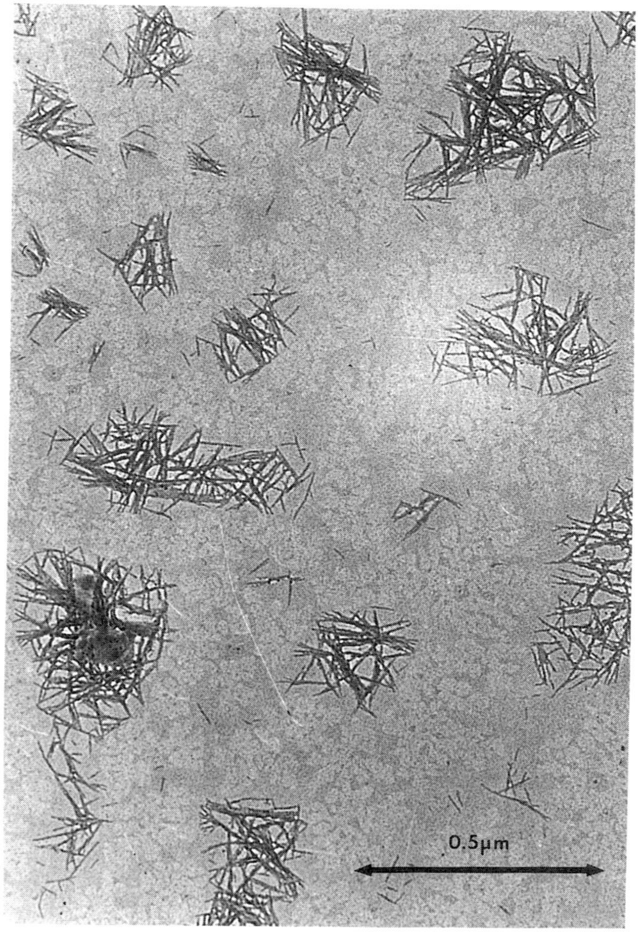

Figure 4. TEM micrograph with droplets showing a variation in the extent of crystallinity with droplet thickness. There is evidence for amorphous material in the thinnest regions of the film.

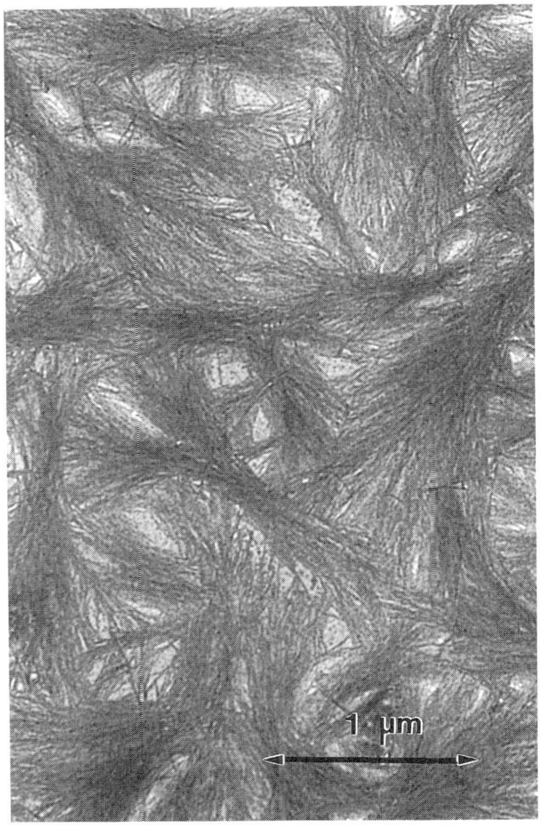

Figure 5. TEM micrograph showing spherulitic structure of SLPF similar to that commonly seen in thin films of semi-crystalline synthetic polymers.

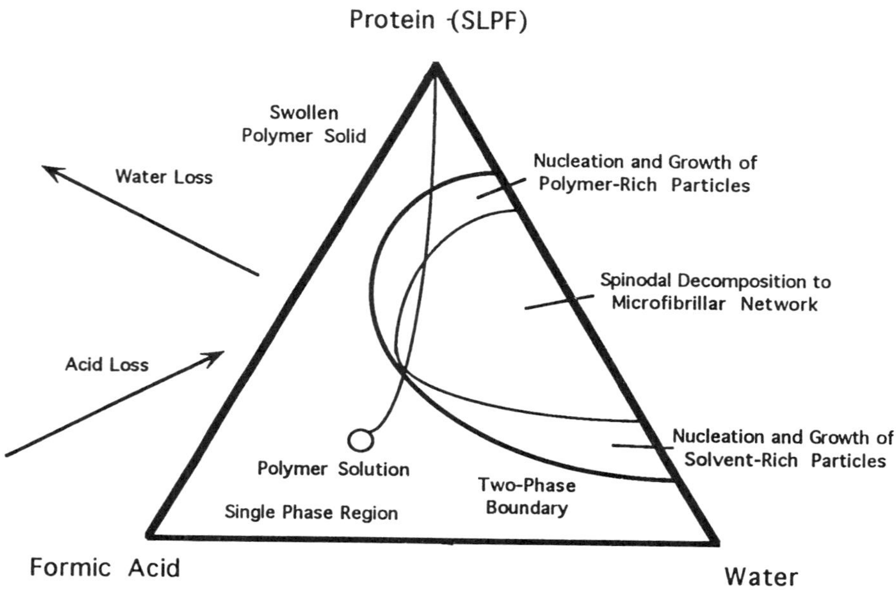

Figure 6. Ternary phase diagram between protein (SLPF), solvent (formic acid) and non-solvent (water). The processing pathway starts with a solution (circle) and ends with solid protein.

region present in the diagram which is located between at least two one-phase regions corresponding to polymer in solution and polymer swollen with solvent. Although regions of liquid crystalline mesophase stability have been observed in natural silk *(16)*, to date there is no evidence for liquid crystalline structure in SLPF-formic acid-water or SLPF-LiBr-water mixtures at any composition.

The processing of solid protein film or fiber from solution involves the loss of both solvating component (formic acid) and non-solvating component (water). The specific pathway taken depends on the relative concentration of these components in the solution during drying to solid. It is expected that there should be three regimes associated with the phase transition which would occur upon entering the two-phase stability region: (a) nucleation of discrete droplets of a polymer rich phase, (b) nucleation of discrete droplets of a solvent rich phase, or (c) the precipitation of a bicontinuous polymer-solvent microfibrillar network. It is known that the formation of a microfibrillar network is an important aspect of fiber formation during processing of lyotropic liquid crystalline polymers *(17)*, it is therefore reasonable to presume that similar concerns may be important in the processing of protein fibers. The optimal processing routes may require passing through the critical point, and therefore it will be necessary to know and control the pathway taken by the solution during film formation. This could be done by systematic variations in the relative amount of solvating component (formic acid or salt) and water during drying.

To determine the position of the two-phase boundary and the corresponding transitions in structure we examined the swelling of SLPF in saturated atmospheres of formic acid/water of various composition. Figure 7 shows the ratio of final weight to initial weight for SLPF after 24 hours exposure of formic acid/water as a function of the formic acid composition. It is evident that there is a significant increase in the polymer swelling above compositions of approximately 70% formic acid.

WAXS data from dry Silk I powder is shown in Figure 8. The data shows several distinct Bragg reflections, with two predominant, intense peaks at d-spacings of

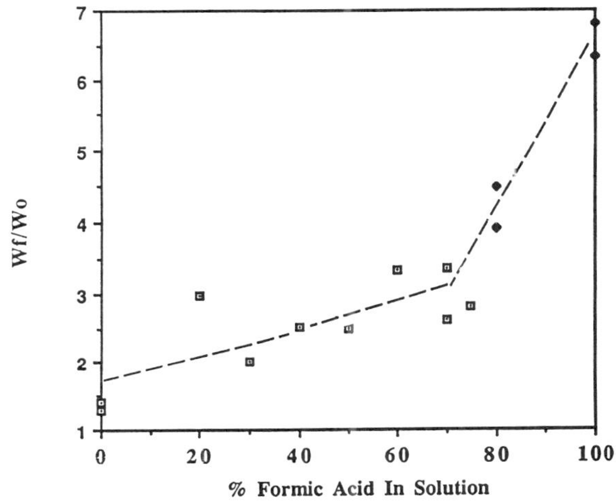

Figure 7. Weight gain of SLPF powder after 24 hours exposure to formic acid water mixtures as a function of concentration of formic acid. There is a significant increase in the amount of solvent absorbed above 70% formic acid.

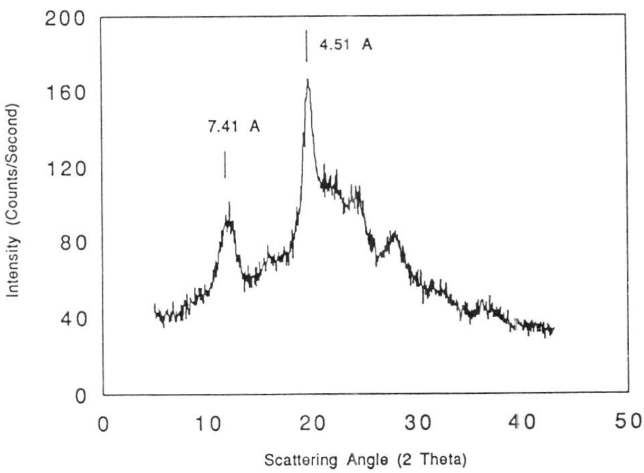

Figure 8. WAXS pattern of dry SLPF powder. There are several distinct Bragg reflections, with two predominant, intense peaks at d-spacings of 0.74 nm and 0.45 nm. The 0.74 nm peak corresponds to the (020) spacing between hydrogen bonded sheets in the crankshaft model for Silk I (12) (Figure 2). Hydrogen bonding contributes to the sharper (210) spacing at 0.45 nm.

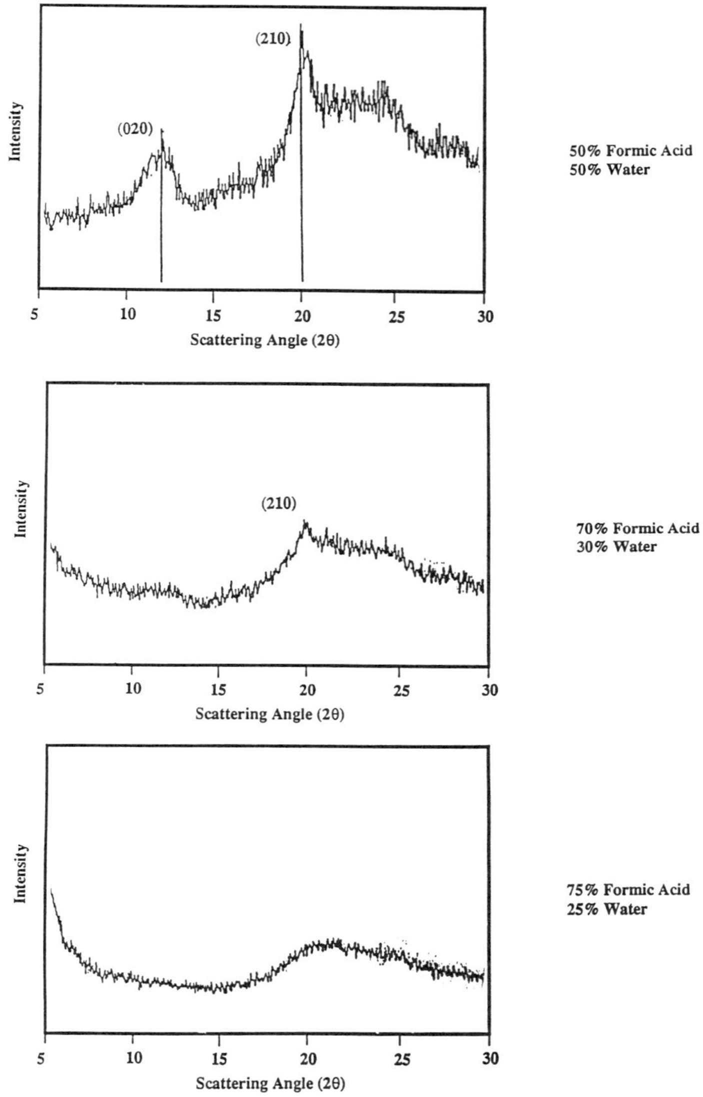

Figure 9. WAXS data of SLPF powder in saturated atmosphere after equilibrium swelling. For 50/50 formic acid / water mixtures and below the pattern remains similar to that for the dry powder (Figure 7). SLPF loses crystalline reflections after exposure to 75/25 formic acid/water. In the SLPF WAXS pattern after exposure to 70/30 formic acid/water the 0.74 nm (020) reflection corresponding to the sheet-to-sheet packing dimension of the crankshaft form for Silk I has nearly disappeared, whereas the 0.45 nm (210) peak still remains. This data indicates that when SLPF is denatured the sheet-to-sheet correlations are lost before the hydrogen bonding within the sheet is disrupted.

0.74 nm and 0.45 nm. The 0.74 nm peak corresponds to the (020) spacing between sheets in the crankshaft model for Silk I (12) (Figure 2). Hydrogen bonding contributes to the sharper (210) spacing at 0.45 nm.

The WAXS analysis of solvated SLPF samples, again as a function of the concentration of the solvent mixture is shown in Figure 9. Below 70% formic acid, the sample retains the WAXS pattern consistent with Silk I crystal structure and is nearly identical to the WAXS pattern from dry powder SLPF. At 75% formic acid the crystalline peaks have disappeared. However, at 70% formic acid the 0.74 nm peak (corresponding to the (020) non-hydrogen bonded planes) has nearly disappeared, whereas the 0.45 nm (210) peak (which involves the hydrogen bonding direction) still remains. Interpreting this data in terms of the Lotz and Keith crankshaft model suggests that the dissolution of crystalline SLPF into solution occurs in steps, with the sheet stacking disrupted first followed by the breakup of the hydrogen bonded sheets. This model is consistent with the fact that the hydrogen bonds securing the sheet are stronger than the van der Waals forces responsible for sheet stacking.

This information suggests that the reverse process of forming SLPF solid films from solution, like the droplets in TEM images, may also occur in stages. The formation of hydrogen bonded planes of molecular aggregates apparently nucleates the whisker crystallites by promoting the stacking of sheets. The whiskers then aggregate into sheaf structures. The amorphous material present may have precipitated out of solution by solvent evaporation before it had time to organize.

Acknowledgments

The authors would like to thank Dr. Joseph Cappello of Protein Polymer Technologies, Inc. for supplying the SLPF polymer and for technical and financial assistance. DCM also acknowledges partial support from the National Young Investigator program of the National Science Foundation.

Literature Cited

(*1*) Cappello, J. *MRS Bulletin,* **1992,** *17,* 48.
(*2*) Cappello, J.; Crissman, W.; Dorman, M.; Mikolajczak, M.; Textor, G.; Marquet, M.; Ferrari, F. A. *Mat. Res. Soc. Symp. Proc.* **1990,** *174,* 267.
(*3*) Tirrell, D. A.; Fournier, M. J.; Mason, T. L. *Curr. Opin. Struct. Biol.* **1991,** *1,* 638.
(*4*) Masilamani, D.; Goldberg, I.; Salerno, A. J.; Oleksiak, M. A.; Unger, P. D; Piascik, D. A.; Bhattacharjee, H. R. In *Biotechnology and Polymers,* Gebelein, C. G., Ed.; Plenum Press: New York, 1991.
(*5*) Stryer, L. *Biochemistry, 3rd ed.;* W.H. Freeman: New York, 1988.
(*6*) Esty, A. *Biomedical Products* **1992,** *16,* 76.
(*7*) Esty, A. *American Biotechnology Laboratory* **1991,** *9,* 44.
(*8*) Shimizu, M. *Bull. Imp. Sericult. Expt. Sta. Japan* **1941,** *10,* 475.
(*9*) Kratky, O.; Schauenstein, E.; Sekora, A. *Nature* **1950,** *165,* 319.
(*10*) Fraser, R. D. B.; McRae, T. P. *Conformation in Fibrous Protiens and Related Synthetic Polypeptiedes;* Academic Press: New York, 1973.
(*11*) Brill, R. *Naturwiss,* **1930,** *18,* 622.
(*12*) Lotz, B.; Keith, H.D. *J. Mol. Biol.* **1971,** *61,* 201.
(*13*) Anderson, J. P.; Martin, D. C.; Cappello, J. *submitted to Biopolymers.*
(*14*) Geil, P. H. *Polymer Single Crystals;* R.E. Krieger Pub. Co.: Huntington, NY, 1973.
(*15*) Coleman, J. E.; Allan, B. J.; Vallee, B. L. *Science* **1960,** *131,* 352.
(*16*) Kerkam, K.; Viney, C.; Kaplan, D.; Lombardi, S. *Nature* **1991,** *349,* 566.
(*17*) Cohen, Y.; Thomas, E. L. *Macromolecules* **1988,** *21,* 433.

RECEIVED May 4, 1993

Chapter 13

NMR Characterization of Silk Proteins

Tetsuo Asakura[1], Makoto Demura[1], Atsuo Uyama[2], Katsuaki Ogawa[3], Keiichi Komatsu[4], Linda K. Nicholson[5], and Timothy A. Cross[5]

[1]Department of Biotechnology, Tokyo University of Agriculture and Technology, Koganei, Tokyo 184, Japan
[2]Department of Bio Research and Development, Tokyo Rikakikai Company Ltd., Ina-machi, Saitama 362, Japan
[3]Research Institute for Biological Science, Katakura Company Ltd., 4–5–25 Chuo, Matsumoto, Nagano 390, Japan
[4]Sericultural Science Research Institute, Shinjuku-ku, Tokyo 109, Japan
[5]Department of Chemistry, Florida State University, Tallahassee, FL 32306

Structures of *Bombyx mori* silk fibroin have been studied in solution, in silkworm and in the solid state by means of solution and solid ^{13}C and ^{15}N NMR spectroscopies. The silk fibroin yields very sharp ^{13}C NMR signals in aqueous solution and in silkworm, indicating the fast segmental motion of the main chain in spite of a fairly high molecular weight, 3 x 10^5. This makes detailed sequential and conformational analyses of the silk fibroin possible. The structure of the silk fiber in the solid state was studied with ^{15}N CP NMR and ^{15}N isotope-labeled silk fibroins on the basis of the chemical shift tensors in detail. The torsion angles of the glycine, alanine and tyrosine residues were determined.

In the class Insecta, many species produce long silken filaments (1). Among them, a well-known silk from *Bombyx mori* has excellent properties which are suitable for clothing and recently, for carrier of the immobilization of enzymes (2).

In this paper, we will describe the structure of silk fibroin, mainly *B.mori* silk fibroin in aqueous solution, in silkworm and in the solid state studied with ^{13}C and ^{15}N NMR spectroscopies. Especially, a new method is developed in which solid state ^{15}N NMR spectra obtained from uniaxially aligned molecules placed with the axis of alignment both parallel and perpendicular to the applied magnetic field (3). Results from the application of this method to the *B.mori* silk fibroin provide structural detail that is consistent with currently accepted structural models based on fiber diffraction studies.

Experimental Materials

B.mori and a wild silkworm, *Philosamia cynthia ricini* larvae were reared with an artificial diet in our laboratory . The isotope labeling of *B. mori* silk fibroin was achieved biosynthetically through the use of an artificial diet supplemented with the isotope enriched amino acids (4) or by the cultivation of the posterior silk glands

0097–6156/94/0544–0148$06.00/0

extracted from the larvae in the culture medium containing the isotope enriched amino acids (5). The highly isotope labeled silks were obtained by the latter method.

The aqueous solution of regenerated silk fibroin from *B.mori* cocoon was prepared as follows. The cocoons were first degummed twice with 0.5% Marselles soap solution at 100°C for 30 min. to remove another silk protein, silk sericin and then washed with distilled water. The silk was then dissolved in 9 M LiBr at 40°C. After dialysis against distilled water for 4 days, the solution was clarified by spinning in a centrifuge at 10,000 rpm for 30 min. The supernatant was collected and brought gently up to the desired concentration with an electric fan.

The powder sample was obtained by freeze-dry treatment of the aqueous solution of silk and then stirred in methanol for 3-4 days to form the Silk II (anti parallel β sheet) conformation. The powder sample of the model compound Boc (t-butoxy carbonyl)-[1-^{13}C]Ala-[^{15}N]Gly-Ala-Gly-OPac (phenyl ester) was prepared by liquid phase synthesis.

The thin sheets of the silk fibroin fibers aligned uniformly were produced with quick-setting bonding reagent and were cut into 8 mm x 12 mm pieces, stacked together and fixed with the bonding reagent to form an 8 mm x 8 mm x 12 mm block. The block is used for the quantitative analytical approach to utilize distinct features of the chemical shift and the line shapes obtained from oriented fibers placed parallel and perpendicular to the magnetic field for restriction of the number of possible orientations of an individual amide plane (3).

NMR Measurements

JEOL GX 400 and EX 270 spectrometers were used for ^{13}C CP, ^{13}C CP/MAS, ^{15}N CP and ^{15}N CP/MAS NMR measurements of silk fibroins in the solid state. A home built solid state NMR spectrometer assembled around an Oxford Instruments 400/89 magnet and a Chemagnetics data acquisition system at Florida State University was also used for the ^{15}N CP NMR measurement of the silk. JEOL GX 270, FX 200 and FX 90Q spectrometers were used for ^{13}C and ^{15}N NMR observations of the silk fibroin in solution and in silkworm. Experimental conditions were described elsewhere (3-12).

Results and Discussion

In vivo and Solution ^{13}C NMR

Using ^{13}C high resolution solution NMR, we can easily observe the silk protein directly in silkworm under non-destructive condition as well as the aqueous solution of the silk (Figure 1). Surprisingly, high resolution ^{13}C NMR spectra are observed and are assigned to triglyceride, trehalose and silk protein in the silkworm. High mobility of these components gives high resolution NMR, but other components do not give any NMR spectra because of their low mobility. The NMR spectrum of the pupa shows the peaks due to triglyceride and trehalose, and therefore the peaks of the amino acid residues of the silk protein are easily assigned in the spectra of both silkworms . In the NMR spectrum of *B.mori* silkworm, the spectral pattern is different from that of *P.c.ricini* silkworm, mainly in the resonance region of the carbons of the Ala residue. The shapes of the peaks from the Ala residue from *P.c.ricini* silkworm are doublets or asymmetric, indicating the presence of both α-helical and random coil conformations in the silk fibroin as reported previously (6,7). However, the corresponding Ala peaks of *B.mori* are all singlets and the peak positions coincide with those of the low-field component of the Ala C$^{\beta}$ and of the high-field component of the Ala C$^{\alpha}$ and C=O of *P.c.ricini*. These data indicate that there is no α-helical portion in the silk fibroin stored in the silk gland of *B.mori* and the conformation is essentially random coil. Thus, it is possible to obtain structural information of the silk fibroin from the ^{13}C NMR directly.

The aqueous solution of the silk fibroin prepared from *B.mori* cocoon also gives high resolution ^{13}C NMR spectrum, indicating very fast segmental motion of the main chain characterized by a very small correlation time, on the order of 10^{-10} s. at room temperature (8). The information of the amino acid sequence of the silk fibroin can be obtained from the carbonyl region (4) . Sixteen peaks are observed (Figure 2). The assignment is performed using [1-^{13}C]Ala , [1-^{13}C]Gly or [^{15}N]Gly labeled silk fibroins, from a comparison of the silk spectrum with those of the crystalline fraction after chymotrypsin hydrolysis and of the model polypeptides such as (Ala-Ala-Gly $)_n$ and (Ala-Gly $)_n$. We assigned these peaks at the pentapeptide sequence level as shown in Figure 2. The relative intensities of the main six peaks are nearly equal, indicating that main sequence of silk fibroin is Gly-Ala-Gly-Ala-Gly-Ser, that is, the alternative copolypeptides of Gly. The content of this sequence is about 70% of *B.mori* silk fibroin.

Solid State NMR

Recent advances in the application of solid state NMR spectroscopy to uniformly aligned biopolymers have opened a window through which to view the detailed structure of biological macromolecules. Atomic resolution structural details are obtained from solid state NMR data in the form of bond orientations, which yield the relative positions of specific atoms within the molecule. A new method is developed in which solid state ^{15}N NMR spectra obtained from uniaxially aligned molecules placed with the axis of alignment both parallel and perpendicular to the magnetic field are analyzed to yield the orientations of specific molecular bonds (3).

A diagram of the scheme used to analyze spectral data to obtain high resolution structural details is shown in Figure 3. Analytical expressions are derived which utilize spectral features read from ^{15}N chemical shift anisotropy lineshapes to calculate a discrete number of possible orientations for a specific site. Figure 4 shows observed and simulated ^{15}N CP NMR spectra of a block of oriented [^{15}N] Ala silk fibroin fibers placed both parallel and perpendicular to the applied magnetic field along with the powder pattern of the silk fibroin. From the simulation, the Euler angles, α_F and β_F, which express the relative orientations of the ^{15}N chemical shift anisotropy (CSA) principal axis system (PAS) and the fiber axis system (FAS) frames of reference were determined. In addition, the angles $\alpha_{DNC'}$ and $\beta_{DNC'}$ for the model compound, Boc-[1-^{13}C]Ala-[^{15}N]Gly-Ala-Gly-OPac , were experimentally determined where $\alpha_{DNC'}$ and $\beta_{DNC'}$ are the Eular angles which express the relative orientations of the ^{15}N CSA PAS and the N-C' bond direction (Molecular Symmetry Axis, MSA system frames of reference) (3). Using these Eular angles, the angles, θ_{NH} and $\theta_{NC'}$, between the NH and NC' bond direction and the fiber axis, respectively, are obtained as follows.

$$\cos\theta_{NX} = \cos\beta_F\cos\beta_{DNX}+\sin\beta_F\cos\alpha_F\cos\alpha_{DNX}\sin\beta_{DNX} \\ +\sin\beta_F\sin\alpha_F\sin\alpha_{DNX}\sin\beta_{DNX}$$

where α_{DNX} and β_{DNX} are the Eular angles for transformation from CSA PAS to MAS system and X is H or C'. A similar ^{15}N solid state NMR analysis was applied to [^{15}N]Gly and [^{15}N]Tyr silk fibroin fibers in order to determine the angles,θ_{NH} and $\theta_{NC'}$. Details have been reported for the former [^{15}N]Gly silk sample (3).

Determination of the torsion angles, ϕ and ψ, of the Gly, Ala and Tyr residues of

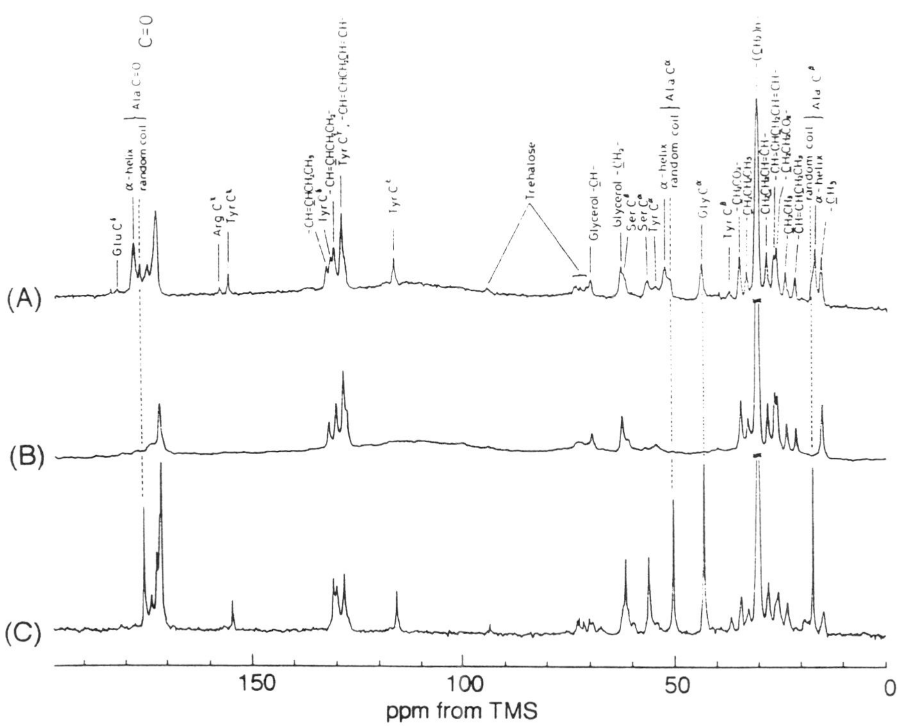

Figure 1. ^{13}C NMR spectra of the silk gland portion of the mature larva (A) and of the abdomen of the pupa (B) of *P.c.ricini*. The spectrum of the mature larva of *B.mori* is also shown (C).

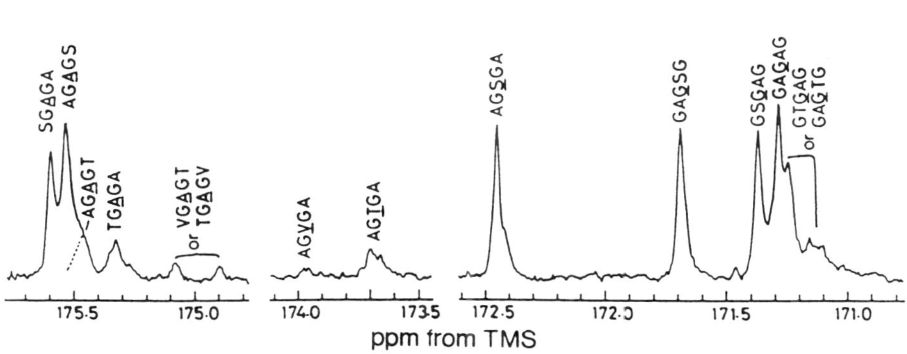

Figure 2. Assignment of the carbonyl region in *B.mori* silk fibroin spectrum.

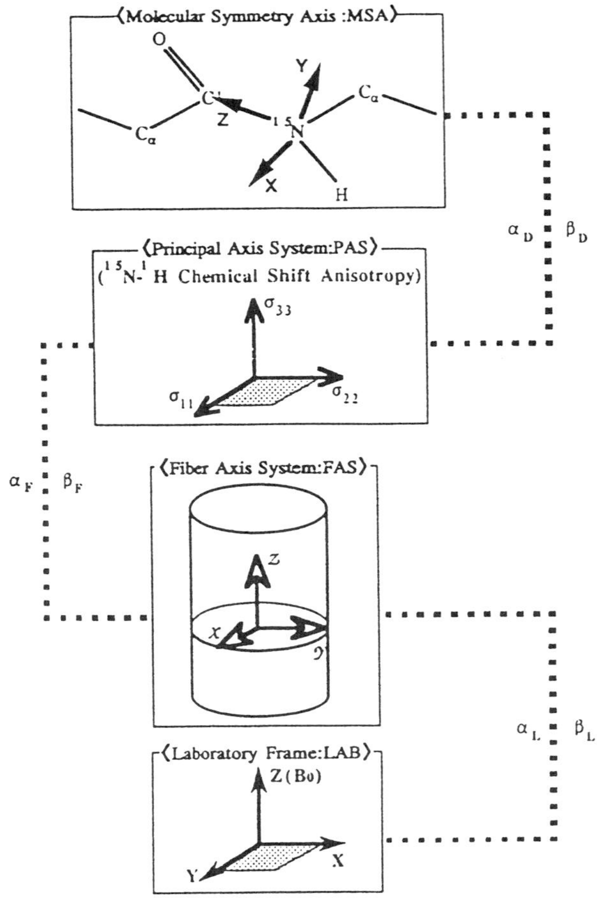

Figure 3. Diagram of the scheme used to analyze spectral data to obtain high resolution structural details

B.mori silk fibroin is possible with two sets of θ_{NH} and $\theta_{NC'}$ values obtained for successive ^{15}N nuclei in the chain, where ϕ describes the rotation about the N-C^{α} bond while ψ describes the rotation about the C^{α}-C' bond. Figure 5 shows the relationship between ϕ or ψ and the orientations of the relevant bonds. The definitions of the symbols used here are described in the reference 9. Several numbers of possible torsion angle solutions are obtained. The ^{13}C CP/MAS NMR (10,11) and ^{15}N CP/MAS NMR data (12) of the silk fibroin can be used to reduce the solutions. The final ϕ and ψ values obtained here are -141 ± 5° and 147 ± 5° for the Gly residue, -139 ± 5° and 146 ± 5° for the Ala residue , and -139 ± 5° and 147 ± 5° for the Tyr residue, respectively.

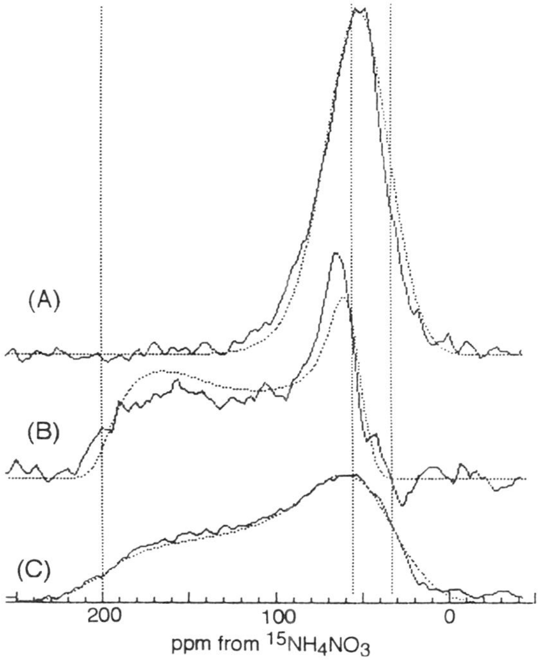

Figure 4. Observed and simulated ^{15}N CP NMR spectra of a block of oriented [^{15}N] Ala silk fibroin fibers placed both parallel (A) and perpendicular (B) to the applied magnetic field along with the powder pattern of the silk fibroin (C).

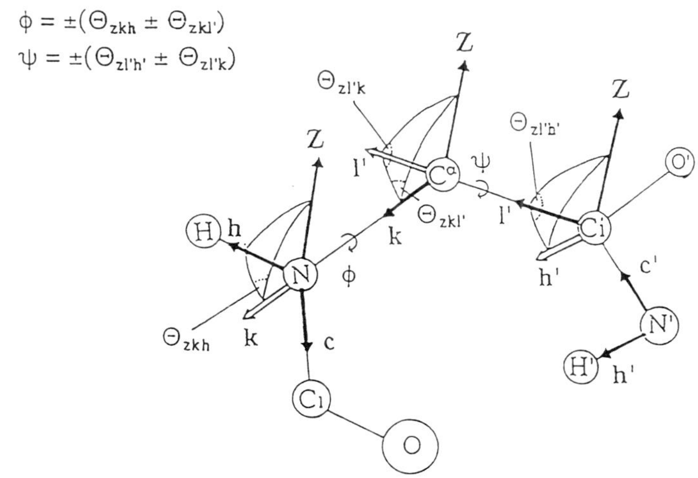

Figure 5. Relationship between ϕ or ψ and the orientation of the relevant bonds.

Acknowledgments
This work has been funded, in part, through the Grant-in-Aid of the Ministry of Education, Science, and Culture of Japan (grant no. 04556007). T.A. gratefully acknowledges support of the Yamada Science Fundation,Japan. T.A. also extends thanks to Toyo Seikan Co., R. & D. Group,Yokohama, Japan for use of their JEOL EX-270 spectrometer.

References and Notes

1.Asakura T. and Kaplan D.L., *Silk Production and Processing*, Encyclopedia of Agricultural Science, Ed. by Arntzen C.J. Academic Press in press.
2.Asakura T., Kitaguchi M., Demura M., Sakai H.and Komatsu K., *J. Appl. Polym. Sci.*, **46**, 49 (1992). and references therein.
3.Nicholson L.K., Asakura T., Demura M.and Cross T.A., *Biopolymers* in press.
4.Asakura T., Watanabe Y. and Itoh T., *Macromolocules* **17**, 2421 (1984).
5.Asakura T., Sakaguchi R., Demura M., Manabe T., Uyama A., Ogawa K. and Osanai M., *Biotechnol. Bioeng.*, **41**, 245 (1993) and references therein.
6.Asakura T., Suzuki H. and Watanabe Y., *Macromolecules* **16**, 1024 (1983).
7.Asakura T., Kashiba H. and Yoshimizu H., *Macromolecules* **21**, 644 (1988).
8.Asakura T., Watanabe Y., Uchida A. and Minagawa H., *Macromolecules* **17**,1075 (1984).
9.Teng.Q., Nicholson L.K. and Cross T.A., *J. Mol. Biol.* **218**, 607 (1991).
10.Saito H., Tabeta R., Asakura T., Iwanaga Y., Shoji A., Ozaki T. and Ando I., *Macromolecules* **17**,1405 (1984).
11.Asakura T., Kuzuhara A., Tabeta R. and Saito H., *Macromolecules* **18**, 1841 (1985).
12.Asakura T., Miyashita N., Demura M., Yoshimizu H., Ando I. and Shiibashi T., *Rept. Prog. Polym. Phys. Japan* . **33**, 633 (1990).

RECEIVED May 4, 1993

Chapter 14

Raman Spectroscopic Analysis of the Secondary Structure of Spider Silk Fiber

D. Blake Gillespie, Christopher Viney, and Paul Yager

Molecular Bioengineering Program, Center of Bioengineering, FL–20, University of Washington, Seattle, WA 98195

As a first step in monitoring molecular and microstructural changes that accompany the spinning of silk, we have employed Raman spectroscopy to determine the secondary structure of the protein in single silk fibers. The relatively well-studed silk of the silkworm *Bombyx mori* was compared to the dragline silk of the spider *Nephila clavipes*. Spectra were analyzed using Williams' secondary structural analysis routines based on the shape of the amide I band (*1, 2*). The proportions of secondary structure in the two types of fibers examined was found to be surprisingly similar. The content of β–sheet in both natural *B. mori* and *N. clavipes* silk fibers is 56 ± 5%, and another 22 ± 5% is β–turn. Oriented spectra also suggest that both the β–sheet and a small proportion of α–helix are highly oriented along the fiber axis, whereas the disordered polypeptide chains are not preferentially aligned. The α–helix is more highly aligned in *N. clavipes* dragline silk.

Although both industry and nature utilize polymer processing methods that allow nearly optimal exploitation of the mechanical properties of the polymer chains, nature employs simpler processing conditions that are chemically and physically far milder than those used for man-made high strength polymers. An understanding of natural polymer processing could be used to develop superior materials, whether the polymers of which they consist of man-made or of natural origin. Silks provide an obvious starting point for analysis of biological polymer-processing systems, representing one of nature's most versatile polymers. Recent demonstrations of cloning and expression of silk fibroins including spider dragline silk (*3-5*) represent exciting scientific opportunities and technological challenges. If one is to make commercial use of the biosynthetic dragline silk, one must develop processing methods for fibroin solutions that equal or improve upon spinning by spiders. To do this, one must first develop a quantitative understanding of the spinning of silks by insects and spiders.

Despite great efforts in the last decade to relate the molecular structure of silks to their mechanical and physical properties (*3*), we are still far from a comprehensive understanding. Two contrasting silks that have been studied in depth are the familiar silk of the cocoon of the silkworm *Bombyx mori*, and the dragline of spiders such as

0097–6156/94/0544–0155$06.00/0

Nephila clavipes. Of these two, spider dragline is the lesser understood, and also the more promising, as it is much tougher than silkworm silk. A popular working model for the microstructure of spider dragline silk is that of Gosline *et al.* (*6*). They view the silk fiber as a microcomposite of small, rigid, inextensible crystallites of β–sheet embedded in a matrix of amorphous rubbery polypeptide. The crystalline domains are presumed to occupy about 30% of the volume of the dry silk; the rest is believed to be occupied by polypeptide segments of about 15 amino acids that are glassy when dry and rubbery when swollen in water. The fibroin that may be the only polypeptide in the dragline silk of *N. clavipes* has a molecular weight of about 320,000. Recently a partial sequence of this fibroin was reported by Xu *et al.*(*7*); this sequence is quite different from that of *B. mori* silk. While the fibroins in both types of silk consist of an alternating sequence of segments likely to form β–sheets and segments likely to remain in random coil or form α–helical structures, the lengths of these segments may be much shorter in spider silk. Spider silk fibroin appears to be much less regular than silkworm silk, and has a repeat of Gly-Gly-X, rather than the Ala-Gly-Ala-Gly-Ser-Gly motif found in *B. mori* silk. The 34-amino acid repeat found by Xu and Lewis in spider dragline (*7*) includes regions high in alanine that may correspond to the amorphous regions suggested by Gosline.

We know that spinning increases the β–sheet content of fibroin solutions, and have a general knowledge of the overall secondary structural compositions and orientations of the pre- and post-spinning materials. However, there is little consensus as to the proportions of secondary structural types. Estimates for *B. mori* fibers range from 74% β–sheet derived from solid state NMR predictions (*8*) to Iizuka's crystallographic work predicting the β–sheet content to be between 36% and 48% (*9*). No direct determination of proportions of secondary structure has been made in spider silks to date. If we are to understand the relationship between structure and function in silks, we must first determine the structure of fibers that are the endpoint of the spinning process. We have collected data on the somewhat better-understood *B. mori* as a control for the first measurements on silk from *N. clavipes*.

Raman Spectroscopy of Proteins

Among the best ways to study protein secondary structure in solution is Raman spectroscopy. Changes in ϕ and ψ angles, in hydrogen bonding, and electrostatics of the environment all affect the vibrations of the peptide bonds, giving rise to a sensitivity of the amide normal modes to the secondary structure of the protein (*10*). In addition, subtle changes in sidechain vibrations can also be correlated with differences in conformation or solvent. Both Fourier transform infrared spectroscopy (FTIR) and Raman scattering (both conventional and NIR-FTIR Raman) have strengths and weaknesses in studying polypeptide structure (*11*) but recent technical advances have greatly enhanced the utility of Raman spectroscopy. CCD detectors have excellent response in the far red, great dynamic range, and, when cooled to liquid nitrogen temperatures, better quantum efficiency than cooled PMTs and almost no noise. Spectra that may have taken 12 hours to collect a decade ago may now be collected in seconds, and smaller samples can be used at lower laser powers (*12*).

Raman spectroscopy has been used to monitor changes in *B. mori* fibroin secondary structure that occur on spinning (*13*), and some excellent spectra have been recorded of both solubilized fibroin and fibers drawn from gland secretion (*14*). The spectra of pre- and post-spinning fibroin are strikingly different, but as yet only qualitative assessments of the dominant type of secondary structure present have been made of such spectra. While monitoring of the peak frequency of the amide I region continues to be adequate for qualitative differentiation between predominantly α–helical structure and predominantly β–sheet or random structure (*11, 15*), the bands strongly overlap, so more complex mixtures of secondary structure produce complex amide I bands that require more sophisticated analysis. Recently, the application of mathematical methods such as factor analysis and partial least-squares have been applied to quantification of protein secondary structure. The methods are similar to ones applied to analysis of circular dichroism spectra, but rely instead on the amide I region, and often the amide II and III regions as well.

Fibers are anisotropic, and, as the laser is polarized, comparison of Raman scattering parallel and perpendicular to the laser polarization can be used to determine the orientation of the groups (such as the amide bonds) responsible for observed scattering (see Figure 1). In the case of highly oriented α–helical fibers and films such as can be formed from poly-γ-benzyl-L-glutamate, it is possible to determine the angle of the amide bond relative to the fiber axis by analysis of the polarization of the amide I band in various fiber orientations relative to the beam propagation and polarization directions (*16, 17*). Not only can the orientation of the peptide backbone be determined, but also that of specific side chains with distinctive vibrational bands, such as the aromatic amino acids. Several technical problems must be overcome, however. The first is collection of sufficient numbers of photons from a very small sample such as a single silk fiber. The signal-to-noise ratio (SNR) is also dependent on adequate rejection of the Rayleigh scattering from the fiber, and the presence of relatively low levels of luminescence (fluorescence and phosphorescence). While fibroin itself should not luminesce when excited at 5145Å, natural products are frequently colored, and often luminescent as well.

Materials and Methods

Preparation of Silks. Raw silkworm cocoons from *B. mori,* of various strains were provided by Marian Goldsmith at the University of Rhode Island, Kingston, RI. Cocoons were degummed by boiling in 0.1% NaOH or a dilute soap solution (*18*) using the commercial detergent, Meri-Suds (Seaport Chemicals, Seattle, WA), followed by an extensive deionized water rinse. The amount of sericin removed in this way was commonly ~25% of the total mass of the raw cocoons (*3*). *N. clavipes* were obtained from Angela Choate at the University of Florida, Gainesville, FL. They were subsequently raised on a diet of *Drosophila* or on crickets (Fluker's Cricket Farm, Baton Rouge, LA). Devices used to acquire *N. clavipes* dragline silk over a range of linear haul-off velocities were modeled on existing designs (*19, 20*). All data presented herein were collected from *N. clavipes* dragline silk spun from the major ampullate gland at approximately 1 cm sec^{-1}, which is similar to the natural spinning rate.

Luminescence of fibers is a major obstacle to collecting silk Raman spectra. Attempts were made to reduce this background signal by dissolving silkworm fibers in concentrated LiBr, followed by dialysis against tap water, as in Yamaura, *et al.* (*21*). The hope was that the contaminant might be water soluble, and so be washed away during dialysis. Not all of the silk could be dissolved, and the residue was slightly colored, resulting in samples with at least as much luminescence as the original material. Discoloration of silk fibers by dietary pigments is also common in *N. clavipes* dragline silk samples. As a result, it was very difficult to obtain spectra from these specimens; the color gives rise to an increased background luminescence as well as increased absorption. The absorption, when found, resulted in heating and subsequent breakage of the fibers in the laser beam. Therefore, spider fibers that were obviously yellow were treated by gentle heating in 0.1% NaOH for 5 -10 minutes. Such treated fibers did not break, and their spectra demonstrated no signs of structural deterioration.

Silk fibers were observed by optical microscopy using a Zeiss ICM 405 inverted microscope in transmission mode, with and without crossed polars. Images were recorded with an MTI DAGE model 66 SIT camera and digitized by a Macintosh

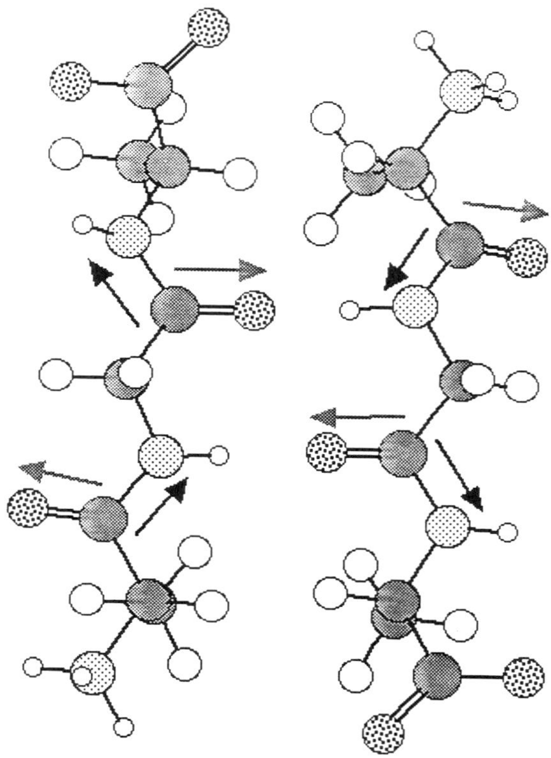

Figure 1. Two identical tripeptides in antiparallel β–sheet configuration. The vertical is both the chain axis and the putative fiber axis in silks. Carbonyls are identified by double bonds, which, in the β–sheet, are oriented normal to the fiber axis. To a first approximation, the amide I mode, being primarily C=O in nature (gray arrows), will scatter most strongly when excited perpendicular to the chain axis (with the fiber oriented along the laboratory y or z axes as defined in Figure 2). The amide III mode, having a great deal of character from the C-N stretch (black arrows), scatters most strongly when excited parallel to the chain axis (with the fiber aligned along the x axis).

II computer equipped with a frame grabber (Data Translation QuickCapture board). Images were processed using the public domain Image software package (Version 1.23h) from NIH. Magnification was verified using a calibrated microscope reticle.

Raman system. Raman spectroscopy was performed with an instrument based on the SPEX 500M 0.5 meter single monochromator (SPEX Corp., Metuchen, NJ), which has moderate resolution but high throughput. The monochromator is coupled to a SPEX Spectrum 1,385 by 578 pixel Astromed MPP CCD detector (0.5 inch chip) operated at 140°K. The detector's low background of 1 photoelectron per pixel per hour allows rapid collection of extremely low signals; spectral resolution is quite adequate for all biologically relevant condensed phase samples. The instrument is controlled by SPEX DM 3000 software running on an IBM 386 clone computer, and coupled to an HP 8-pen plotter (Hewlett Packard Corp., Model #7440A). The system is operated in a manner in which the slit image is projected along the short axis of the detector (vertical), and the grating disperses the spectrum horizontally. As the CCD area can currently be electronically sampled as if it were 1 to 4 independent horizontal detectors, it is possible to simultaneously collect four independent spectra from segments at different heights along the slit axis. Using this CCD collection of spectra from a single row of pixels was feasible, but the S/N ratio was greater if more rows were employed. Because there is only one slit in the instrument, a filter to remove Rayleigh-scattered light is required for scattering samples such as fibers. The holographic filter employed (RHE 632.1D, POC Corporation, Torrance, CA) cuts out scattering at less than 800 cm^{-1}, but has very flat transmission to the limits of detector sensitivity, retaining the important reference bands provided by water Raman scattering at ~3300 cm^{-1}. The excitation source is a tunable argon ion laser (Lexel Laser, Inc., Model #95, Fremont, CA) operated at 5145Å, and the equipment is mounted on a 4'x6' floating optical table (Newport Corp., Irvine, CA).

Single silk fibers were suspended in the laser beam by a metal frame, around which the ends of the fibers were wound. The environment of the fibers was ambient laboratory air at 20°C; no attempt was made to control the relative humidity. The scattering geometry used in the collection of all spectra was 90° as shown in Figure 2. Fibers could be oriented along the laboratory x, y or z axes. Spectra were collected at laser powers from 50mW to 1W from single silk fibers, and occasionally from fiber bundles. However, use of multiple fibers tended to increase background scattering at the expense of signal, as did use of a pellet of finely chopped silk. The time allowed for photobleaching of background varied from sample to sample, but no more than five minutes of exposure was needed before acquisition of a spectrum. Photobleaching was essentially complete after ten minutes. A single fiber could withstand high laser powers for the duration of an experiment with no signs of degradation apparent in the spectrum.

To determine the compounds responsible for luminescence of fibroin samples, a Perkin Elmer LS-5B (Norwalk, CT) fluorescence and phosphorescence spectrophotometer was employed. This instrument was equipped with a printer-plotter and thermostattable cell holders, and was interfaced to an XT clone microcomputer. It was used for fluorescence measurements of liquid and solid samples, using either normal 1cm x 1cm cuvettes, or a microsample cuvette (General Atomics, San Diego, CA) holding as little as 3 µl. To date, no fluorophores have been identified by this method aside from silk's own aromatic side chains.

Data Analysis. Raw spectra must be processed prior to secondary structural analysis. Collection of spectra in several windows of 500 cm^{-1}, and their subsequent concatenation into a single data file were carried out using SPEX DM3000 software, version 3.12. Baseline subtraction, normalization, peak-fitting, and removal of side chain Raman peaks near the amide I band were performed using the Macintosh

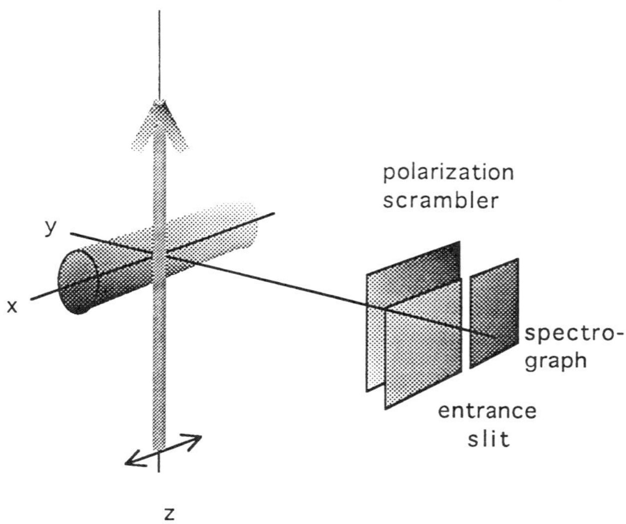

Figure 2. A schematic diagram of the orientation of the experimental axes as described in the text. The laser propagates upwards along the z axis, and is polarized along the x axis. The fiber, which is shown as a cylinder lying along the x axis, is rotated around the z axis for collection of the y axis data. A camera lens (not shown) images the illuminated portion of the fiber onto the spectrograph entrance slit.

software package Igor (WaveMetrics, Lake Oswego, OR) on a Macintosh II microcomputer.

Methods have been developed for the interpretation of infrared (*22-24*) and Raman (*1, 2, 25*) amide I regions, and comparable accuracies are obtained under ideal conditions with both spectroscopic techniques. To quantify the fibroin secondary structure using Raman spectra, we have used the method developed by Prof. R.W. Williams at USUHS in Bethesda (*1, 2*). His technique relies on assuming that the amide I region of the spectrum of the protein in question can be treated as a linear combination of spectra from their component secondary structures. Raman data from a 300 cm^{-1} wide region of the spectrum is collected, luminescence background and sidechain peaks are subtracted; and the remaining amide I band is normalized and used as input for a non-negative least squares routine, code for which was provided for our use by Dr. Williams.

This non-negative least-squares fitting incorporates a basis set of 15 proteins whose secondary structure is known from crystallographic studies. Specifically determined are the percentages of α–helix ends and their ordered cores, β–sheet, β–turn, and disordered structure in the unknown protein; error for each category is limited to about ±5%. The basis set proteins include tropomyosin, lactate dehydrogenase, insulin, hemoglobin, BPTI, melittin, concanavalin, alcohol dehydrogenase, lysozyme, fd phage, crambin, thermolysin, ribonuclease A, carboxypeptidase A, calcium binding protein, erabutoxin B and adenylate kinase. The original FORTRAN code was modified to run on a local VAX mainframe.

Each Raman system is optically unique, so to use Williams' basis set we had to determine what convolutions of our spectra were required in order to make them compatible with the stored basis spectra. A comparison of lysozyme spectra demonstrated that a spectral shift of + 0.8 cm^{-1} was necessary to align our spectrum with that in the basis set. After subtraction of the positively sloping baseline, a Gaussian fit was made to the aromatic side chain peak at 1615 cm^{-1}, which was then subtracted, according to the method of Williams (*1*). This process is illustrated in Figure 3. Spectra were normalized to the ring breathing modes in the 900-1000 cm^{-1} region and to the methylene deformation mode at 1450 cm^{-1}. Williams' routine then calls for the extraction of 20 points along the amide I band, every five cm^{-1} from 1615 to 1710 cm^{-1}. These values are summed, each intensity is divided by this sum and is multiplied by 10^4. Any new spectrum collected is normalized in this way before being input into the prediction program.

Results and Discussion

Optical microscopy of silk fibers from *B. mori* cocoons and spun from live *N. clavipes* indicated:

1. *B. mori* fibers were generally 20 µm in diameter, were not perfectly cylindrical, and were highly birefringent.
2. The diameter of *N. clavipes* dragline fibers varied from 6 to about 20 µm in diameter, depending on the specimen and method of silk collection. The fibers were moderately birefringent, and swelled up to 50% in diameter upon immersion in water (see Figure 4). The narrower fibers were used for Raman spectroscopy.

At the 200 µm slit widths that we had been using for capillary samples, the lateral position of the illuminated portion of the silk fibers determined the Δ cm^{-1} shift at which the Raman bands appeared. The pixels of the CCD detector in the Raman system are spaced at 20 µm intervals, so setting the entrance slit at 20 µm provides the maximum possible spectral resolution for the system. With our f/1.2 lens collecting the scattered light, the sample is magnified about three-fold onto the slit, so

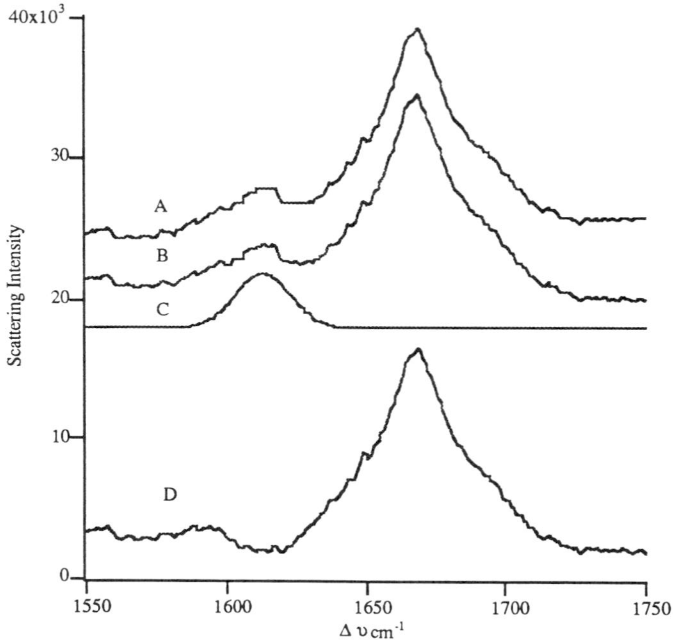

Figure 3. A sample of the preparation of amide I Raman data for secondary structural analysis. **A)** A portion of a typical raw spectrum from the SPEX system, imported to Igor; **B)** the spectrum after subtraction of a positively sloped baseline; **C)** a Gaussian fit to the aromatic side chain band at 1615 cm^{-1}; and **D)** the spectrum after subtraction of **C.**

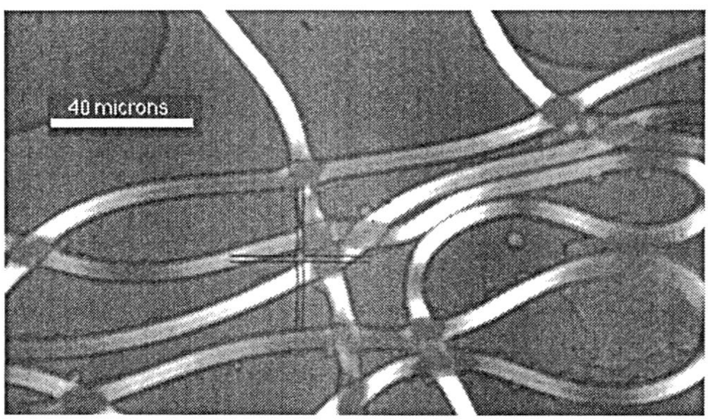

Figure 4. An optical micrograph of fibers of *N. clavipes* dragline silk under water and between crossed polars. Note the birefringence, indicating alignment of the polymer within the fibers.

the silk fibers at least filled the 20 μm opening. This fortuitously provided an optimal combination of resolution and light collection.

The Williams routine's basis set is constituted of spectra collected from isotropic protein samples. The ideal sample for comparison with this basis set would be an isotropic sample of spun silk. Unfortunately it is nearly impossible to prepare such a sample, as the fibers are intrinsically oriented; minced fibers were an inadequate solution as luminescence increased greatly in proportion to signal. Anisotropic orientation of the sample enhances the intensity of bands that are derived from Raman tensors that are aligned along the excitation light polarization axis. We therefore prepared "pseudoisotropic" spectra. If the fiber axis is denoted by *Z*, and the two axes perpendicular to the fiber as *X* and *Y*, it is immediately obvious that *X* and *Y* are equivalent, but that both their contributions must be taken into account. The laser beam is polarized along the x-axis (see Figure 2). We created the pseudoisotropic spectra by adding spectra in which the laser polarization axis lay along the two differentiable fiber axes (*X*/*Y* and *Z*). To collect data with the laser polarization along the fiber *Z*-axis, the fiber is aligned along the experimental x-axis. To excite the fiber *X*/*Y*-axes, the fiber can either be aligned along the laboratory y- or z-axes. At first the fibers were positioned so that they were near-parallel to the incident beam (laboratory z-axis), but we found that the polarization of the laser was rotated as light was refracted into the fiber. Aligning the fiber along the laboratory y-axis ensured that the excitation light was polarized only perpendicular to the fiber *Z*-axis. The "pseudoisotropic" spectra for the two types of silks were formed by normalizing the x- and y-axis spectra, and adding two times each y-axis spectrum (for the two principal components of this orientation) to one times the x-axis spectrum. These data were collected for silk from both *B. mori* (Figure 5) and *N. clavipes* (Figure 6).

The results of the secondary structural analysis of the amide I bands of the six spectra are summarized in Table I. Secondary structural predictions of silkworm and spider silks based on the amide I band of the "pseudoisotropic" spectra are within the range of values seen in the literature (*3*). We estimate the antiparallel β–sheet content for both degummed *B. mori* cocoon silk and *N. clavipes* dragline to be surprisingly similar, at 56 ±5%. None of the analyses yielded a prediction of even a minor component of parallel β–sheet for either silkworm or spider silk. The least-squares fitting routine is constrained to return only positive values, so a content of exactly 0.0% reflects the fact that the routine estimated a negative value for the percentage parallel β–sheet. The Williams routine predictions of parallel β–sheet is less accurate than that for other types of secondary structure because parallel β–sheet is underrepresented in the basis set. The helical segments present must be relatively short, based on the comparable amounts of ordered helix and helix ends. The high proportion of turns found in both silks (~22%) suggests either that the stretches of β–strand in crystallites are not extremely long, or, if the strands are long, that the turns are not evenly distributed along the β-strands. If one assumes a structure for regions of β–sheet that is analogous to the platelet structure of polyethylene, and four residues are present in each turn, then the runs of straight β–strand are approximately 12 residues long.

While the "pseudoisotropic" spectral data are the most directly comparable to that of other proteins, the ratio of the x- and y-axis values in Table I give a clear sense of the orientation of secondary structure in the fibers (See Figure 1). The x-axis fiber orientation (normal to the incident beam) will result in the preferential excitation of the C-N stretch (amide III). Conversely, the z-axis fiber orientation maximally excites the carbonyl-dominated amide I mode. The predicted sensitivity of the amide I and III bands to the fiber geometry is immediately apparent in Figures 5 and 6. This is consistent with the having 80% of the protein in β–sheet (60% sheet plus the associated 21% β–turn). The α–helical structure that *is* present in both types of fiber

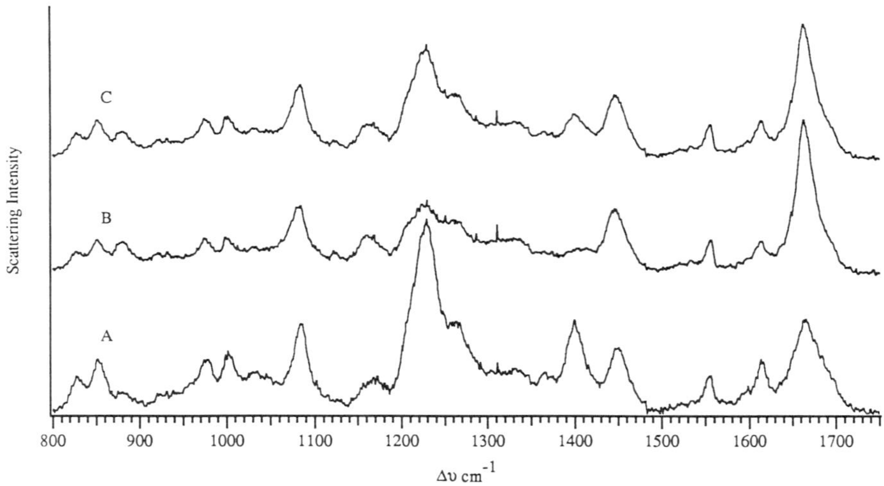

Figure 5. Spectra of one *B. mori* degummed silk fiber taken in the x (A) and y (B) fiber orientations, overlaid upon a "pseudoisotropic" spectrum (C) created by adding two times the y-orientation spectrum to one times the x spectrum. Note differences in the x- and y-oriented spectra, reflecting the alignment of β–sheet regions parallel to the fiber axis. The typical signal-to-noise ratio can be seen more clearly in the enlargement of the amide I region in Figure 3.

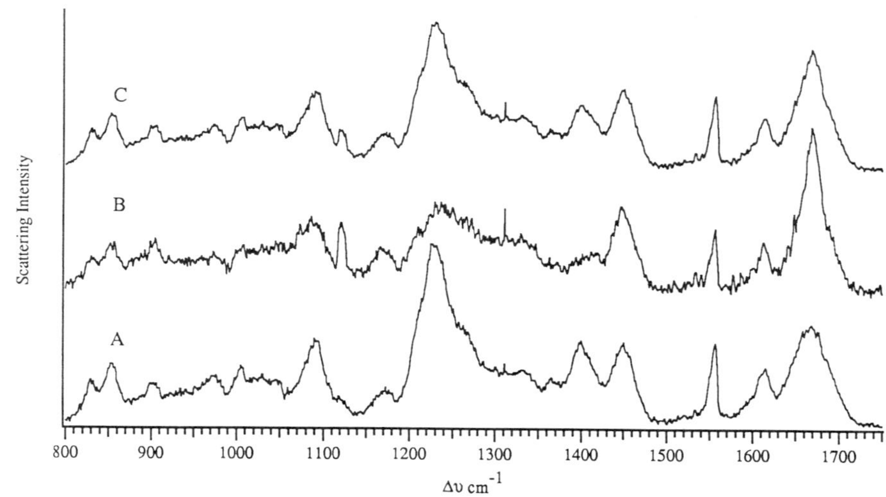

Figure 6. Spectra of one *N. clavipes* dragline silk fiber taken in the x-axis (A) and y-axis (B) fiber orientations, overlaid upon a "pseudoisotropic" spectrum (C) created by adding two times the y-orientation spectrum to one times the x spectrum. As with *B. mori*, note the differences in the x- and y-oriented spectra, reflecting the alignment of β–sheet regions parallel to the fiber axis.

Table I. Proportions of Secondary Structure Determined (± 5%)

fiber	orientation	% ordered α–helix	% helix ends	% anti-parallel β–sheet	% turn	% dis-ordered	% total helix	% total sheet	total
Bombyx mori	x	8.4	10.2	48.6	20.9	10.2	18.6	48.6	0.982
	y	5.5	7.0	58.1	23.6	11.1	12.5	58.1	1.050
	"pseudo-isotropic"	5.6	7.0	56.5	23.1	10.9	12.6	56.5	1.030
Nephila clavipes	x	13.8	4.8	48.8	18.2	13.0	18.6	48.8	.986
	y	0.0	1.9	63.5	21.6	11.9	1.9	63,5	.982
	"pseudo-isotropic"	5.0	6.0	56.4	21.0	10.0	11.0	56.4	.976

appears to be highly aligned along the fiber axis, particularly in *N. clavipes*. One can use the proportion of disordered structure as an internal reference for the success of the Williams method. If this component is truly disordered, one would expect to find the same proportion of random structure in both orientations, which is precisely what is found in both types of silk. While this does not prove the accuracy of the secondary structure determinations, it is strongly supportive.

A key question in the application of this secondary structural analysis method for the first time to fibrous proteins is the adequacy of the basis set. To accurately predict secondary structure of an unknown protein using any of the so-called chemometric techniques, one must have a basis set that includes all the types of protein conformation found in the unknown sample. All of the 15 basis set proteins (with the exception of poly-L-lysine) are globular proteins, which have an outer surface that is hydrated. Silk fibers have relatively low water content, and therefore the fibroin molecules may be in environments that are similar to the interiors of globular proteins, but the fibroin will have far less hydrated outer surface. Consequently, in the absence of absolutely trustworthy literature values even for *B. mori* silk, we do not have independent confirmation of the accuracy of our secondary structural values. However, it has been known for some time that a chymotryptic digest of solubilized *B. mori* silk produces a β–sheet-forming precipitate that represents approximately 60% of the original dry mass of the washed silk (*26*). Furthermore, recent electron diffraction studies of *N. clavipes* dragline fibers in our laboratory have demonstrated that approximately 50% of volume of the fibers consists of β–sheet crystallites (Thiel *et al.*, unpublished data).

Using the green 5145 Å laser line available to us, luminescence was observed in all cases. This technique is fine for studying the structure of static silk strands, but clearly prevents the study of moving silk strands that may stay in the laser beam for a few seconds or less during spinning. As there is nothing intrinsically luminescent in the fibroin protein, we conclude that the fluorophores are either contaminants, or ingested pigments that pass into the silk gland. Fluorescent contaminants may also be present in synthetic fibroin, but these solutions can be purified using chromatographic techniques. Unfortunately, chromatography is not applicable to live spiders. To avoid luminescence the only possibilities are varying the diet of the spiders (which we have not found to be altogether successful) or using a redder laser. This latter approach has the potential to solve the luminescence problem for both natural and artificial silk, and so will be pursued.

This CCD-based Raman system allows collection of high S/N ratio spectra from extremely small biological samples in conveniently short periods of time. The demonstration that high-quality spectra can be obtained from single fibers opens up the use of this technique for the study of processing of silk, as well as for other areas requiring extreme sensitivity. We hope that the successful use of sophisticated secondary structural analysis techniques such as the Williams method will broaden the appeal of the use of Raman spectroscopy for the study of a wide range of protein-based biomaterials. However, it should be emphasized that this work represents only a preliminary investigation of the properties of these silks. The structural basis set will be expanded to include additional fibrous proteins. The final fiber structure's dependence on draw rate, relative humidity, and temperature remain to be explored.

We have only a rudimentary understanding of how the fibroin molecules self-organize to form a high strength polymer, and how they manage to do so under the mildest of conditions. If we can understand how the spinning process works, we can use this knowledge to improve processing in existing man-made polymers, allow utilization of fermentation-derived silks, and extend the use of silks and silk-like polymers to structures other than fibers.

Acknowledgments

This work was supported by NSF grant # BCS-9202007 to C.V. and P.Y., and by a grant from the Whitaker Foundation to the Center for Bioengineering. We gratefully thank Prof. Robert W. Williams of the Uniformed Services University of the Health Sciences for providing us with his FORTRAN code for protein secondary structure analysis, and both Christie Beacham and Christopher Bryden for assistance in modification of the FORTRAN code for use on our local computers. Also thanks to Anne Huber, Keith Guess, Dwayne Dunaway and Keven Kerkam for preparation of spider silk for this study, and to Hillary MacDonald for constant help in maintaining the Raman system.

References

1. Williams, R. W. *Meth. Enzym.* **1986,** *130,* 311-331.
2. Williams, R. W. *J. Mol. Biol.* **1983,** *166,* 581-603.
3. Kaplan, D. L.; Lombardi, S. J.; Muller, W. S.; Fossey, S. A. In *Biomaterials: Novel Biological Processes;* Byrom, D., Ed.; Stockton Press: New York, 1991, pp 3-53.
4. Lombardi, S. J.; Kaplan, D. L. *Polym. Preprints.* **1990,** *31 (1),* 195-196.
5. Stinson, S. C. *C&EN.* **1990,** July 16, 26-32.
6. Gosline, J. M.; DeMont, M. E.; Denny, M. W. *Endeavour.* **1986,** *10,* 37-43.
7. Xu, M.; Lewis, R. V. *Proc. Natl. Acad. Sci. USA.* **1990,** *87,* 7120-7124.
8. Asakura, T.; Kuzuhara, A.; Tabeta, R.; Saito, H. *Macromol.* **1985,** *18,* 1841-1845.
9. Iizuka, E. *J. Appl. Polym. Sci.: Appl Polym. Symp.* **1985,** *41,* 173-185.
10. Spiro, T. G.; Gaber, B. P. *Ann. Rev. Biochem.* **1977,** *46,* 553-572.
11. Rath, P.; Bousche, O.; Merrill, A. R.; Cramer, W. A.; Rothschild, K. J. *Biophys. J.* **1991,** *59,* 516-22.
12. Archibald, D. D., Structural studies of high aspect-ratio self-assembled lipid microstructures with the use of microscopy and FT-NIR-Raman spectroscopy, Ph.D. Thesis, University of Washington, 1990
13. Zheng, S.; Li, G.; Yao, W.; Yu, T. *Appl. Spect.* **1989,** *43,* 1269-1272.
14. Magoshi, J.; Magoshi, Y.; Nakamura, S. *Poly. Commun.* **1985,** *26,* 309-311.
15. Jackson, M.; Haris, P. I.; Chapman, D. *Biochim. Biophys. Acta.* **1989,** *998,* 75-79.
16. Wilser, W. T.; Fitchen, D. B. *Journal of Chemical Physics.* **1975,** *62(2),* 720-724.
17. Tsuboi, M.; Ikeda, T.; Ueda, T.; Okawa, M.; Hidema, K. In *Proceedings of the 12th International Conference on Raman Spectroscopy;* Durig, J. R.; Sullivan, J. F., Ed.; John Wiley & Sons: Columbia, South Carolina, 1990, pp 726-727.
18. Levin, M.; Pearce, E. *Fiber Chemistry; Marcel Dekker, New York, NY,* 1985
19. Work, R. W.; Emerson, P. D. *Journal of Arachnology.* **1982,** *10(1),* 1-10.
20. Wilson, R. S. *Quart. J. Micr. Sci.* **1962,** *104,* 557-571.
21. Yamaura, K.; Okumura, Y.; Ozaki, A.; Matsuzawa, S. *J. Appl. Polym. Sci: Appl. Polym. Symp.* **1985,** *41,* 205-220.
22. Dousseau, F.; Pézolet, M. *Biochemistry.* **1990,** *29,* 8771-9.
23. Lee, D. C.; Haris, P. I.; Chapman, D.; Mitchell, R. C. *Biochemistry.* **1990,** *29,* 9185-93.
24. Sarver, R. W.; Krueger, W. C. *Anal. Biochem.* **1991,** *194,* 89-100.
25. Bussian, B. M.; Sander, C. *Biochemistry.* **1989,** *28,* 4271-7.
26. Drucker, B.; Hainsworth, R.; Smith, S. G. *Journal of the Textile Institute.* **1953,** *43,* T420-T435.

RECEIVED August 26, 1993

Chapter 15

Crystal Structure of Silk of *Bombyx mori*

Yasuhiro Takahashi

Department of Macromolecular Science, Faculty of Science, Osaka University, Toyonaka, Osaka 560, Japan

The crystal structure of silk fibroin (*Bombyx mori*) was reexamined by using newly collected intensity data. Four molecules pass through a rectangular unit cell with parameters, a = 9.38 A, b = 9.49 A, and c (fiber axis) = 6.98 A, and the space group $P2_1\text{-}C_2^2$. The sheet formed by hydrogen bond is arranged parallel to the ac-plane. The molecular conformation is essentially the same pleated-sheet conformation as Marsh, Corey, and Pauling (1) proposed, but the sheet formed by hydrogen bonds assumes the antipolar-antiparallel structure in which the methyl groups of the alanine residues alternately point to both sides of the sheet along the hydrogen bonding direction, differing from the polar-antiparallel structure proposed by Marsh, Corey, and Pauling in which the methyl groups are all on the one side of the sheet. In the crystal structure of silk, two antipolar-antiparallel sheets with different orientations statistically occupied a crystal site with different probabilities . The crystalline region of silk is composed of rather irregular stacking of the antipolar-antiparallel sheets with different orientations.

The structural study of silk (*Bombyx mori*)can be said to start from the work of Nishikawa and Ono in 1913, in which they pointed out the characteristic feature of the so-called "fiber diagram" (2). In 1955, the crystal structure of silk fibroin was proposed by Marsh, Corey, and Pauling (1), which has been accepted so far. However, their crystal structure model is based on the quantitative intensity estimation of six equatorial reflections (discrepancy factor R = 37 %) and the qualitative comparison between the calculated and observed intensities on the layer lines. Judging from the present level of the crystal structure analysis of the fibrous sample, it is insufficient to be accepted. The purpose of

0097–6156/94/0544–0168$06.00/0

the present study is to reexamine the crystal structure. The results are partly given in the note (3).

The cocoon fibers were rolled up on a metal holder in a shape of cylinder with 1 mm diameter after purification according to the usual procedure. The specimen used for Weissenberg photograph was prepared by cutting it off 1 mm long after sticking by Aron Alpha. The doubly oriented sample was prepared by rolling the silk gland from the lumen. X-ray measurements were carried out by CuKα radiation monochromatized by a pyrolyzed graphite. The fiber diagram of silk fibroin is shown in Figure 1. The intensities of four strong reflections were measured by a PSPC system. The integrated intensities of 26 reflections were measured by a drum scan densitometer and five weak reflections were visually estimated, after fiber and Weissenberg photographs were taken by the multiple film method. Some overlapped reflections are separated by the least-squares method under the assumption of Gaussian profile. Thus, the intensities of 35 independent reflections were collected.

The protein of silk fibroin is composed of L- and H-chains linked by a disulfide bond, the molecular weights of which are ca. 25,000 and ca. 350,000, respectively. The amino acid sequences of the L-chain and a part of the H-chain are clarified (4 - 7). It is not considered that the L-chain is associated with the crystalline region judging from the amino acid sequence. The regular sequence of the H-chain is associated with the crystalline region(7, 8). The amino acid sequence in the crystalline region is considered as $(-\text{GAGAGS}-)_n$ where G, A, and S denote glycine, alanine, and serine, respectively.(6, 7) As the first approximation, the amino acid sequence is considered to be $(-\text{GA}-)_n$. The fiber identity period was estimated as 6.98 A, although the very weak streak layers corresponding to the sequence $(-\text{GAGAGS}-)_n$ are observed between the equator and the first layer line. From the fiber identity period, the molecular conformation is considered to be essentially the pleated-sheet structure as Marsh et al.(1) proposed. The observed reflections can be indexed by the rectangular unit cell with parameters, a = 9.38 A, b = 9.49 A, and c (fiber axis) = 6.98 A and the space group $P2_1-C_2^2$, which are essentially the same as those reported by Marsh et al.(1) The unit cell contains four molecular chains, a pair of which forms a sheet structure by hydrogen bonds. From the space group $P2_1$, two molecules forming the sheet in the unit cell are symmetrically independent and two sheets are related by two-fold screw axis. Four models can be constructed for the hydrogen bonding sheet, polar-antiparallel sheet (PA), polar-parallel sheet (PP), antipolar-antiparallel sheet (AA), antipolar-parallel sheet (AP) models (Figure 2). The molecules in the parallel sheet models assume the same direction while, in the antiparallel sheet models, the up- and down-molecules alternate. In the polar models, the methyl groups of alanine residues are all on one side of the sheet, while in the antipolar models, the methyl groups alternately point to both sides of the sheet along the hydrogen bonding direction. The crystal structure proposed by

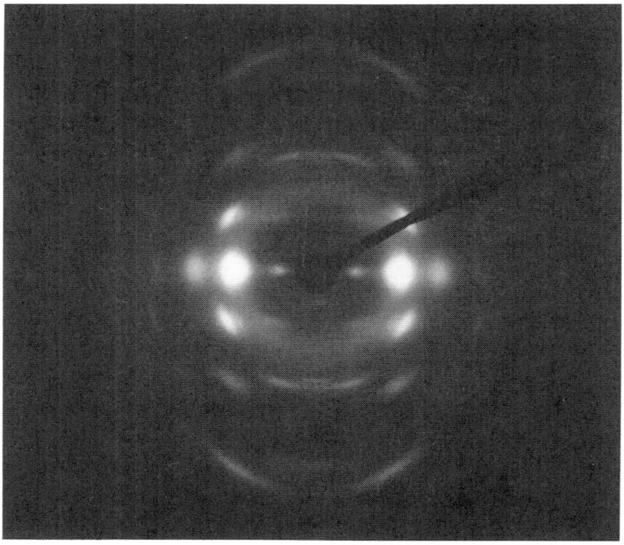

Figure 1. X-ray fiber diagram of silk (Bombyx mori).

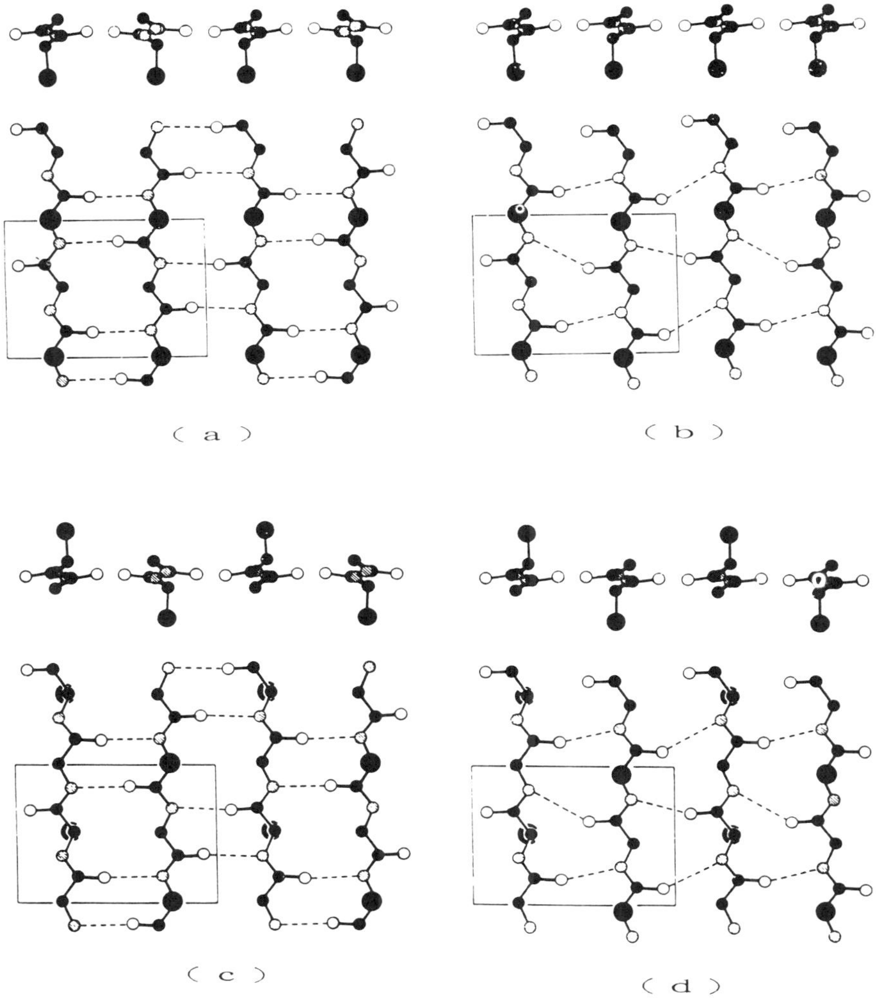

Figure 2. Four kinds of sheet structure models.
(a) Polar-antiparallel model which Marsh et al (1) proposed, (b) polar-parallel model, (c) antipolar-antiparallel model, and (d) antipolar-parallel model.

Marsh et al. corresponds to the regular arrangement of the polar-antiparallel sheets (1).

The constrained least-squares refinement (9, 10) was carried out by using the accepted bond lengths and bond angles, where the same conformation was assumed for the symmetrically independent molecules and the anisotropic temperature factor $\exp[-(ha^*/2)^2B_x + (kb^*/2)^2B_y + (lc^*/2)^2B_z)]$ was adopted by taking into account the anisotropy in vibration and disorder. The refinement was first made for the models based on the regular arrangement of the sheets. R-factors converged to 17.9 % for PA, 19.8 % for PP, 24.5 % for AA, and 18.3 % for AP. The PA model gives the lowest R-factor, but these values are not necessarily sufficient to distinguish which model is correct. Many researchers pointed out the inter- and intra-sheet disorder in the crystalline region of silk and the related polypeptides (11 - 13). The anisotropy in the reflection broadening observed on X-ray diffraction pattern, especially on the doubly oriented sample, suggests that the silk includes the disorder in the direction perpendicular to the hydrogen bonding sheets, i. e., the stacking disorder of the hydrogen bonding sheet (intersheet disorder). The disorder was taken into account by the statistical coexistence of the four sheets with different orientations at a crystal site with different probabilities. This type of disorder was found in the crystal structure of poly(vinylidene fluoride) form II (14, 15) and native cellulose (16). Each sheet models are constructed by allotting the same probability to the molecules forming the sheet at two symmetrically independent crystal sites. The refinement gave far better agreement between the observed and calculated structure factors: R-factors, 13.7 % for PA, 13.8 % for PP, 9.7 % for AA, and 11.6 % for AP. AA model gave the lowest R-values 9.7 %, in spite of the fact that, in the regular arrangement, AA model gave the largest R-value 24.5 %.

Finally, serine residue were taken into consideration by replacing the 1/3 of alanine by the 1/3 of serine in weight. R-value did not reduce in the cases of PA and PP models, but when the internal rotation angle C(=O)-C(α)-C(β)-O(H) assumes gauche conformation, AA model gave the lowest value 8.5 %, while PA model gives 10.7 % when the bond assumes minus gauche. Therefore, it is concluded that the AA model is most probable. The probabilities of the molecules with different orientations occupied a crystal site are 72 %, 24 %, 1 %, and 4 %. Accordingly, within the accuracy of the present analysis, it can be said that a crystal site is occupied in the ratio 3 : 1 by two sheets (AA) related by two-fold rotation axis parallel to the b-axis. In other words, the crystalline region of silk is constructed by stacking of two kinds of sheets with different orientations. The crystal structure (AA model) is shown in Figure 3 and the molecular structure is shown in Figure 4 in comparison with the pleated sheet model reported by Marsh et al. A sheet structure (AA) formed by hydrogen bonds is shown in Figure 5.

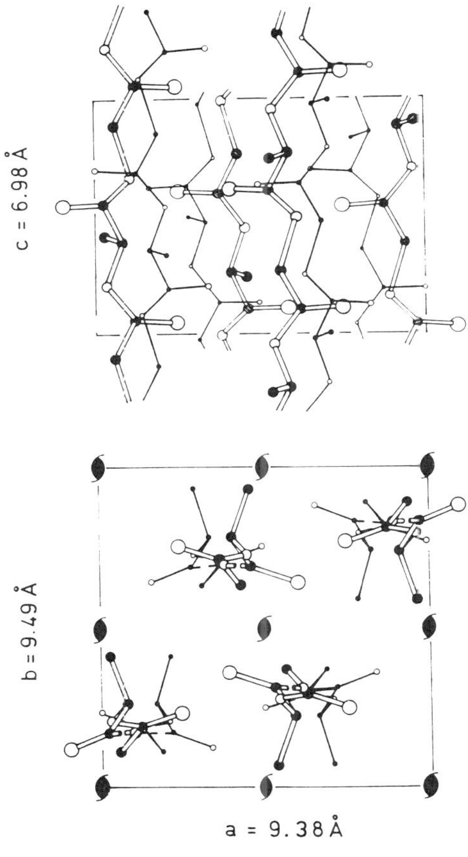

Figure 3. Crystal structure of silk.

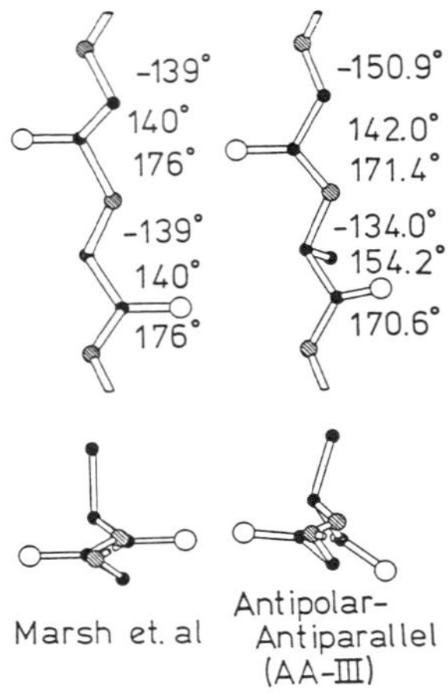

Figure 4. Molecular structure of silk.

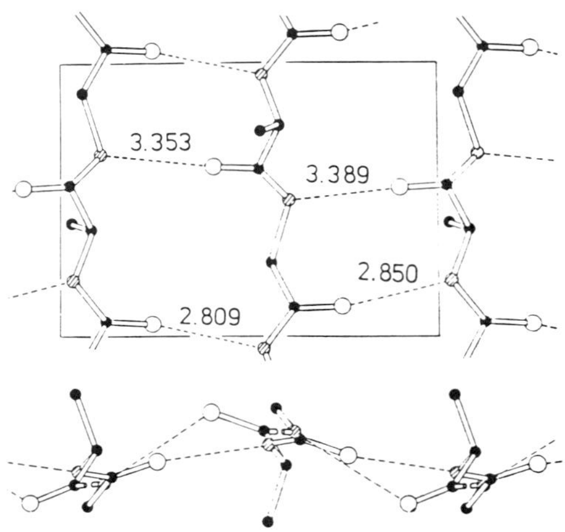

Figure 5. Sheet structure formed by hydrogen bonds of silk. Figures are given in A unit.

Literature Cited

1. Marsh, R. E.; Corey, R. B.; Pauling, L., *Biochim. biophys. Acta*, **1955**, 16, 1.
2. Nishikawa, S.; Ono, S., *Proc. Tokyo Math. Phys. Soc.*, **1913**, 7, 131.
3. Takahashi, Y.; Gehoh, M.; Yuzuriha, K., *J. Polym. Sci. Polym. Phys. Ed.*, **1991**, 29, 889.
4. Shimura, K., *Experimentia*, **1983**, 39, 455.
5. Yamaguchi, K.; Kikuchi, T.; Takagi, A.; Kikuchi, F.; Oyama, F.; Shimura, K.; Mizuno, S., *J. Mol. Biol.*, **1989**, 210, 127.
6. Tsujimoto, Y.; Suzuki, Y., *Cell.*, **1979**, 18, 591.
7. Lucas, F.; Shaw, J. T. B.; Smith, S. G., *Biochem. J.*, **1957**, 66, 468.
8. Warwicker, J. O., *Acta Cryst.*, **1954**, 7, 565.
9. Takahashi, Y.; Sato, T.; Tadokoro, H.; Tanaka, Y., *J. Polym. Sci. Polym. Phys. Ed.*, **1973**, 11, 233.
10. Takahashi, Y., *Rep. Progr. Polym Phys. JPN.*, **1992**, 35, 231.
11. Arnott, S; Dover, S. D.; Elliot, A., , **1967**, 30, 201.
12. Fraser, R. D. B.; MacRae, T. P.; Parry, D. A. D.; Suzuki, E., *Polymer*, **1969**, 10, 810.
13. Lotz, B.; Brack, A.; Spach, G., *J. Mol. Biol.*, **1974**, 87, 193.
14. Takahashi, Y.; Matsubara, Y.; Tadokoro, H., *Macromolecules*, **1983**, 16, 1588.
15. Takahashi, Y; Tadokoro, H., *Macromolecules*, **1983**, 16, 1880.
16. Takahashi, Y.; Matsunaga, H., *Macromolecules*, **1991**, 24, 3968.

RECEIVED July 7, 1993

Chapter 16

Toward Single-Fiber Diffraction of Spider Dragline Silk from *Nephila clavipes*

S. G. McNamee[1], C. K. Ober[1], L. W. Jelinski[2], E. Ray[2], Y. Xia[2], and D. T. Grubb[1]

[1]Department of Materials Science and Engineering and [2]Biotechnology Program, Cornell University, Ithaca, NY 14853

The results of a preliminary synchrotron X-ray study of the spider dragline silk from *Nephila clavipes* are presented. Comparisons of two different detector systems are made. The diffraction from 100-fiber bundles was observed as a function of environment and load on the samples. While the introduction of a dehydrating helium environment affected the tension of the samples, no concurrent change in the scattering behavior was observed. Mechanical loading of the bundles resulted in an increase in the orientation of the scattering centers within the fibers. Ideas for potential experimental protocols for single fiber diffraction are discussed.

The dragline silk from the major ampullate gland of the golden orb weaver spider, *Nephila clavipes*, has excellent structural and mechanical properties. For example, the specific modulus of dragline silk is greater than that of steel and the ability of the silk to absorb energy is superior to that of Kevlar (*1*). Although research on this biomaterial has been ongoing for many years, it has generated renewed interest in view of the potential for modern biotechnology to produce engineered proteins. The physical and mechanical properties of silks in general continue to intrigue numerous investigators, as demonstrated by the breadth of topics covered at this workshop. For a good introduction to the subject, the reader who is not familiar with this biomaterial or the arachnid that produces it is referred to references *1-4* and the references contained therein, as well as other chapters of this volume.

The fact that spider silk fiber is drawn from a liquid crystalline solution under mild conditions (*5*) is of great interest to those who wish

0097–6156/94/0544–0176$06.00/0

to spin high-performance fibers from a liquid crystalline melt or solution. There are numerous factors (rheological, chemical, and thermal) that combine in the processing of spider silk to produce practically defect-free fibers. It is known that the nascent silk in the gland is transformed from an isotropic liquid to a liquid crystalline solution (*5*) as water is removed, but the role that water plays in the processing and in the material's structure and performance after the fiber exits the spinneret is not clear. If water is lost or gained as the fiber is spun from the spider, we may expect the microstructure of the fiber, that is the combination of β-sheet and amorphous domains, to change as the silk travels from the spider during the spinning process. Oriented fibers, crystalline domains and liquid crystalline phases all possess distinct X-ray diffraction characteristics that we would like to use to monitor microstructural changes upon spinning. The manner in which the microstructure of the fibers responds to post-spinning mechanical loading is also of interest, since the natural fiber's properties are set without the need for any post-spinning draw, in contrast to conventional synthetic polymer fiber processing.

The use of *in situ*, time resolved, single fiber X-ray diffraction can help us to accomplish the goals of better understanding the processing and resultant microstructures of silk fibers drawn from the liquid crystalline state. This study is particularly timely in view of recent uncertainties in the molecular mechanism of spider silk elasticity (*6,7*). Fourier transform infrared (FTIR) techniques have been used to follow changes in the helical content of the silk as it is stretched (*6*). It is anticipated that X-ray diffraction can provide complementary results, with the added advantage of *in situ* measurements.

This paper describes our preliminary work on spider silk diffraction using a synchrotron X-ray source. While these studies were carried out on multifiber bundles, it is our ultimate goal to investigate a single fiber as it is being spun from the spider and to determine the effect of process conditions on the resulting fiber. This effort to obtain X-ray diffraction information from a single thread of dragline silk is intended to complement the magnetic resonance imaging (MRI) and solid state nuclear magnetic resonance (NMR) work being done on the *in vivo* chemical processing of the fibers. The diffraction experiments provide information primarily about the crystalline components, whereas the magnetic resonance experiments will provide information about the molecular dynamics of the amorphous regions. More generally, the techniques developed will also be useful in ongoing research into synthetic high performance fibers and liquid crystalline polymers.

Experimental

Fiber samples of dragline silk were collected from the spiders of the species *Nephila clavipes* originally obtained in Panama and maintained in individual humidified cages. The spiders were hand-fed crickets three times a week. Fiber samples were obtained by the motor method at a draw rate of 24 cm/sec. Bundles of 100 fibers were mounted on card stock under just enough tension to avoid tangling of the fibers.

The samples were mounted in a tensile apparatus developed for work at the Cornell High Energy Synchrotron Source (CHESS), which is described elsewhere (*8*). Care was taken to clamp the ends of the bundles as securely as possible, using adhesive tape and closed cell foam rubber grips on the tensile clamps. The distance between the clamps was changed by driving a stepping motor until the fibers were held steady. The piezoelectric load cell was then set to zero.

The entire stretching device was mounted on a sensitive, stepping-motor-driven, uniaxial slide to allow micron-scale translation of the sample within the X-ray beam. Two separate detectors were used. The first was a storage phosphor image plate and Kodak scanner system which allowed a very large area of detection with 100 μm resolution. The second was a direct coupled 80 mm diameter image-intensified detector of similar resolution that had a smaller active detector area but a much more easily used data-manipulation system. The interested reader is directed to references *9* and *10*, which describe the two detectors that led to the one that was used in this experiment, and also to reference *11*, which compares various quantitative X-ray imagers. The monochromated beam (hutch A1, CHESS, λ = 1.565 Å) was collimated to a diameter of 300 μm when using the 80 mm detector and 500 μm when using the storage phosphor.

In the initial experiment using the storage phosphor, a poly(ethylene) bag was used to enclose the sample and the image plate in a helium atmosphere. Relative humidity within the bag was monitored as was the change in the tension of the fiber sample while the helium environment was being introduced. For the second experiment using the 80 mm detector, a stream of helium was passed over the fiber sample while the change in the tension of the sample was monitored.

Exposure times were about 5 minutes with a sample to detector distance of 58 mm using the 80 mm detector, and 10 minutes with a sample to detector distance of 102 mm using the storage phosphors. No visual evidence of beam damage to the fibers was observed after any of the scattering experiments.

Results and Discussion

During the initial experiment using the storage phosphor detector, it was observed that the tension on the fiber bundle increased when the sample was placed in a helium atmosphere, where the relative humidity was 30%. The helium atmosphere was introduced in an effort to reduce the air scatter that was fogging the image plate and obscuring the diffracted image. The air scatter was only slightly diminished and the use of the bag was discontinued when subsequent images were not improved sufficiently to warrant the inconvenience of the bag. It is important to note that careful background subtraction was necessary for observation of the fiber diffraction pattern.

Figure 1 shows equatorial scans through the diffracted images as captured by the two detectors employed in this study. The measured **d** spacing of the 120/210 reflections agree with the literature (*12*). The S/N for the storage phosphor is clearly better than that of the image intensifier and the area of the storage phosphor was much greater than the 80 mm detector. However, the data handling and manipulation was much more facile with the smaller area detector and allowed for much faster data collection. Further discussion of the data is limited to that which was taken using the image intensifier. The slight asymmetry of the image intensifier data is due to a slight skew in the face of the detector at the time of the experiment.

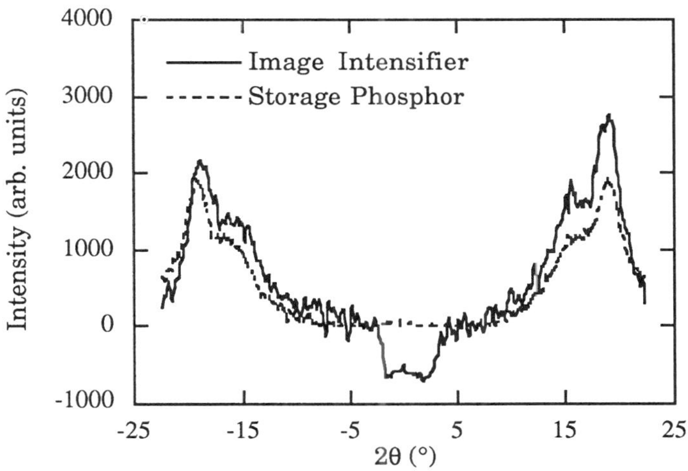

Figure 1. Comparison of the equatorial scans through the primary layer reflections as monitored by both detectors. 100-Fiber bundles under comparable loads of ca. 2 GPa.

Figure 2 shows that no structural change could be detected as a result of the flow of helium over the bundle despite the 46% increase in stress. Prior to exposing the bundle to the helium, the load on the bundle corresponded to a calculated stress of 0.358 GPa. The stress increased to 0.524 GPa after the fiber tried to shrink as a result of the dehydrating flow of helium but was constrained by the tensile device. Scanning parallel to the fiber direction through the layer lines shows again no significant change in the crystalline scattering, as seen in Figure 3. In Figure 3, ξ is defined as $\mathbf{arctan(L_n/a)}$ where $\mathbf{L_n}$ is the

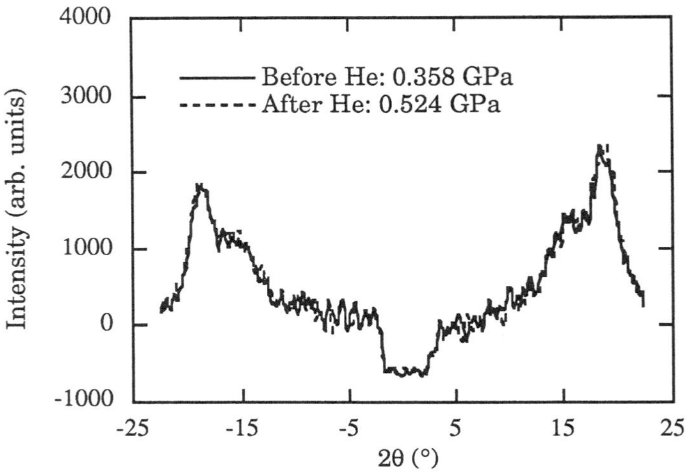

Figure 2. Effect of helium (He) environment on the equatorial scattering from the fiber bundle. Data captured using image intensifier.

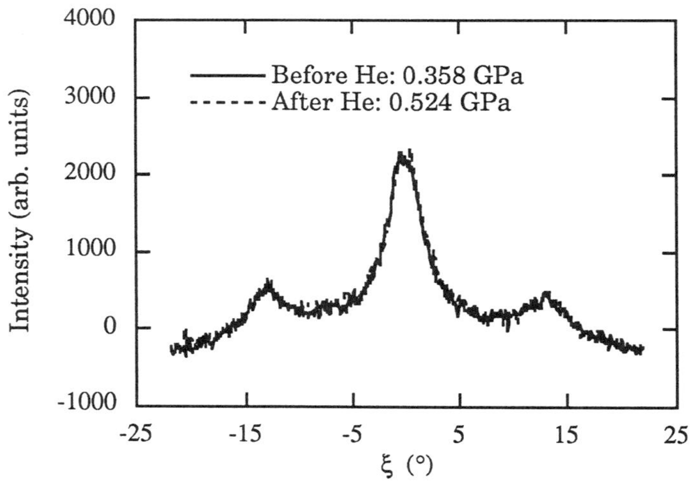

Figure 3. Effect of helium (He) environment on the right-hand primary and secondary layer reflections from the fiber bundle. Data captured using image intensifier.

distance from the center of the 120/210 reflections and **a** is the sample to detector distance. The right side of the image is profiled as it was the more intense side due to the aforementioned skew of the detector. Taken together, the data in Figures 1, 2 and 3 suggest that essentially no change occurs in the crystalline regions of silk upon dehydration and under low amounts of stress. However, the situation differs at greater levels of stress.

Table I shows the stress to which the bundle was subject for each of the X-ray images captured on the 80 mm detector after the bundle had been exposed to the helium and then left to equilibrate in the ambient humidity. The increase in stress for exposure **A** was a result of the attempted shrinkage of the fiber after application of the helium. Subsequent increases in the stress on the sample were a result of mechanically stretching the fiber bundle. The diameter of a single fiber was approximately 4 µm; a 100-fiber bundle would have a cross-sectional area of 1257 μm^2. This value was used for calculating the stress on the bundles while under tension. Strain calculations were not made, given the uncertainty of the differences in the modulus of the fiber before and after being subject to the helium environment. Experiments to compare the effects of environment on the single fiber and bundle moduli are presently underway.

Table I: Mechanical loading of the sample within the X-ray beam

Exposure reference	Load(gm)	Stress(GPa)	% increase in stress
A	67.3	0.524	46
B	159.1	1.24	136
C	246.3	1.92	55
D	321.3	2.51	30

One can see in Figure 4 that the intensity of the equatorial reflections increases slightly with increasing load, but that the position of the peaks remains constant. Analysis of the diffraction peak shapes showed a slight decrease in the FWHM with increasing load. This indicates an increase in the orientation of the scattering centers. Our results show very little change in the crystalline component under stress, suggesting that the amorphous component must participate in the distribution of load in the fiber. The data are in qualitative agreement with FTIR studies (*6*), and would further suggest that spider silk's excellent mechanical properties depend greatly on the amorphous regions' response to stresses. A slight rotation of the 120/210 reflections about the axis of the X-ray beam is due to a straightening out of the fiber as it is loaded, a problem that the next generation of tensile tester will address.

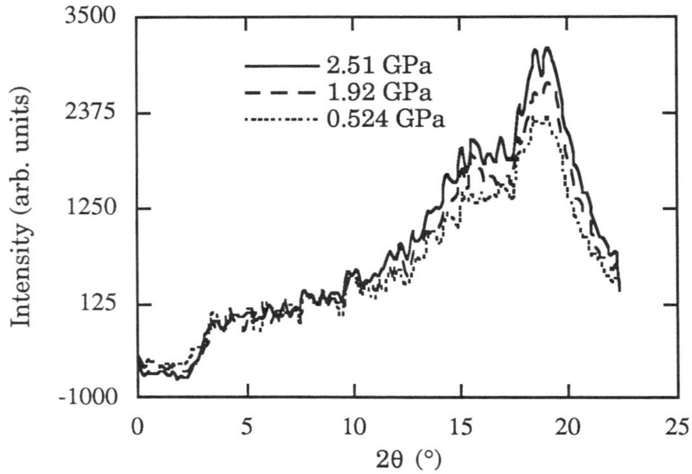

Figure 4. Effect of stress on equatorial scatter from the fiber bundle. Only the right-hand reflection is shown.

Although we have previously been successful in scattering from single fiber synthetic polymer samples (*8*), we were unable to capture a signal from a single fiber of dragline silk. We attribute this to several factors. Because the volume of material was small and the inherent cross-sectional scattering areas of the atoms that make up the silk are small as well, sufficient exposure times for the X-ray fluxes would have been prohibitively long. Furthermore, the S/N from a single fiber was simply too small for the detectors that were available. By using a bundle of 100 fibers, the S/N problem was overcome in the present study, but that, of course, precluded probing the silk as it was being spun from the spider. However, the data reported here enable us to calculate the feasibility of *in situ* scattering from a single fiber. Our calculations suggest that it will be necessary to either develop new detectors or increase the flux of X-rays scattering from the fiber. Work on the former is progressing, but it is the increase in flux that holds particular promise at this time.

The sample of interest is only a few microns wide, so much of the 500 μm-wide X-ray beam is wasted. Very precise X-ray slits may be used, but the problem of aligning the fiber sample to the beam becomes more difficult. Furthermore, in order to probe the fiber at different points along its length, the slits would have to be so short that the primary beam would be attenuated to the same extent as with present cylindrical collimators. A solution to this dilemma may emerge from the research being done at CHESS in the area of tapered glass capillaries to funnel a 500 µm beam of X-rays to a submicron diameter beam of very high brightness (*13*). Using tapered capillaries, one can obtain an increase in flux on the order of 1000-fold. Development of such a system for use with single polymer fibers, including dragline silk, is currently underway.

This work has shown the feasibility of probing the microstructure of spider silk fiber bundles under various conditions of stress and environment with a sufficiently bright X-ray source. As the technology of micron-sized X-ray beams is developed, the possibility of scattering from a single fiber of spider silk becomes greater. Within such an experimental protocol, one may scan across a single fiber to profile the domains of β-sheet packing vs. other regions in the material. It may be possible to follow the transition of the non-β-sheet regions as they transform due to water adsorption or absorption, due to applied mechanical strain, or due to other changes in the environment. Should the development of the fiber from a liquid crystalline solution happen as the silk emerges from the spinneret, it may be possible to follow the concurrent change in the microstructure as the single fiber is being drawn from the spider, profiling the fiber at different draw rates or at different distances from the spinneret.

Acknowledgments

The authors wish to thank E. Eikenberry and S. Gruner for the very generous use of their 80 mm detector. Funding by the NSF and the ACS/PRF is gratefully acknowledged. The authors appreciate the use of the facilities of CHESS and Cornell's Materials Science Center. S. Mc. thanks the Department of Education for a fellowship. Other people who have aided in this study include M. Went, C. Trojan, and Z. Li.

Literature Cited

1. Gosline, J. M.; DeMont, M. E.; Denny, M. W. *Endeavour*. **1986,** *10*, 37-43.
2. Vollrath, F. *Scientific American*, **1992,** *Mar.*,70-76.
3. Lewis, R. V. *Accounts of Chemical Research*, **1992,** *25*, 392-398.
4. Kaplan, D. L.; Lombardi, S. J.; Muller, W. S.; and Fossey, S. A. In *Biomaterials: Novel Materials from Biological Sources;* Byrom, D., Ed.; Stockton Press: NY, NY, 1991; pp 3-53.
5. Kerkam, K.; Kaplan, D.; Lombardi, S.; Viney, C. *Materials Synthesis Based on Biological Processes.* MRS Proceedings, 1991, Vol. 218; pp 239-244.
6. Dong, Z.; Lewis, R.V.; Middaugh, C.R. *Arch. of Biochem. and Biophys.* **1991,** *284*, 53-57.

7. Hinman, M.B.; Lewis, R.V. *J. Biol. Chem.* **1992,** *267*,19320-19324.
8. Prasad, K.; Grubb, D. *J. of Polym. Sci.: Part B: Polym. Phys.* **1990** *28*, 2199-2212.
9. Templer, R.H.; Gruner, S.M.; Eikenberry, E.F. *Advances in Electronics and Electron Physics* Academic Press: London, 1988; pp 275-283.
10. Eikenberry, E.F.; Tate, M.W.; Belmonte, A.; Lowrance, J.; Bilderback, D.; Gruner, S.M. *IEEE Trans. Nucl. Sci.*, **1991,** *38*, 110-118.
11. Eikenberry, E.F.; Tate, M.W.; Bilderback, D.H.; Gruner, S.M. In *Photoelectronic Image Devices, 1991;* Morgan, B.L., Ed.; Institute of Physics; Bristol, 1992; pp 273-280.
12. Work, R.W.; Morosoff, N. *Text. Res. Jour.* **1982,** *52*, 349-356.
13. Hoffman, S.A.; Thiel, D.J.; Bilderback, D.H. In *Optics for High-Brightness Synchrotron Radiation Beamlines;* SPIE: 1992;Vol. 1740, pp 252-257.

RECEIVED May 4, 1993

Chapter 17

X-ray Moduli of Silk Fibers from *Nephila clavipes* and *Bombyx mori*

Mary A. Becker[1], D. V. Mahoney[1], P. G. Lenhert[2], R. K. Eby[1], David Kaplan[3], and W. Wade Adams[4]

[1]Department of Polymer Science and Maurice Morton Institute of Polymer Science, The University of Akron, Akron, OH 44325–3909
[2]Department of Physics, Vanderbilt University, Nashville, TN 37235
[3]Biotechnology Division, Natick Research, Development, and Engineering Center, U.S. Army, Natick, MA 01769–5020
[4]Materials Directorate, Wright Laboratory, WL/MLPJ, Wright-Patterson AFB, OH 45433–7702

X-ray moduli were determined for silk fibers from the golden orb-weaver spider *Nephila clavipes* and the domesticated silkworm *Bombyx mori* to provide insight into their physical properties. Differences between the respective moduli, 17 and 29 GPa, are discussed in terms of the lack of scaling with the basal area per molecule, the effect of the amorphous content and the stress history dependence. Both annealed and irradiated *B. mori* fibers yielded decreased X-ray moduli. All these results as well as the considerable difference between the experimental and computer modeled moduli raise questions regarding the applicability of the uniform stress assumption for the X-ray modulus calculation at room temperature conditions.

Many biological materials exhibit extraordinary physical properties considering that they are assembled *in vivo* from abundant natural resources. The silk dragline fibers (major ampullate gland) produced by *Nephila clavipes* provide structural support for orb-shaped webs and dissipate large quantities of energy when prey fly into the snare (*1, 2*). It is the combination of high strength, modulus and extensibility that yields the latter property and makes spider silk unique among the extracellular fibrous proteins. Their properties and the idea of obtaining silk fibers directly from a live creature, as opposed to the tedious task of unwinding cocoons, even lead to attempts at product development over a period of time going back at least to the beginning of the 18th century (*3-5*). Some unusual products, such as smothering caps of felted *Nephila* silk employed in the suffocation of people, dooming bags, fishing nets, optical cross hairs, strings for musical instruments, etc., have been reported (*5, 6*). Knowledge of the physical and chemical structures/properties is required to understand the unusual characteristics of silk produced by the spider.

0097–6156/94/0544–0185$06.00/0

Silk's physical properties are a consequence of both the composition and molecular architecture of a fibrous protein, fibroin. Amino acid compositions comprising a large percentage of glycine and alanine are characteristic of fibroins, and in general, of all extracellular matrix proteins. For the extruded fibroins, the conformation of the polypeptide is influenced by the amino acid sequence as well as by the shear stresses imposed on the fibroin molecules during the extrusion process. It is the shear stresses which effectively orient and extend the polypeptide chains (*7, 8*). The high percentage of small side chain amino acids allows for close packing of the extended molecules and facilitates crystal formation.

The composition of *Bombyx mori* silk, the most extensively studied fibroin, has been investigated since the middle of the 19th century (*9*), however, the exact amino acid sequence of this highly repetitive molecule has been difficult to obtain. The sequence of the crystalline precipitate remaining after digestion of fibroin with chymotrypsin, the C_p fraction, is (*10, 11*)

$$\text{Gly-Ala-Gly-Ala-Gly-Ser-Gly-Ala-Ala-Gly-[Ser-Gly-(Ala-Gly)}_n]_8\text{-Tyr}$$

More recently, *B. mori* fibroin has been found to consist of a light (L-) and a heavy (H-) chain linked by a disulfide bond (*12*). The amino acid sequence of the L-chain (25 kd) has been determined by cDNA, while only the initial region of the repetitious H-chain (350 kd) has been reported (*13-15*).

The amino acid composition of *N. clavipes* silk has the characteristically high glycine and alanine content consistent with other fibroins (*16, 17*). The repeats -(Gly-Gly-Xaa)$_n$- and -(Tyr-Gly-Gly-Leu-Gly-Ser-Gln-Gly-Ala-Gly-Arg-Gly-Gly)-, however, have been proposed as part of *N. clavipes*' major ampullate gland silk sequence, replacing the classic repeat -(Gly-Ala-Gly-Xaa)$_n$- arrangement found in *B. mori* fibroin (*18, 19*).

The first scattering of X-rays by *B. mori* fibers was reported in 1913, but it wasn't until much later that a pseudo-unit cell was determined to be orthorhombic with monoclinic symmetry (*20, 21*). In 1960, Warwicker published a comparative study of the X-ray patterns from seventy different silks produced by species in the classes Insecta and Arachnida (*22*). This led to the silks being classified into five groups based on their X-ray patterns, specifically the two principle equatorial reflections (*22*). Some of the five groups were subdivided into "a", those which exhibited equivalent intensities of the principle equatorial reflections and "b", those with different intensities. A common **c** dimension of 6.95 ± 0.05 Å along the fiber axis was determined for all the fibroins (*22*). A **b** dimension of 9.44 Å, the direction of the hydrogen bonding within the anti-parallel β-pleated sheets, was also the same for all five X-ray groups (*22*). The variation in the **a** dimension, intersheet spacing, from 9.3 Å to 15.7 Å, was attributed to the different amino acid compositions and sequences of the fibroins (*22*). While the five X-ray groups are not identical to the taxonomical classification of the species, they do indicate a similarity in the basic structure and provide a basis for further comparative studies.

The distribution and size of the crystalline and amorphous regions of the extruded proteinaceous fibers influence the physical properties. One means of investigating the crystalline regions utilizes X-ray diffraction measurements to determine the lattice strain in the direction of the applied macroscopic stress (*23, 24*). The ratio of the applied stress to lattice strain is commonly referred to as the "X-ray modulus", "crystal modulus", or "lattice modulus" and, subject to certain assumptions, is the ultimate specimen modulus.

The lattice strain is determined by expressing the change in the position of a diffraction peak compared to the initial position as a fractional change in d-spacing with tension (*23, 24*). The calculation of the X-ray modulus assumes a uniform stress distribution throughout the material (*23, 24*). Previous workers have argued for this assumption on the basis that the X-ray moduli of PVA and PE fibers are a function of only the lattice spacing and independent of overall crystal orientation and crystallinity (*25*). The fact that the same moduli were also determined for both dry and wet conditions of hygroscopic PVA fibers was also cited as indicating that the uniform stress assumption is valid (*26*).

Three conditions have been proposed as requirements for a high X-ray modulus in fibers (*27*). First, the molecular axis should be parallel to the applied stress. Second, the molecular conformation and configuration should allow a close-packing arrangement of the molecules. Third, the force constant for deformation of the molecule should be large. Deformation by bond stretching rather than bond angle bending or torsion generally carries a higher force constant. Helical polymers, for example, usually deform by changes in torsion and therefore have lower moduli than fully extended chain polymers.

The dimensions of the pseudo-unit cell proposed for *B. mori* are **a** = 9.29 Å, **b** = 9.44 Å and the fiber axis **c** = 6.95 ± 0.05 Å corresponding to Group 1 in Warwicker's classification (*22*). The similarity in the X-ray diffraction by the fiber and the C_p fraction is interpreted to mean the sequence is representative of the crystalline regions of the fiber (*28*). The -(Gly-Ala-Gly-Xaa)- repeat in the crystalline regions allows for the closest packing of all the fibroins reported to date (*22, 28*). Therefore, Group 1 fibroins would be expected to have the highest X-ray moduli of all the β-pleated sheet proteins.

The similarity of the amino acid composition of *N. clavipes* to that of *Nephila madagascariensis* suggests that similar unit cell dimensions might be anticipated (*16, 29*). The pseudo-unit cell of *N. madagascariensis* is classified as Group 3b and has dimensions of **a** = 10.6 Å, **b** = 9.44 Å, and **c** = 6.95 ± 0.05 Å (*22*). These dimensions are comparable to those of Tussah silk (*Antheraea pernyi*), also Group 3b, which is described as having an -(L-Ala)$_n$- sequence in the crystals (*28*). The amino acid residues in the unit cells of *N. clavipes*, *N. madagascariensis* and *A. pernyi* have not been determined explicitly. Nevertheless, the similarity of their pseudo-unit cell dimensions and β-pleated sheet conformation provides a basis for a comparison of their X-ray moduli.

In this paper, measurements and computations of the X-ray moduli for the silks of *B. mori* and *N. clavipes* are presented. The results are discussed in the context of the validity of the assumption of uniform stress.

Materials and Methods

The female *N. clavipes* spiders were obtained from Panama. Dragline fibers were collected in a manner similar to a recently published method as well as to one described as early as the 18th century (*30, 31*). The silk was collected at a rate of approximately 1.1 cm/s. A bundle of fibers (approximately 5100) was prepared, mounted in a sample holder and secured with room-temperature curing epoxy. The density was taken as 1.35 (± 0.005) g/cm^3 (*32*). An effective cross-sectional area of a fiber bundle was determined from the mass of a specific length and the density. This yielded a sample cross-sectional area of 9.5 x 10^{-8} m^2.

Commercially degummed *B. mori* yarns were bundled in groups of 40 or 100 yarns, mounted in sample holders and secured using room-temperature curing epoxy (*33*). The silk yarns had approximately 8 twists/cm which is calculated to cause a very small error in the X-ray modulus (see Appendix). The density was taken as 1.353 g/cm^3 (*34*). This yielded an effective cross-sectional area of 9.8 x 10^{-8} m^2. Two additional *B. mori* samples were prepared from artificially-aged fibers. The first was exposed to 500 kJ/m^2 (0.75 W/m^2) of simulated sunlight-through-glass in an Atlas Ci-65 Weather-ometer. The irradiated fibers were then mounted in sample holders as described above. The second sample consisted of fibers mounted in the sample holders, secured with epoxy and then annealed at 200 °C for one hour at a fixed length. Both samples exhibited a discoloration after the aging.

A Rigaku Denki RU-200 rotating anode was the source of nickel-filtered Cu radiation used for all experiments. X-ray data for the N. *clavipes* sample was collected using a Siemens General Area Detector with stress applied by a dead weight. Data for *B. mori* samples were collected for both increasing and decreasing tension as well as various orientations of the tensioning device according to a published procedure (*33*). The stress relaxation of the *B mori* sample after the initial application of strain was compensated for by a further increase in applied strain before X-ray measurements were made. (There is a related creep effect in the method used for the *N. clavipes* sample.) The 002 diffraction peak was used for calculation of the X-ray moduli. For comparative purposes, the 006 diffraction peak of *B. mori* was also used. The relatively small effects of changes in relative humidity and temperature on density during the experiments were not investigated (*34-36*).

Molecular modeling was done on a Silicon Graphics system using the Sybyl 5.5 software from Tripos Associates, Inc. Arrays of nineteen peptides, each ten alanine residues in length and terminated with blocking groups, were constructed in the anti-parallel β conformation to represent Group 3b. Various packing geometries and deformation protocols were applied to these arrays in order to obtain the energy of the central molecule as a function of strain. The second derivative of this energy with respect to strain was used together with a correction for end effects, the initial length and the basal area per molecule to obtain the modulus (*37*).

Results and Discussion

The 002 diffraction peak from the *N. clavipes* fiber bundle gave a d-spacing of 3.47 Å or a pseudo-unit cell **c** dimension of 6.94 Å. This agrees with the pseudo-unit cell **c** dimension, 6.95 ± 0.05 Å, for all fibroins in the β-pleated sheet conformation (*22*). The pseudo-unit cell dimensions perpendicular to the chain axis conformed with the published values for Group 3b, for which **a** = 10.6 Å and **b** = 9.44 Å (*22*). Presumably the increased spacing in the **a** direction is the result of bulkier side groups incorporated into the crystals (*22*). The X-ray modulus of the dragline silk was 16.7 ± 0.6 GPa (002) at 27 °C and approximately 15% relative humidity (Fig. 1).

The 002 diffraction peak from the *B. mori* sample gave a d-spacing of 3.50 Å or a pseudo-unit cell **c** dimension of 7.00 Å. This also corresponds with the characteristic **c** dimension of all fibroins in the β-pleated sheet conformation (*22*). Scans of the layer lines 0 to 6 showed only minor differences, after background corrections, between the peak positions and the cell corresponding to Group 1, **a** = 9.29 Å and **b** = 9.44 Å (*22*).

The X-ray modulus determined as a function of increasing strain for the crystals in *B. mori* silk fibers was 28.7 ± 0.2 GPa (002) at 20 °C and a relative humidity ranging between 40% and 60%. The modulus determined as a function of decreasing strain was a higher value, 30.5 ± 0.5 GPa (002). The difference in the data (Fig. 2) is related to the stress relaxation observed immediately after the strain was altered and to a hysteresis reported earlier (*1, 38*). Note also that the value of the modulus obtained depends somewhat upon the time elapsed between the application of the strain and the taking of the X-ray data. Effects relating to relaxation phenomena in silk have been known for a long time (*39*). The values of the modulus obtained for increasing and decreasing strains can be averaged to yield 29.6 GPa (002).

For *B. mori* fibers, the 002 reflection gave a lower X-ray modulus than the 006 reflection. The latter gave a value of 35.3 ± 0.2 GPa for increasing strain and 36.4 ± 0.8 GPa for decreasing strain with an average of 35.8 GPa (006). One possible cause of the difference between the (002) and the (006) modulus values is that the overlap of the amorphous halo and the 002 peak interferes with the baseline subtraction. If this overlap changes with the stress, it would introduce error to the change in the d-spacing. The 006 peak does not overlap the amorphous halo as much as the 002 peak and thus could yield a more accurate X-ray modulus. Another possibility, known to occur in some other polymers, is a change in the Laue lattice factor $|G|^2$ as crystallite size increases in the chain direction with applied tensile stress (*40-42*). However, the change in crystallite size with applied stress is negligible for *B. mori* fibers (*43*). Thus, the source of the discrepancy is probably in the overlap with amorphous scattering.

The artificially-aged *B. mori* fiber bundles had substantially reduced X-ray moduli. Those of silk irradiated with simulated sunlight-through-glass were 20.5 ± 0.4 GPa (002) and 24.8 ± 0.5 GPa (006). Those of the annealed silk were 22.3 ± 0.2 GPa (002) and 25.5 ± 0.3 GPa (006). A comparable X-ray modulus of 23 GPa has been reported for *B. mori* silk fibers which also had been annealed at 200 °C for one hour at constant length (*43*). Biochemical analyses have shown that the amorphous regions are affected initially by degradative photochemical mechanisms (*44*). Similarly, it might be anticipated that the oxidative degradation of the annealed fibers would primarily affect the amorphous regions which would be more accessible to oxygen. This has been concluded on the basis of other work (*45*). Changes in optical, mechanical and physical properties were interpreted in terms of changes in the amorphous regions upon heat treatment to about 260 °C (*45*). The present X-ray results agree with those in that work (*45*). That is, qualitative comparison of the diffraction patterns obtained before and after annealing to 200 °C indicates a small increase in crystallinity with the crystal structure remaining unchanged. Thus, the decreased X-ray modulus for the artificially-aged silk could be the result of changes in the amorphous regions. This suggests that the connectivity of the crystalline and

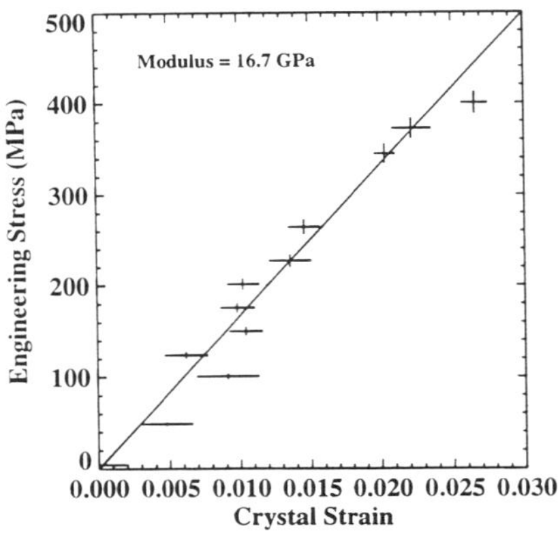

Figure 1. The stress-strain curve based on x-ray diffraction by the (002) planes of crystals in *Nephila clavipes* fibroin for increasing strain. The line is a weighted fit to the data which are represented by crosses giving the uncertainty.

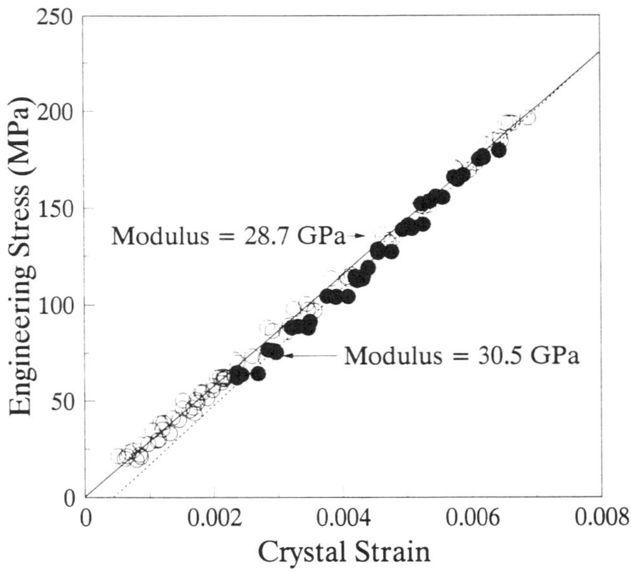

Figure 2. The stress-strain curve based on x-ray diffraction by the (002) planes of crystals in *Bombyx mori* fibroin for increasing (-O-) and decreasing strain (··●··). The lines are least-squares fits to the data which were obtained with one sample using four different sets of measurements, each using a different orientation of the tensioning device.

amorphous regions affects the X-ray modulus measurements and might not correspond to the proposed uniform stress model (*43*). The measured X-ray moduli of pitch-based carbon and poly(paraphenylene benzobisthiazole) fibers are known to vary with changes in structure and morphology (*46, 47*).

The difference between the moduli of *N. clavipes* and *B. mori* could be the result of increased spacing in the **a** direction due to bulkier side groups in the crystals of *N. clavipes*. The difference between the moduli of *B. mori* and *A. pernyi* silk have been accounted for by such a change (*48*). When the modulus of *N. clavipes* is normalized to the pseudo-unit cell dimension of *B. mori* in the **a** direction, the value is increased only slightly to 19 GPa, compared to the 29 GPa given above for *B. mori*. Thus, the change in cross-sectional area of the crystals does not fully account for the differing measured moduli of these two fibers. Since *N. clavipes* is more amorphous than *B. mori* silk (*1*), this difference in the moduli might result from the effects of overlap of the amorphous halo and the 002 peak as well as from differences in amorphous content per se. Further, there might be differences arising from the fact that *N. clavipes* was measured under fixed load and *B. mori* under fixed strain. These results raise the issue of the validity of the uniform stress assumption and the influence of the connectivity as well as stress relaxation on that assumption. It should be noted that the assumption of uniform stress yields the lower limit of the modulus, while the assumption of uniform strain yields the higher limit (*46, 47, 49*). The X-ray modulus can also be calculated using the assumption of uniform strain by making use of elastic constants and crystal orientation (*46, 47*). However, the elastic constants are unknown for β-pleated fibroin crystals.

Molecular modeling yielded a modulus of about 150 GPa for the arrays of molecules of -(L-Ala)$_{10}$-. This computational modulus for agrees fairly well with the value of 155 GPa calculated by Treloar's method for the pseudo-unit cell *B. mori* crystals (*50*). It has been argued that the latter result is in error because it neglects the low stiffness for torsion about the backbone bonds (*43*). However, intersheet restrictions on the torsion due to side chain interactions as well as intramolecular restrictions are known (*28, 51, 52*). It is our observation that there are also intrasheet restrictions due to the methyl branch interactions. All would add significantly to the effective torsional stiffness.

While poly-(L-alanine) may not explicitly correspond to the crystalline regions of *N. clavipes*, it is considered a model for the pseudo-unit cell of *A. pernyi* and both species are classified in Warwicker's Group 3b (22). In this sense, the computed modulus of 150 GPa for poly-(L-alanine) can be compared to the X-ray moduli of *N. clavipes*, 17 GPa, and *A. pernyi*, 20 GPa (*48*). Furthermore, computational moduli for single chain homopolymers of alanine, glycine and serine suggest that the molecular stiffness changes relatively little among homopolymers containing of the smaller side chain amino acids. This implies that the computed modulus can also be compared to the X-ray modulus of 29.6 GPa for *B. mori*. In contrast to these results, test calculations for polyethylene using similar modeling approaches, however, yielded a computational modulus much closer to that experimentally observed. Thus, the computational results for fibroins suggest that the measured values of X-ray moduli based on the proposed use of the assumption of uniform stress are too small (*43*).

The large disparity between the computational and the experimental X-ray moduli of the fibroins cannot simply be accounted for by various arrays and protocols of the modeling program. A discrepancy between the theoretical and experimental modulus values of nylon has been noted by others (*40, 53*). However, simple explanations, such as amide chain shortening for the case of nylon-6, do not correct for the discrepancy in moduli values for fibroins. While the assumption of a uniform stress distribution has been verified by others (*26, 43, 48, 54*), X-ray diffraction is known to yield lower experimental moduli than spectroscopic techniques such as Raman and coherent inelastic neutron scattering (*55*). The approximate 110 GPa difference between the computed modulus of poly-(L-alanine) and the X-ray moduli of *N. clavipes*, *A. pernyi*, and *B. mori* silk cannot be accounted for. It does raise questions regarding the validity of the uniform stress assumption for fibroins at room temperature conditions.

Conclusions

The X-ray moduli (002) are 17 GPa for *N. clavipes* and 29 GPa for *B. mori*. These values do not scale according to the difference between the basal area per molecule in the two crystals. Further, they are much less than the values calculated by computational modeling for poly-(L-alanine) and by Treloar's method for the fibroin pseudo-unit cell (*48*). The modulus value for *B. mori* is altered by an observed stress relaxation phenomena and lowered by degradation effects in the amorphous components of the fiber.

All the discrepancies among the various X-ray and computational moduli determined for *N. clavipes* and *B. mori* raise questions regarding the effect of relaxation phenomena and crystalline-amorphous connectivity on X-ray modulus measurements for these silks. These observations, in turn, question the validity of the assumption of uniform stress used in the analysis of the experimental data.

Appendix: Effect of Yarn Twist on Modulus Calculations

Consider the yarn to fiber modulus relationship using idealized helical geometry analogous to the single filament to X-ray modulus relationship. Based on the twist per unit length and the idealized helical yarn geometry, the fibers subtend an average angle θ of less than 5°, with respect to the yarn axis (*57*). The idealized helical geometry predicts the yarn modulus E_y is related to the fiber modulus E_f by a factor of $\cos^2 \theta$, $E_y = E_f \cos^2 \theta$. Assuming the analogous situation for the relationship between the measured crystal modulus of the yarn and the crystal or X-ray modulus of the fiber, the X-ray modulus would be altered less than 1%. This alteration is comparable to the experimental error of the X-ray modulus measurements.

Acknowledgements

The helpful suggestions and assistance of Dr. Hao Jiang are gratefully acknowledged. Portions of this work were supported by the Wright Laboratory through contracts with Systran Corp. (M.A.B., P.G.L., R.K.E.) and Lawrence Associates, Inc. (D.V.M, M.A.B., R.K.E.). Portions of this work were reported previously (*38, 56*).

Literature Cited

1. Gosline, J.M.; DeMont, M.E.; Denny, M.W. *Endeavor* **1986**, *10*, pp 37-43. See also Gosline, J.M.; Denny, M.W.; DeMont, M.E. *Nature*, **1984**, *309*, pp 551-552.
2. Vollrath, F. *Scientific American* **1992**, *3*, pp 70-76.
3. Wilder, B.G. *The Atlantic Monthly* **1866**, *18 (106)*, pp 129-145.
4. Dusuzeau, J. *Laboratorie d'etudes de la Soie Industrie,"Rapport presente a la Chambre de Lyons por Commerce d'Industrie"*, **1892**, pp 163-171 plus plates, refers to the work of Bon in 1708.
5. Mariox, M.J. *Laboratorie d'etudes de la Soie Industrie, "Rapport presente a la Chambre de commerce de Lyons por commerce et d'Industrie"*; **1899**, pp 23-34. See also Ref. 31.
6. McKeown, K.C. *Spider Wonders of Australia*; Angus & Robertson Ltd.: Sydney, 1936; pp 233-234.
7. Komatsu, K. *J. A. R. Q.* **1979**, *13(1)*, pp .64-72.
8. Magoshi, J.; Magoshi, Y.; Nakamure, S. In *Proc. of the 7th International Wool Textile Research Conference,* Tokyo, 1985, Vol.1. pp 329-336.
9. Cramer E. *Journal fur Praktische Chemie*, **1865**, *96*, pp 76-98; See also Roard, *Ann. de Chim.* **1807**, *65*, p 44.
10. Lucas, F.; Shaw, J.T.B.; Smith, S.G. *Biochem. J.* **1957**, *66*, pp 468-479.
11. Strydom, D.J.; Haylett, T.; Stead, R.H. *Biochem. Biophys. Res. Comm.* **1977**, *79(3)*, pp 932-938.
12. Sasaki, T,; Noda, H. *Biochim. Biophys. Acta* **1973**, *310*, pp 91-103.
13. Yamaguchi, K.; Kikuchi, Y.; Tagaki, T.; Kikuchi, A.; Oyama, F.; Shimura, K.; Mizuno, S. *J. Mol. Biol.* **1989**, *210*, pp 127-139.
14. Tsujimoto, Y.; Suzuki, Y. *Cell*, **1979**, *16*, pp 425-436.
15. Tsujimoto, Y.; Suzuki, Y. *Cell*, **1979**, *18*, pp 591-600.
16. Work, R.W.; Young, C.T. *J. Arachnol.* **1987**, *15*, pp 65-80.
17. Lombardi, S.J.; Kaplan, D.L. *J. Arachnol.* **1990**, *18*, pp 297-306.
18. Mello, C.; Senecal, K.; Young, B.; Vouros, P.; Kaplan, D.L. this monograph.
19. Lewis, R.V. *Acc. Chem. Res.* **1992**, *25*, pp 392-398.
20. Nishikawa, S.; Ono, S. *Proc. Math. Phys. Soc. Tokyo* **1913**, *7*, pp 131-138, plates.
21. Marsh, R.E.; Corey, R.B.; Pauling, L. *Biochim. Biophys. Acta* **1955**, *16*, pp 1-34.
22. Warwicker, J.O. *J. Mol. Biol.* **1960**, *2*, pp 350-362.
23. Sakurada, I.; Nukushina, Y.; Ito, T. J. Poly. Sci. **1962**, *57*, pp 651-660.
24. Dulmage, W.J.; Contois, L.E. *J. Poly. Sci.* **1958**, *28*, pp 275-284; also see Refs. *33, 46, 47.*
25. Sakurada, I.; Kaji, K.; Nakamae, K.; Wadano, S. *Bulletin of the Institute for Chemical Research, Kyoto University*, **1966**, *44*, pp 168-182.
26. Sakurada, I.; Ito, T.; Nakamae, K. *Die Makromolekulare Chemie* **1964**, *75*, pp 1-10.
27. Nakamae, K.; Nishino, T. In *Integration of Fundamental Polymer Science and Technology-5*; Lemstra, P.; Kleintjens, L.; Eds., Elsevier: New York, 1991, pp 121-130.

28. Marsh, R.E.; Corey, R.B.; Pauling, L. *Acta Cryst.* **1955**, *8*, pp 710-715.
29. Lucas F.; Shaw, J.T.B.; Smith S.G. *J.Mol. Biol.* **1960**, *2*, pp 339-349.
30. Work, R.W.; Emerson, P.D. *J. Arachnol.* **1982**, *10*, pp 1-10.
31. Anonymous *Seta-Bollettino di Sericoltura* **1957**, *18*, pp 17-20, refers to de Termeyer, R.M. *Opuscoli scientifici di Entomologia di Fisica e d'Agricoltura*, Vol 1, Milan, 1807 which reports work from the previous 30 years.
32. Zemlin, J.C. Technical Report 69-29-CM, 1968, U.S. Army Natick Laboratories, Natick, MA.
33. Lenhert, P.G.; Adams, W.W. In *The Materials Science and Engineering of Rigid-Rod Polymers*; Adams, W.W.; Eby, R.K.; McLemore, D.E., Eds.; Materials Research Society Symposium Proceedings; Materials Research Society: Pittsburgh, PA, 1989, Vol. 134; pp 329-340. See also Lenhert, P.G.; Adams, W.W. *Tensile Modulus by X-ray Diffraction; Instrument and Method*; Wright Laboratory, Materials Directorate, WPAFB, OH 45433, July 1990; WRDC-TR-90-407.
34. Goodings, A.C.; Turl, L.H. *J. Textile Inst.* **1940**, *31*, pp T69-80.
35. McLaren, A.D.; Rowen, J.W. *J.Poly. Sci.* **1951**, *7(2/3)*, pp 289-324.
36. Wright, B.A. *Nature*, **1948**, *162*, p 23.
37. Klunzinger, P.E.; Eby, R.K.; Adams, W.W. In *Hierarchically Structured Materials;* Aksay, I.; Baer, E.; Sarikaya, M.; Tirrell, D., Eds.; Materials Research Society Symposium Proceedings; Materials Research Society: Pittsburgh, PA, Vol. 255, 1992, pp 119-128.
38. Lenhert, P.G.; Becker, M.A.; Eby, R.K Presented at the Fiber Society 50th Technical Conference, Princeton, NJ, August 19-23, 1990.
39. Weber, W.; *Poggendorf's Ann. der Physik Chemie*, **1835**, *34*, pp 247-257.
40. T. Kaji, K.; Sakurada, I. *Makromol. Chem.* **1978**, *179*, pp 209-217.
41. Kaji, K. *Makromol. Chem.* **1974**, *175*, pp 311-325.
42. Nishino, T.; Tada, K.; Nakamae, K. *Polymer*, **1992**, *33(4)*, pp 736-743.
43. Nakamae, K.; Nishino, T.; Ohkubo, H. *Polymer*, **1989**, *80*, pp 1243-1246.
44. Becker, M.A. Ph.D. Dissertation, The Johns Hopkins University, 1993.
45. Tsukada, M.; Freddi, G.; Nagura,M.; Ishikawa, H.; Kasai, N. *J. Appl. Poly. Sci.* **1992**, *46*, pp 1945-1953. See also Warwicker, J.O. *Biochim. Biophys. Acta*, **1961**, *52*, pp 319-328.
46. Arsenovic, P.; Jiang, H.; Eby, R.K.; Adams. W.W.; Liu, J.M. in *Carbon '88*; McEnaney, B; Mays, T.V., Eds.; IOP Publishing Ltd., Bristol, 1988; pp 485-487.
47. Jiang, H.; Eby, R.K.; Adams, W.W.; Lenhert, P.G. in *The Materials Science and Engineering of Rigid-Rod Polymers*; Adams, W.W.; Eby, R.K.; McLemore, D., Eds.; Materials Research Society Symposium Proceedings; Materials Research Society: Pittsburgh, PA, 1989, Vol. 134; pp 341-350. See also Jiang, H. Ph.D. Dissertation, The Johns Hopkins University, 1989.
48. Nishino,T.; Nakamae, K. *Polymer*, **1992**, *33*, pp 1328-1329.
49. Johnson, W. In *Strong Fibers, Vol 1*; Watt, W.; Perov, B.V. Eds.; Elsevier Science, New York, 1988; Chapter X.
50. Iizuka, E. *Biorheology*, **1965**, *3*, pp 1-8.
51. Lotz, B.; Colonna Cesari, F. *Biochimie*, **1979**, *61*, pp 205-214.

52. Oka, M.; Baba, Y.; Kagemoto, A.; Nakajima, A. *Polymer Journal*, **1990**, *22(5)*, pp 416-425.
53. Dumbleton, J.H.; Buchanan, D.R. *Polymer*, **1968**, *9*, pp 601-602.
54. Nakamae, K.; Nishino, T.; Ohkubo, H. *J. Macromol. Sci-Phys.*, **1991**, *B30(1&2)*, pp 1-23.
55. Fanconi, B.; Rabolt, J.F. *J. Polymer Sci.-Phys. Ed.*, **1985**, *23*, pp 1201-1215.
56. Mahoney, D.; Eby, R.K.; Kaplan, D.; Adams, W. *Bull. Am. Phys. Soc.* **1992**, *37(1)*, Abstract K'34 10.
57. Hearle, J.W.S. In *Structural Mechanics of Fibers, Yarns, and Fabrics*; Hearle, J.W.S.; Grosberg, P.; Backer,S. Eds.; Wiley-Interscience, New York, 1969; Chapter 4.

RECEIVED July 9, 1993

Chapter 18

Aspects of the Morphology of Dragline Silk of *Nephila clavipes*

D. V. Mahoney[1], D. L. Vezie[2], R. K. Eby[1], W. Wade Adams[3], and David Kaplan[4]

[1]Department of Polymer Science and Maurice Morton Institute of Polymer Science, The University of Akron, Akron, OH 44325–3909
[2]Department of Materials Science and Engineering, Massachusetts Institute of Technology, Cambridge, MA 02139
[3]Materials Directorate, Wright Laboratory, WL/MLPJ, Wright-Patterson AFB, OH 45433–7702
[4]Biotechnology Division, Natick Research, Development, and Engineering Center, U.S. Army, Natick, MA 01760–5020

Samples of golden orb-weaver dragline silk were observed to undergo large tensile and compressive deformations (~40%) without evidence of fracture in tension or kinking in compression. The absence of kinking is consistent with the absence of a fibrillar structure within the fiber such as those microfibrils observed in high performance polymer fibers. Therefore, the lack of failure prompted an examination of the morphology of the silk. Low voltage high resolution scanning electron microscopy (LVHRSEM) and atomic force microscopy (AFM) were used to examine as-spun, fractured and abraded fibers. The images yielded a rich variety of features. These were examined by the original height profiles of the AFM images and by Fourier transforms of the profiles. A broad range of dimensions of features is present. These features occur on the surface of as-spun silk as well as on the surfaces exposed through abrasion and liquid nitrogen fracture.

Silk is an extracellular fibrous protein which has been genetically tailored to exhibit specific properties. Spiders spin a variety of silk fibers for a number of purposes with each silk having a different composition and being synthesized in a different gland. The uses include immobilizing prey, protecting offspring, fabricating structures, and sifting the air for food. Because the spider must expend energy to manufacture silk, the task must be accomplished with as little material as necessary. For this, the material must have properties designed specifically for the task. One example is spider dragline silk which is manufactured in the major ampullate gland. The spider uses this silk as the structural foundation of the web and as a lifeline. The material has unusual characteristics of strength, modulus and elasticity (*1, 2*). Some spiders

0097–6156/94/0544–0196$06.00/0

produce silk with a strength similar to some synthetic high-performance fibers and a specific energy to break which is an order of magnitude greater (*2*). This fiber is spun from solution with water as the solvent in a process that is a subject under much scrutiny (*3, 4*). We are investigating the structural/morphological basis of the mechanical properties of the dragline silk of *Nephila clavipes*.

One limiting factor for many high performance fibers is their lower material properties in compression relative to those in extension (*5, 6*). The compressive strength of many of these fibers is approximately an order of magnitude lower than the tensile strength (*5*). The cause of this disparity is attributed to the presence of microfibrillar structures within the fiber (*7-9*) which fail by kink band formation at relatively low loads (*10, 11*). In synthetic fibers, such fibrils occur as a consequence of the kinetics of fiber processing. The formation of silk fibers is a process which is not fully understood. However, the fibers are formed from a liquid crystalline phase with the solvent (water) being removed in the gland (*3*). The apparent absence of a fibrillar structure in the silk might indicate that the compressive properties will be favorable, therefore the structure/morphology of the fibers is being investigated. There have been some microscopic investigations of the morphology of the naturally spun silk fibers of *Bombyx mori* (*12-15*). Electron microscopy studies were made of stained or shadowed samples obtained by digestion and/or mechanical fragmentation of the fibers. These revealed ribbon-like crystals approximately 10 nm in lateral extent and very much longer. Fewer studies have been made on naturally spun fibers from spiders and these have been confined primarily to optical microscopy of the outer surface (*1, 2, 16, 17*). In this note, we present the initial results of our investigation.

Experimental Methods and Materials

The female *N. clavipes* spiders were obtained from Panama. The dragline silk was obtained from live specimens through mechanical silking similar to published techniques (*18, 19*). The spiders were anesthetized with CO_2. With the aid of an optical microscope, the fibers were located coming from the anterior pair of spinnerets and were drawn at a rate of ~1.1 cm/s over a rotating drum. Some samples of silk were abraded with a glass slide and some were fractured by tensile deformation in liquid nitrogen prior to examination. The fibers used for study by atomic force microscopy (AFM) of the unabraded surface were obtained by pulling the strands from the spider by hand so that they did not come into contact with any surface prior to mounting on the sample holder.

The first method employed to examine the fibers was low voltage, high resolution scanning electron microscopy using a Hitachi S-900 field emission gun, immersion lens SEM. Due to its ability to yield high resolution images with a low voltage, only a very thin conducting coating was required. The samples were adhered to the sample holder using double stick tape and sputter coated with a thin film of AuPd prior to being imaged at 1.0 keV.

The second technique used was AFM (*20*) for which no surface modification of the sample was necessary. The fibers were mounted on holder discs with a thermoplastic adhesive which becomes tacky at ~40 °C and melts at ~120 °C. It was melted, applied to the disk and then cooled to the tacky state for mounting. The mounted fibers were examined with an optical microscope to verify adhesion and that they were primarily in the plane of the adhesive. The optical lever type AFM, TopoMetrix 2000, was used in the repulsive mode under ambient conditions. The scanning tip was silicon nitride with pyramidal geometry. The aspect ratio was 1 to 1 and the tip radius was less than 50 nm. Two scanning modes were used and both yield calibrated lateral displacements of the sample by the piezoelectric scanner. In the variable height mode, the scanner displaces vertically to maintain a constant force on the tip during scanning. In this mode, calibrated vertical displacement information may be obtained. In the constant height mode the piezoelectric scanner does not displace vertically during scanning and the image is generated with vertical dimensions given in units of feedback current which exaggerate contour variations.

Results

The dragline silk from *N. clavipes* appears smooth when viewed using the LVHRSEM at lower magnification (Figure 1). Longitudinal striation marks, which are usually seen on the surface of extruded synthetic fibers, were not observed at this magnification. The loop of the fiber with a small radius of curvature prompted an examination of fibers undergoing further deformation. The knot shown in Figure 2 was tied by hand. A looped portion near the center of the figure appears to lie nearly perpendicular to the direction of view. In Figure 3, it can be seen that the surface of this loop shows a variety of surface features (as well as some dirt). At high magnifications such features are evident on all the fibers studied. A simple calculation of the surface strain (tension and compression) in the bend of the fiber yields a value of approximately 40% (*21*). This value is close to but within the upper limit of the maximum strains reported for these fibers in tension (*22, 23*). There is no evidence of failure either by kinking at the inner radius of the bend, where the fiber is in a state of compression, or by breaking at the outer radius of the bend, where the fiber is in a state of extension. High performance fibers of polyethylene, Kevlar, carbon, poly(p-phenylenebenzobisthiazole), and poly(p-phenylenebenzobisoxazole) all fail compressively by kink band formation (*24*) which is not seen in the silk fibers. The lack of kinking is consistent with the absence of fibrils or with the presence of fibrils in which the lateral interactions between them are just as strong as the interactions parallel to them. In the present case, the ability to undergo large compressive deformations is probably facilitated by a rapid stress relaxation as is observed in tension for *B. mori* (*25, 26*) and *N. clavipes*. The absence of kinks for uniaxial compression was also verified to a strain of 5% by using an axial compression apparatus (*27*).

Because the fibers did not fail compressively by kinking, the presence or absence of microfibrils became a subject of interest. Therefore, the inner morphology of the fibers was also examined. An example of the images observed with abraded fibers is shown in Figure 4. It reveals a striated interior morphology. In order to reduce the possibility that these features are artifacts of deformation during the abrading, fibers that were fractured in tension at the temperature of liquid nitrogen were also examined. Figure 5 shows a fracture that has propagated along the length of the fiber. The surface observed in this image is similar to that observed for the fibers abraded at room temperature. In the absence of kink formation, it seems likely that the striations do not represent fibrils but are the result of fracturing a highly oriented, non-fibrillar structure. Such features can be seen in fractured, extended chain lamellae of bulk polyethylene and polytetrafluoroethylene (*28, 29*).

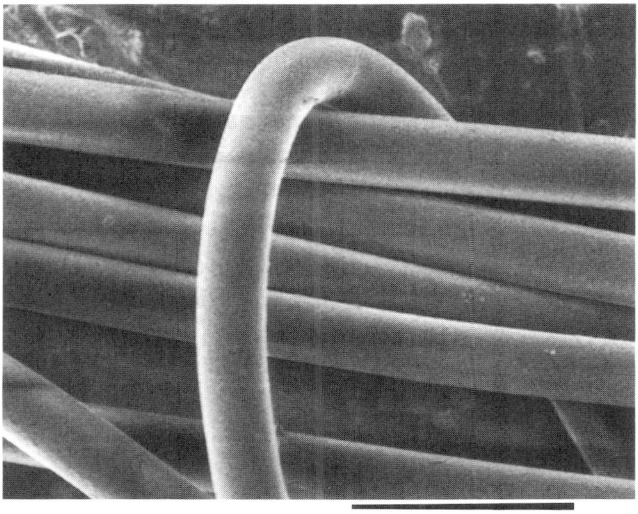

Figure 1 LVHRSEM micrograph of *Nephila clavipes* dragline silk fibers. The scale bar is 12,000 nm.

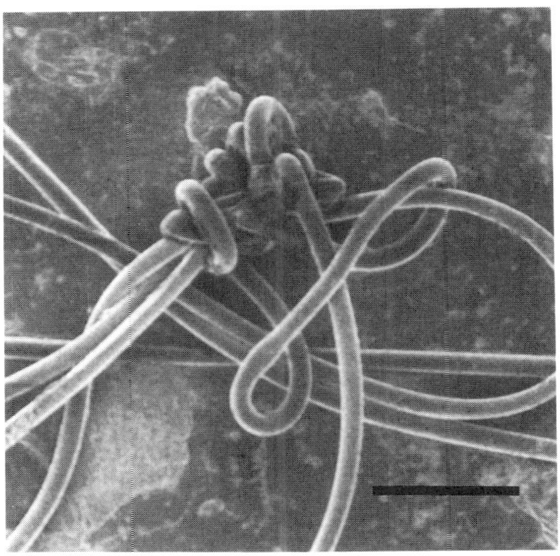

Figure 2 LVHRSEM micrograph of knot tied in dragline silk. The scale bar is 30,000 nm.

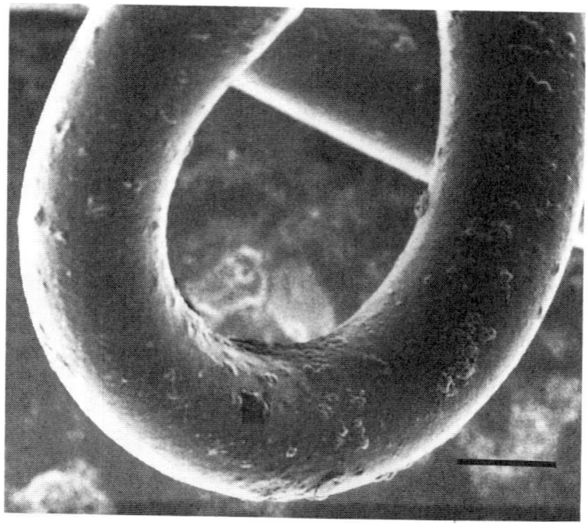

Figure 3 LVHRSEM micrograph of tight bend in knot of dragline silk. The scale bar is 3,000 nm. Surface contamination is apparent in this image. Note the lack of evidence of failure by kinking in the inner compressed portion of the fiber and by fracture in the outer extended portion of the fiber.

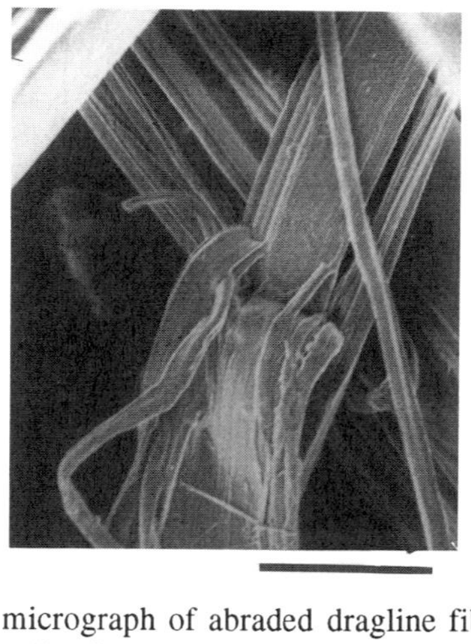

Figure 4 LVHRSEM micrograph of abraded dragline fibers. The scale bar is 3,000 nm. Note the smaller features apparent in this image.

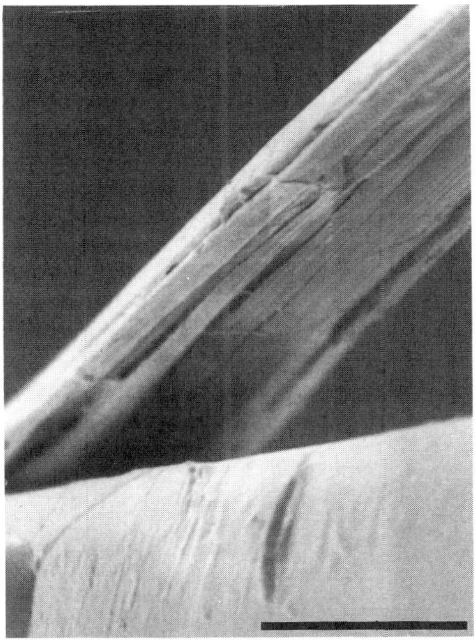

Figure 5 LVHRSEM micrograph of fracture surface of dragline fiber achieved through liquid nitrogen temperature fracture in tension. The scale bar is 3,000 nm.

In order to examine the morphology at a higher magnification, similar investigations were undertaken utilizing the AFM. Figure 6 is an image which reveals that the exterior surface of the fiber has considerable undulation in both the lateral and longitudinal directions. A higher magnification, curvature-corrected image of the exterior surface is shown in Figure 7. It reveals marks similar to extrusion lines in the longitudinal direction. It also can be seen in this image that the channels are not continuous, and that the depths of the channels vary along the length of the fiber. Undulation in the direction of the fiber is also apparent. The possible origin of these features covers a wide range of possibilities involving the material and the spinning process.

Because the contour information is digitized, image analysis may be done. A measure of the surface roughness is the average of the absolute value of the variation from the mean height of the profile over the length of the contour. Of course, a great many areas must be examined to be sure that an accurate sampling of the overall structure of the material has been obtained. Also, considerable caution must be used in interpreting the resulting numbers. Values of roughness were calculated over a number of scan lines perpendicular to the fiber direction for each image area after the images were corrected for fiber curvature. A variety of areas and area sizes were also used. The value for the average roughness of the silk fibers in the transverse direction is 9 ± 6 nm. The variability in the roughness is indicated by the magnitude of the standard deviation. It is due to the considerable variation of the roughness from region to region which persists even for silk obtained from a single spider.

The inner morphology of the abraded fibers was also studied with the AFM. An example of the images obtained is shown in Figure 8. The undulation of the silk both along and perpendicular to the fiber direction is apparent, with the latter being considerably more pronounced in the abraded images than it was on the exterior surface of the fibers. The undulations which run perpendicular to the fiber direction might indicate the presence of a fibrillar structure. However, the depth of the channels between these possible fibrils appears to be too shallow to indicate definitely independent entities. Further, as noted above, a fibrillar structure is not a prerequisite to achieving this type of polymer fracture surface. In fact, the structure that can be seen in the abraded surface images is similar to the fracture surface structure that is observed for fractured extended chain lamellae of a bulk polymer (*28-30*). Material crossing the channels was observed as seen in Figure 8. This material appears to be a contiguous portion of the silk bordering the channel, suggesting the absence of a fibrillar structure prior to abrading the silk. Another example of the type of morphology observed for the abraded silk is shown in Figure 9. The unabraded surface of the fiber is seen in the left portion while the right portion shows an area where the scanning tip has lost contact with the sample. Material crossing the channels is more prevalent in this image. Figure 10 shows a constant height image of abraded silk containing a bit of material crossing the channels. This material appears to be a contiguous portion of the material bordering both sides of the channel. This image also reveals considerably smaller features with dimensions down to tens of nanometers.

In all the images that been have examined, rather large scale undulations with dimensions that vary considerably have been observed. These are persistent both on the outer and the interior surfaces. By examining the surface profiles taken both parallel and perpendicular to the fiber direction, simple measurements were made of the lateral dimensions. On the unmodified fiber surfaces, there is an average

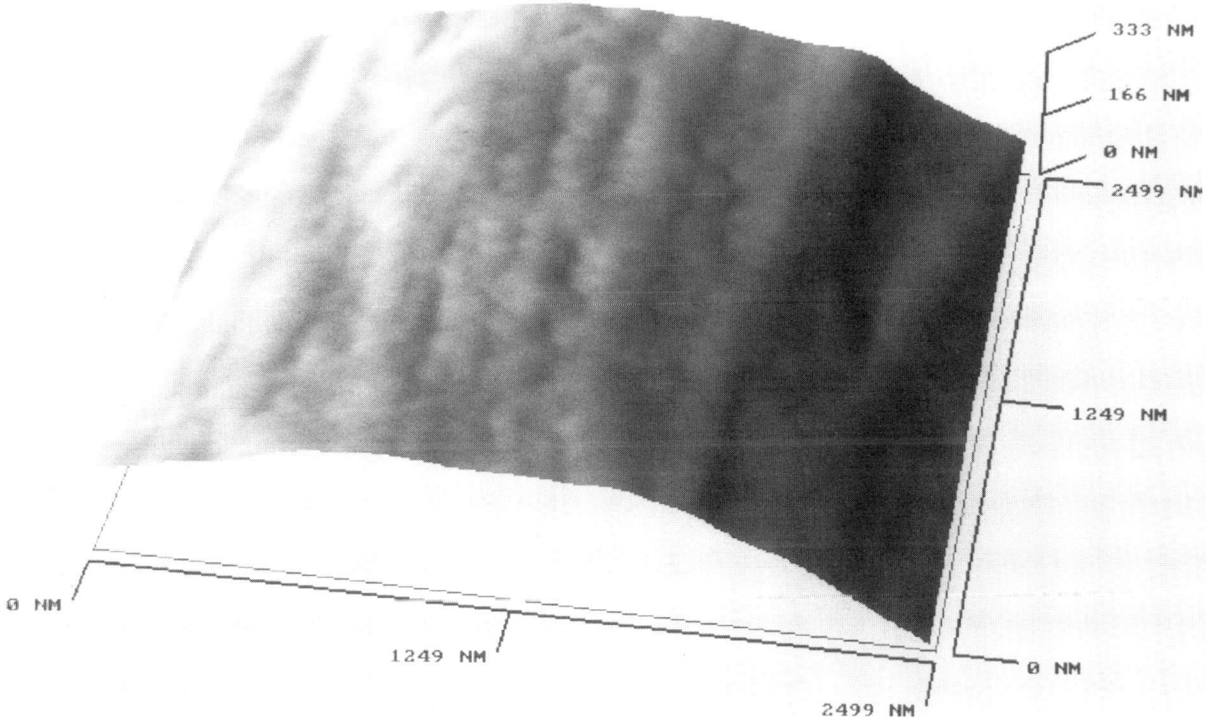

Figure 6 AFM constant force image (200x200 data points) of surface of dragline silk with the fiber axis aligned parallel with the right-hand axis. Note the undulation in both the lateral and longitudinal fiber directions. The image is 2499 nm by 2499 nm and approximately 333 nm in elevation.

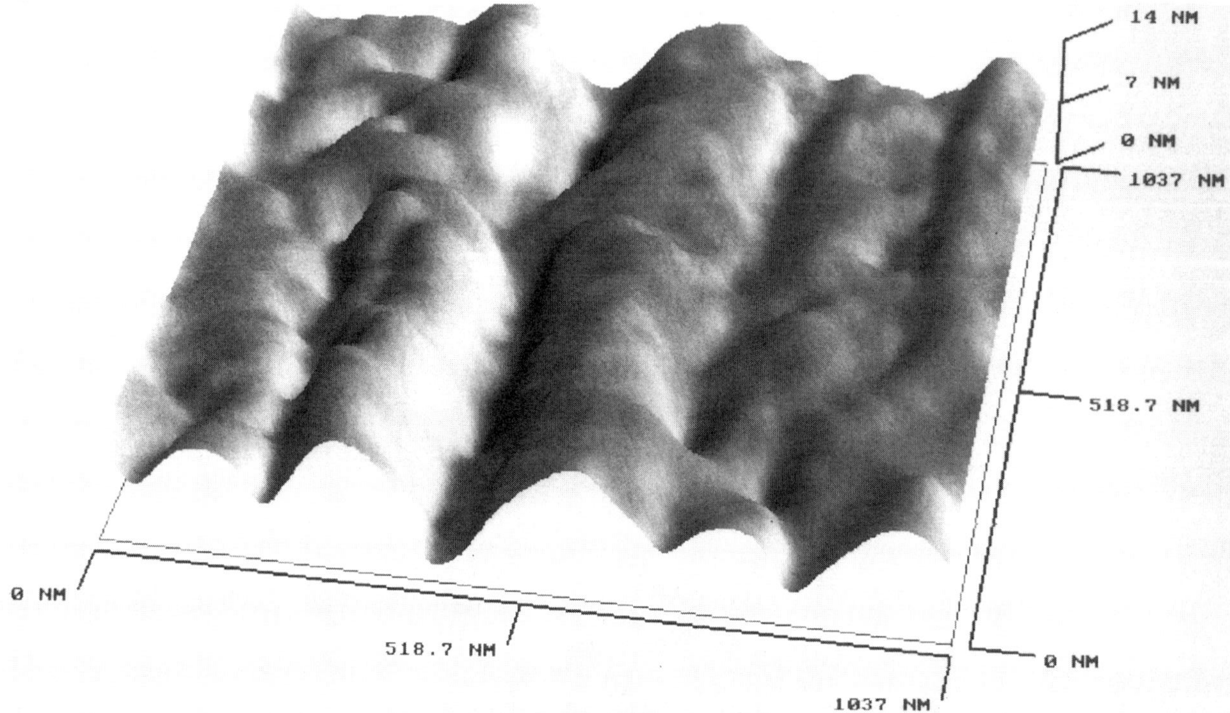

Figure 7 Curvature-corrected AFM constant force image (200x200 data points) of surface of dragline silk with the fiber axis aligned parallel to the right-hand axis. The considerable undulation of the silk is apparent in this image. Note that the extrusion lines are not completely continuous. The image is 1037 nm by 1037 nm and approximately 14 nm in elevation.

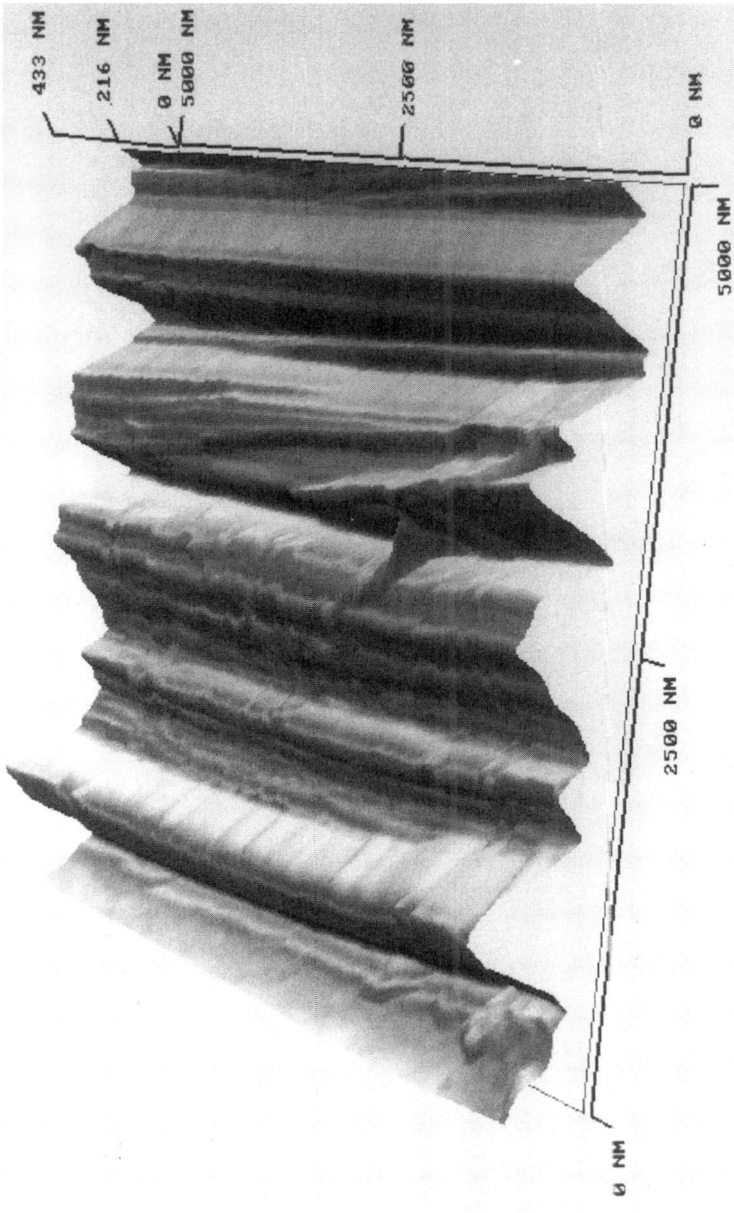

Figure 8 AFM constant force image (400x400 data points) of abraded dragline silk with the fiber axis aligned parallel to the right-hand axis. Note the large lateral undulations in the silk. The image is 5000 nm by 5000 nm and approximately 433 nm in elevation.

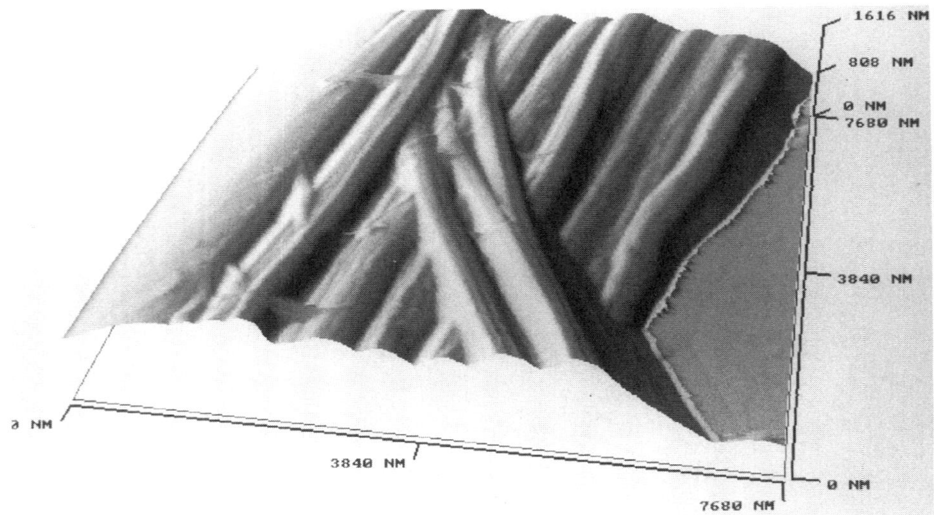

Figure 9 AFM constant force image (400x400 data points) of abraded dragline silk with the fiber axis aligned parallel to the right-hand axis. The unabraded fiber is visible on the left, while the scanning tip has lost contact with the surface on the right. Note the striations and the connecting material crossing them. The image is 7680 nm by 7680 nm and approximately 1616 nm in elevation.

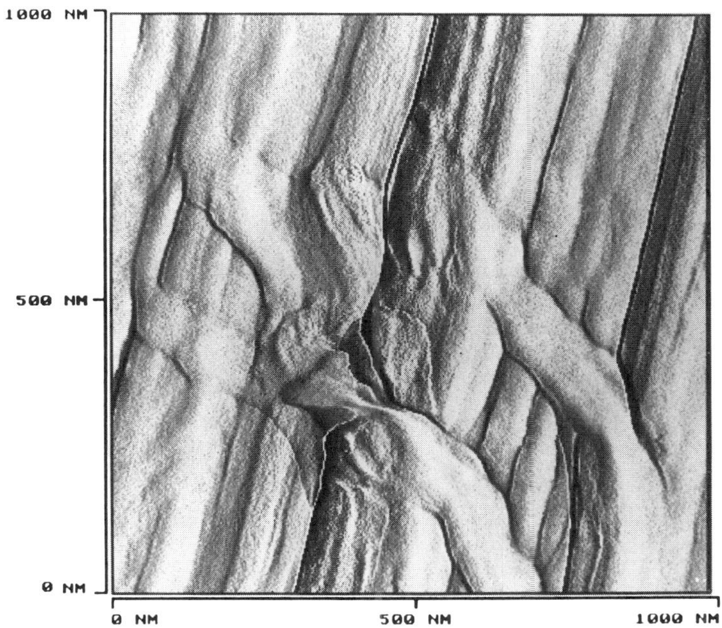

Figure 10 AFM constant height image (400x400 data points) of abraded dragline silk with the fiber axis aligned nearly parallel to the right-hand axis. Note material crossing the channels. Also seen in this image is the resolution of smaller features down to tens of nm. The image is 1000 nm by 1000 nm.

dimension for these large scale features of 113 ± 20 nm and 98 ± 13 nm in the perpendicular and parallel directions to the fiber direction, respectively. Again, some caution must be used in interpreting these numbers. Note that although the dimensions perpendicular and parallel to the fiber direction are similar, they do not arise from identical ensembles of features. The large standard deviations are associated with the large range of feature dimensions. Along with these relatively large scale features is a nearly continuous range of other features having dimensions down to ~40 nm. The latter dimension approaches lateral dimensions of the crystallites found in *B. mori* silk (*12-15*). These smaller features are less pronounced and frequent. Therefore, they are more difficult to measure. The dimensions for the abraded surfaces might reflect the fracture process as well as the inherent morphology and are therefore, not presented. It can be noted, however, that they are of the same magnitude as those given above.

Fast Fourier transforms are another method of examining the contours of the surfaces. By performing two-dimensional fast Fourier transforms of the curvature-corrected image, it was found that the peaks of the intensity spectrum fall primarily on the axes of the transform plane. Therefore, an analysis of the features of the silk was obtained by examining the power spectra of one-dimensional transforms taken parallel and perpendicular to the fiber direction. These were effectively continuous with a peak at longer wavelengths and a tail extending to much shorter ones. The spectra of the unabraded silk samples had a higher intensity tail (to shorter wavelengths) than did those for the abraded silk images. The minimum number of the highest peaks in the transform spectrum required to reproduce the prominent features of the original image in the inverse transform ranged from three to five. The weighted average and the standard deviation of these were determined to be 220 ± 24 nm and 342 ± 40 nm for the unabraded silk perpendicular and parallel to the fiber direction, respectively. These are of the same order as those determined from the raw contour plots, however, the numbers must be interpreted with some caution. (For the reasons given above, the similar numbers for the abraded silk are not reported.)

Discussion/Conclusion

The dragline silk fiber of *N. clavipes* is able to undergo large deformations, both in compression and in extension, without evidence of failure. This is consistent with a material of crystalline and amorphous domains with the amorphous domains able to move in a fashion that permits the stress to be relaxed. There is the appearance of a possible fibrillar structure upon abrading of the silk fibers; however, it probably results from the longitudinal fracture of the oriented fiber. This structure is not as evident on the surface of the unabraded silk and the work performed so far shows no conclusive evidence of a fibrillar structure. The lack of a fibrillar structure is further supported by the absence of kinking of the fibers under high compressive deformation. Kinking occurs in all highly oriented fibers possessing a fibrillar structure. The silk exhibits a rich variety of prominent features on both the outer surface and on the surface exposed through abrading the fibers. These features are measurable through examination of both the raw traces and the Fourier transforms of the surface contours of AFM images. Both measurements show similar average dimensions and are consistent with a wide distribution of feature sizes.

There is a need for much more microscopy and other work. For example, we know of no small angle X-ray scattering performed on the silk to clarify the existence of a fibrillar structure. Such measurements and further microscopic investigations are underway to confirm the apparent absence of a fibrillar structure.

Acknowledgments

Portions of this work were supported by TopoMetrix and by Wright Laboratory, Wright Patterson AFB, through a contract with Lawrence Associates, Inc. (D.V.M., D.L.V., and R.K.E.). We are indebted to I. Jangchud for his valuable guidance in AFM and D. Thomas for assistance with the axial compression device.

Literature Cited

1 Zemlin, J.C. "A Study of the Mechanical Behavior of Spider Silks"; technical report to the US Army Natick Laboratories (report 69-29-CM), Collaborative Research, Inc., Sept. 1968.

2 Gosline, J.M.; DeMont, M.E.; Denny, M.W. *Endeavour* **1986**, *10*, 37-43.

3 Kaplan, D.L.; Fossey, S.; Viney, C.; Muller, W. *Self-Organization (Assembly) in Biosynthesis of Silk Fibers -A Hierarchical Problem*; Materials Research Society: Pittsburgh, PA, **1992**, *255*, 19-29.

4 Lewis, R.V. *Acc. Chem. Res.* **1992**, *25*, 392-398.

5 Kumar, S.; Helminiak, T.E. *Sampe Journal*, **1990**, *26*, 51-61.

6 Dobb, M.G. In *Handbook of Composites, Vol, 1 - Strong Fibers*; Watt, W., Perov, B.V., Eds.; Elsevier, 1985, pp 686-687.

7 Yang, H.H. In *Aromatic High-Strength Fibers*, SPE Monograph Series; 1989 Wiley, New York, 1989; p. 261.

8 Tanner, D.; Fitzgerald, J.A.; Riewald, P.G.; Knoff, W.F. In *High Technology Fibers*; Lewin, M.; Preston, J., Eds.; Marcel Dekker: New York, 1989; Vol. 3(B), Chapter 2.

9 Snétivy, D.; Vancso, G.J.; Rutledge, G.C. *Macromolecules* **1992**, *25*, 7037-7042.

10 Bacon, R. In *Chemistry and Physics of Carbon*; Walker, P.L.; Thrower, P.; Dekker, M.; 1973; Vol. 9, p. 89.

11 Warner, S.B. *J. Mater. Sci.* **1987**, *6*, 951.

12 Lotz, B.; Gonthier-Vassal, A.; Brack, A.; Magoshi, J. *J. Mol. Biol.* **1982**, *156*, 345-357.

13 Dobb, M. G.; Fraser, R. D. B.; Macrae, T.P. *J. of Cell Biol.* **1967**, *32*, 289-295.

14 Mercer, E. H. *Aust. J. Sci. Res. (B)* **1952**, *5*, 366-373.

15 Marsh, R. E.; Corey, R. B.; Pauling, L. *Biochimica et Biophysica Acta.* **1955**, *16*, 1-34.

16 Vollrath, F.; Edmonds, D. T. *Nature* **1989**, *340*, 305-307.

17 Vollrath, F.; Fairbrother, W. J.; Williams, R. J. P.; Tillinghast, E. K.; Bernstein, D. T.; Gallagher, K. S.; Townley, M. A. *Nature* **1990**, *345*, 526-528.

18 Work, R.W.; Emerson, P.D. *J. Arachnol.*, **1982**, *10*, 1-10.

19 Anonymous *Seta-Bollettino di Sericoltura* **1957**, *18*, pp 17-20, refers to DeTermeyer, R. M. *Opuscoli scientifici di Entomologia di Fisica e d' Agricoltura*, Vol 1, Milan, 1807.
20 Magonov, S.N.; Cantow, H.J. *J. Appl. Sci.: Appl. Polym. Symposium*, **1992**, *51*, 3-19.
21 Beer, F.P.; Johnston, Jr., E.R. *Mechanics of Materials*; McGraw-Hill: New York, 1981; p. 155.
22 Butterworth, G. A. M.; Lenz, R. W. "The Tensile Properties of Spider Silk"; final report c65593 to U.S. Army Natick Laboratories; Fabric Research Laboratories, Inc., Dedham, MA, May, 1966.
23 Kaplan, D.L.; Lombardi, S.J.; Muller, W. In *Biomaterials: novel Materials from Biological Sources*; Byrom, D., Ed.; Stockton Press: New York, 1991.
24 Vezie, D. L. Ph.D. Dissertation, MIT, 1993.
25 Weber, W.; *Poggendorf's Ann. der Physik Chemie*, **1835**, *34*, pp 247-257.
26 Lenhert, P. G.; Adams, W. W.; *Tensile Modulus by X-ray Diffraction; Instrument and Method*; Wright Laboratory, Materials Directorate, WPAFB, OH 45433, July 1990; WRDC-TR-90-407.
27 Macturk, K. S.; Eby, R. K. Adams, W. W. *Polymer* **1991**, *32*, 1782-1787.
28 Sakaoku, K.; Miyata, S.; Itoh, T. *Bull. Inst. Chem. Res., Kyoto Univ.* **1977**, 200-205.
29 Bassett, D. C.; Davitt, R. *Polymer* **1974**, *15*, 721-728.
30 Kulawansa, D. M.; Langford, S. C.; Dickinson, J. T. *J. Mater. Res.* **1992**, *7*, 1292-1302.

RECEIVED July 7, 1993

Chapter 19

Thermal Properties of Silk Proteins in Silkworms

Shigeo Nakamura[1], Jun Magoshi[2], and Yoshiko Magoshi[3]

[1]Department of Applied Chemistry, Faculty of Engineering, Kanagawa University, Kanagawa-ku, Yokohama 221, Japan
[2]National Institute of Agrobiological Resources, Tsukuba, Ibaraki 305, Japan
[3]National Institute of Sericultural and Entomological Science, Tsukuba, Ibaraki 305, Japan

The thermal behavior of silk proteins in silkworms has been investigated in detail. As the temperature of the amorphous random-coil fibroin of Bombyx mori is raised, water is evaporated off up to 100° C. Intra- and intermolecular hydrogen bonds are broken between 150 and 180° C. The glass transition is observed at 175° C. The random-coil to β-form transition accompanied by reformation of hydrogen bonds occurs above 180° C. Crystallization to the β-form crystals starts at above 190° C. The α-form and β-form fibroins lose water up to 100° C on heating, and molecular motions in the crystalline regions starts at 175° C. The α-form to β-form transition is induced thermally at 270° C. Silk sericin shows a glass transition at 170° C and crystallizes at 250° C. The glass transition temperature of wild silk fibroins varies with species of silkworm ranging from 160 to 210° C.

It has been observed that, when being heat-treated, silk filament starts to lose its weight gradually at 175° C, and the color changes from white to light yellow, brown and then to black at 250° C (1).

Schwenker and Dusenbury (2) has first recorded DTA curves of silk filament. In the DTA curves of silk filament, a peak due to evaporation of water appears at 108° C and an endothermic peak of thermal degradation occurs at 326° C.

Several studies have been reported on the thermal behavior of silk. Ishikawa et al. (3) found that the DTA peak due to thermal degradation appears at 305, 348 and

0097–6156/94/0544–0211$06.00/0

234°C for domestic silk, tussah silk and silk from anaphe infracta, respectively.

However, systematic research on the thermal behavior of silk, especially fibroin, the main component of silk filament, has not been carried out until the authors report detailed investigation of thermal behavior of silk fibroin (4).

EXPERIMENTAL

Liquid silk used for experiments was taken from the posterior part of the middle division of the silk gland in full-grown larvae (one day before spinning) of silkworm. After the sericin enveloping fibroin as an outer layer was removed by washing thoroughly with deionized water, the liquid silk was diluted with water. The aqueous solution of silk fibroin was dialyzed against the deionized water and diluted to a predetermined concentration. Amorphous film with random-coil conformation was cast from a 0.8% solution onto glass plates.

Differential scanning calorimetry (DSC) curves were obtained on a Perkin-Elmer type II DSC apparatus under nitrogen. Thermal expansion measurement was made using a hand-made apparatus by recording the change in length of the film specimen under a constant tension of less than $250gcm^{-2}$.

The dynamic modulus E' and the loss tangent (tan δ) were measured with Rheovibron model DDV-II at a frequency of 110Hz. Dielectric measurement was made with a 4270A automatic capacitance bridge (Yokogawa-Hewlett Packard) equipped with a vacuum cell. A thin gold layer was evaporated on the surface of the specimen.

X-ray diffraction patterns were obtained with a Rigaku D-3F x-ray apparatus. The temperature of specimen was controlled to ±0.5°C. Infrared spectra were measured on a JASCO model 402G infrared spectrometer. The specimen was placed between two CaF_2 or NaCl plates.

RESULTS AND DISCUSSION

Thermal Behavior of Random-coil Fibroin. Two endothermic peaks appear at about 100 and 280°C in the DSC curve of amorphous silk fibroin of domestic silkworm Bombyx mori with random-coil conformation (Figure 1(a)) (5). An exothermic peak is observed at 212°C and an endothermic shift of the base line occurs at 175°C.

The prominent endothermic peak at about 280°C is attributed to the degradation of silk fibroin. A weight loss starts abruptly at about 250°C in the TG curve. The broad endothermic peak at about 100°C is due to the evaporation of water in the specimen. This peak becomes smaller by drying the specimen prior to measurement (Figure 1(b)). The endothermic shift starting at 175°C is the glass transition of silk fibroin, because it disappears by heat-treating the specimen at 220°C to induce crystallization (Figure 1(C)).

Further detailed investigation is undertaken to clarify the thermal behavior of silk fibroin with random-coil conformation.

In the linear thermal expansion curve of amorphous random-coil fibroin film (Figure 2) (4), the specimen contracts very slowly up to 100° C. This contraction is attributed to the evaporation of water, because the contraction becomes smaller with prior drying at 100° C. The length of specimen begins to increase abruptly at about 165° C. The temperature at which the intersection of the straight line at the lower temperature side with that at the higher temperature side is 175° C, which agrees well with the glass transition temperature observed by DSC.

The dynamic modulus (E') and the loss tangent (tan δ) of amorphous random-coil fibroin film is recorded as a function of temperature (Figure 3) (4). Figure 3 contains also the results for films with different β-form contents or different degrees of crystallinity. These specimens are prepared by immersing the random-coil fibroin film in mixtures of CH_3OH and H_2O with various compositions (6, 7). At lower temperatures the E' of random-coil fibroin hardly decreases with temperature and starts to decrease abruptly at about 170° C. It reaches the minimum at 200° C and then increases. After passing through the maximum at about 220° C, it decreases again. The tan δ begins to increase abruptly at about 170° C and reaches the maximum (α_a dispersion) at 195° C. Then it turns to decrease again and reaches the minimum at 230° C.

The maximum of E' at 220° C increases in height with increasing β-form content or crystallinity, whereas the minimum at 200° C becomes smaller. Only a shoulder is observed for the specimen with 25% crystallinity. For the specimens with 27 and 30% crystallinity, both the maximum and the minimum disappear. The tan δ peak shifts slightly and continuously to higher temperature and decreases in height with increasing β-form content or crystallinity. Only a shoulder appears for the 27 and 30% crystalline specimens. Accordingly, the tan δ peak at 195° C is related to the molecular motions in the amorphous phase, namely which is induced by the segmental motions of fibroin molecules in the amorphous region. The increase in modulus starting at about 200° C is attributed to the crystallization of amorphous fibroin accompanied by the random-coil to β-form conformational transition.

The dielectric constant of regenerated amorphous fibroin films increases slowly with increasing temperature, has the maximum at about 100° C and then starts to increase abruptly at about 150° C (8). The peak at 100° C is due to the evaporation of water in the specimen, because the peak can not be observed after the specimen is heat-treated. Two loss peaks appear in the dielectric loss tangent measurement. The prominent peak at 175° C is independent of frequency. The minor peak is remarkably affected by the frequency and shifts from -40° C at 1kHz to 30° C at 1MHz. The prominent loss peak does not

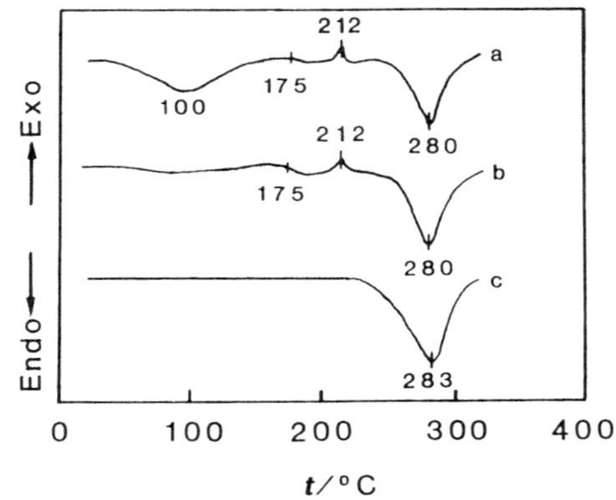

Figure 1. DSC curves of amorphous silk fibroin in the random-coil conformation: (a) untreated, (b) dried prior to measurement, (c) heat-treated at 200°C. Heating rate: $8 K min^{-1}$. Reproduced with permission from Reference 5.

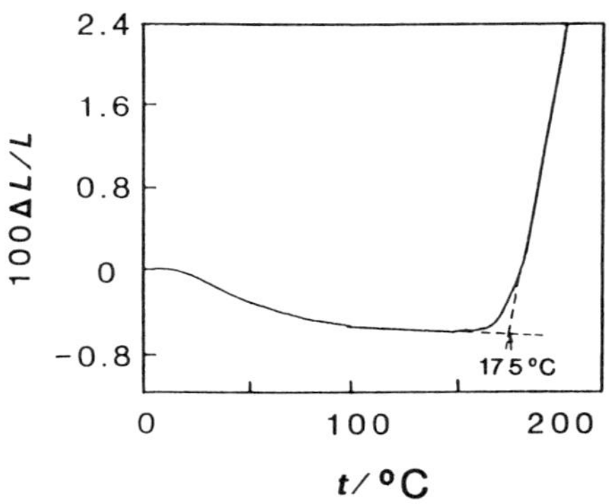

Figure 2. Thermal expansion of amorphous silk fibroin in the random-coil conformation. Heating rate: $3 K min^{-1}$. Reproduced with permission from Reference 4.

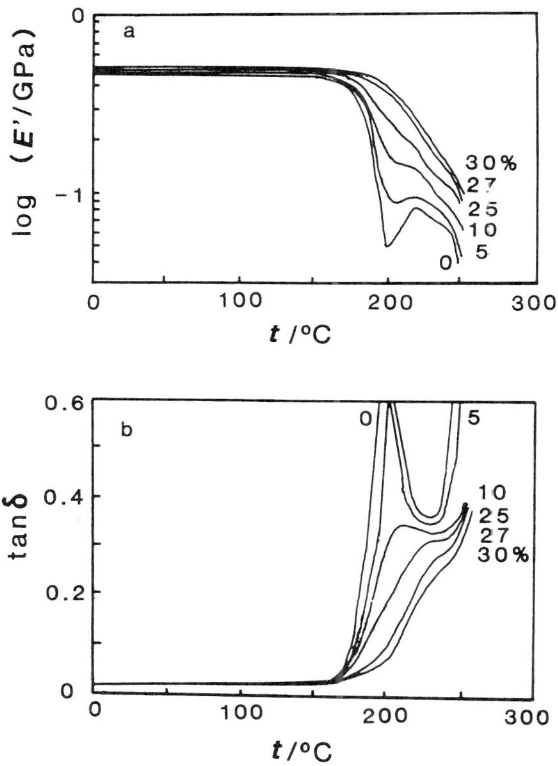

Figure 3. Temperature dependence of the dynamic modulus E' (a) and the dynamic loss tangent tan δ (b) of silk fibroin with different β-form content or different crystallinity. Reproduced with permission from Reference 4.

appear for the heat-treated specimen. This peak is caused by the conformational transition to the β-form which has occurred during heat-treatment. The minor peak is attributed to the local motion (β-dispersion) of the amorphous fibroin with absorbed water since the specimen dried at 100° C in vacuum shows a peak at about 10° C (0.3Hz). Therefore, the peak is supposed to be due to the H_2O-coupled local mode motion of fibroin in the random-coil conformation.

It is also observed from the temperature dependence of the line width of broad-line NMR that the micro-Brownian motion in the amorphous region becomes to be vigorous at about 170° C (9).

X-ray diffraction patterns are recorded for amorphous fibroin film during stepwise increase in temperature (Figure 4) (4). No crystalline peak is observed below 190° C. The amorphous halo slightly shifts to smaller angles owing to the thermal expansion of the specimen. When the temperature is raised to 200° C, the characteristic x-ray diffraction pattern of the β-form crystals appears. Therefore, the exothermic peak of DSC at 212° C is attributed to the crystallization of amorphous random-coil fibroin to the β-form crystals.

Infrared spectra of random-coil silk fibroin are recorded in the range 400-1700cm^{-1} with stepwise increasing temperature (4). The spectrum at room temperature shows bands 1660, 1540, 1235 and 650cm^{-1}. These bands are assigned to the amide I, amide II, amide III and amide V bands of the random-coil conformation. At 180° C, bands caused by the β-form appear at 1630, 1535, 1265, and 700cm^{-1}. However, even at 210° C infrared spectra do not correspond exclusively to the β-form, and bands caused by the random-coil conformation persist.

In the infrared spectra of the random-coil fibroin the absorbance of the amide I, II, and III bands decreases linearly with increasing temperature and an abrupt change in the slope occurs at 180° C (Figure 5) (4). The absorbance of the amide I band of the β-form at 1630cm^{-1} decreases linearly and starts to increase abruptly at 190° C (Figure 6). On cooling, the absorbance of these bands increases linearly with decreasing temperature and no abrupt change in the slope is observed as seen in the broken lines in Figures 5 and 6. This means that the random-coil to β-form transition occurs irreversibly on heating.

The absorbance of the 3450cm^{-1} band due to free (not hydrogen-bonded) OH stretching vibration decreases with temperature up to 150° C, then increases up to 180° C and then again decreases, whereas that of the 3270cm^{-1} bands caused by the hydrogen-bonded NH stretching decreases up to 190° C and then increases. The change of the absorbance of these bands with heating is also irreversible. On cooling, the absorbance of these bands increases monotonously with decreasing temperature. The increase in absorbance of the 3450cm^{-1} band on heating is induced by the breaking of intra- and intermolecular hydrogen bonds and the decrease in absorbance of the 3450cm^{-1} band is

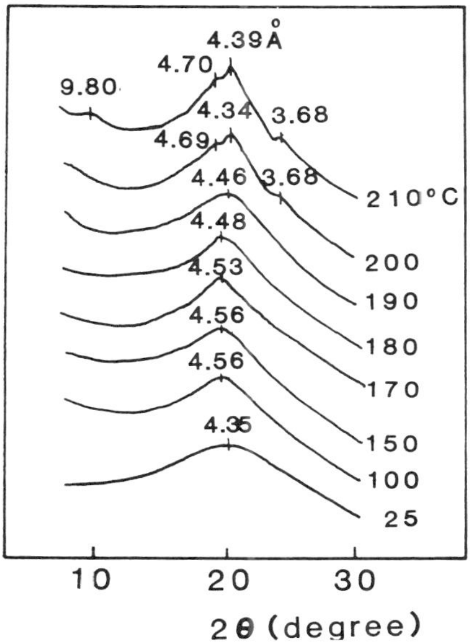

Figure 4. X-ray diffraction patterns of amorphous random-coil fibroin in the course of step-wise increase in temperature.

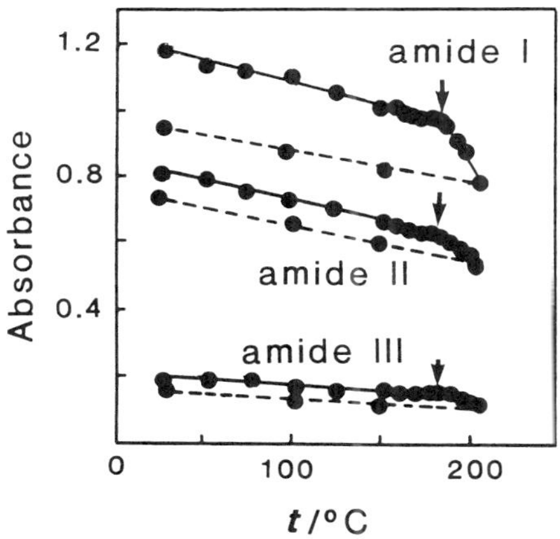

Figure 5. Absorbances of the amide I, II and III bands of the random-coil conformation as a function of temperature: heating (——); cooling (------). Reproduced with permission from Reference 4.

caused by the reformation of the hydrogen bonds. Therefore, decrease in absorbance of the $3450cm^{-1}$ and the increase in absorbance of the $3270cm^{-1}$ band at above 180° C reveal the reformation of hydrogen bonds once broken by heating.

From these results, the thermal behavior of amorphous silk fibroin with random-coil conformation is summarized as follows. As the temperature is increased, water existing in the specimen is evaporated off up to about 100° C. The intra- and intermolecular hydrogen bonds are broken in the temperature range 150-180° C, resulting in the increase in molecular motions. The glass transition occurs at 175° C. The irreversible random-coil to β-form conformational transition takes place above 180° C accompanying the reformation of hydrogen bonds. Thermally induced crystallization to β-form crystals starts at about 190° C.

Thermal Behavior of Crystalline β-Form Fibroin. For the crystalline β-form fibroin, three endothermic DTA peaks are observed at about 100, 252 and 317° C at a heating rate of $1Kmin^{-1}$. A broad peak at 100° C is due to the evaporation of water. Two endothermic peaks at higher temperatures are attributed to the degradation of the β-form silk fibroin. In the DTA curve of the β-form fibroin, a weight loss occurs in two steps. These results indicate that the degradation of the β-form silk fibroin proceeds in two steps.

In the DTA curves of the β-form fibroin at different draw ratios, a broad endothermic peak at about 100° C is due to evaporation of water and an endothermic peak at higher temperature is attributed to the degradation of the β-form fibroin (Figure 7) (10). The latter peak shifts to higher temperatures with increasing draw ratio.

The dynamic modulus E' of the β-form fibroin film increases with draw ratio, but the tan δ decreases (10). The decrease in modulus at about 100° C is not affected by draw ratio and is attributed to evaporation of water. The decrease in modulus and the increase in tan δ starting at 175° C are due to the onset of molecular motions in the crystalline region because an abrupt increase in spacing is observed in x-ray diffraction patterns at 175° C.

Thermal Behavior of Crystalline α-Form Fibroin. The oriented α-form fibroin can not be obtained because the α-form to β-form conformational transition occurs by drawing (11). The E' and the tan δ of the drawn α-form fibroin films behave very similarly to those of the drawn β-form films (Figure 8), because the α-form to β-form transition induced by drawing (10).

At lower draw ratios a small endothermic peak appears at 270° C in DTA curves and is attributed to the thermally-induced α-form to β-form transition, which is supported by x-ray diffraction patterns.

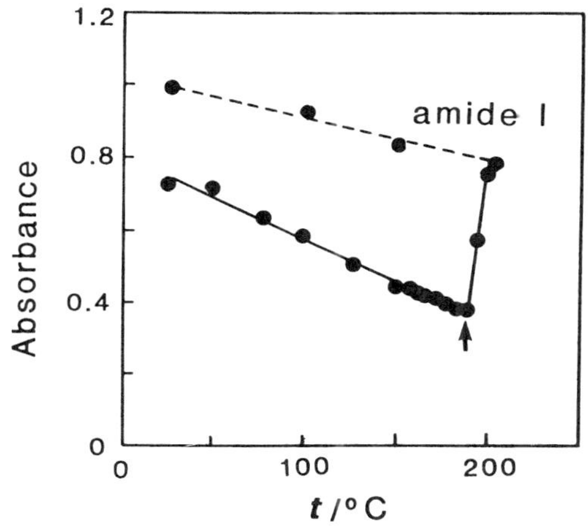

Figure 6. Absorbance of the amide I band of the b-form conformation as a function of temperature: heating (———); cooling (------). Reproduced with permission from Reference 4.

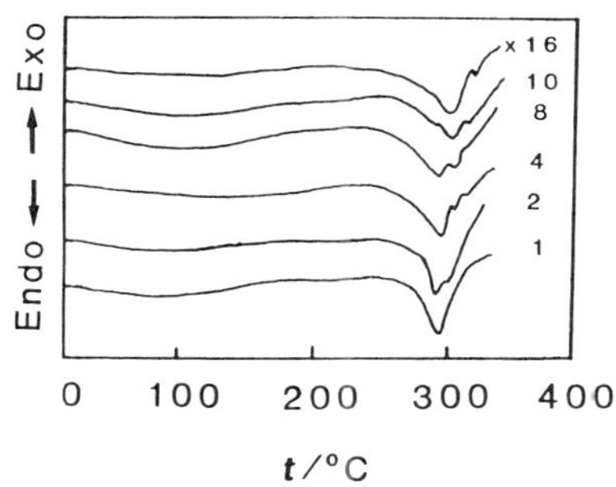

Figure 7. DTA curves of the β-form fibroin at different draw ratios. Heating rate: $20Kmin^{-1}$.

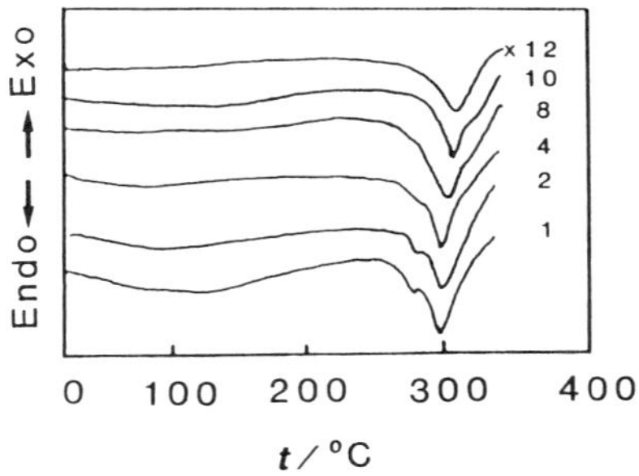

Figure 8. DTA curves of the α-form fibroin at different draw ratios. Heating rate: 20Kmin^{-1}.

Thermal Behavior of Silk Sericin. Two endothermic peaks appears in the DTA curve of sericin (12-14). A peak at 100°C is due to the evaporation of water and a peak due to the thermal degradation is observed at 260 to 270°C. The latter peak is by 10°C higher for sericin A, the more soluble fraction, than for sericin B, the less soluble fraction.

The glass transition of sericin is observed at 170°C for powdered amorphous specimen. An exothermic peak due to crystallization appears at 205°C in the DSC curve of powdered sericin.

Thermal Behavior of Wild Silk Fibroin. The thermal behavior of wild silk fibroins is somewhat different from that of domestic silk fibroin due to the difference in amino acid composition. The glass transition temperature of domestic silk fibroin in Bombyx mori is 175°C, while those of wild silks from sakusan (Antheraea pernyi), kususan (Dictyopica japonica), erisan (Attacus ricini) and tensan (Anteraea yamamai) are 160, 198, 191, and 201°C, respectively (15, 16).

Literature Cited

1. Shirakashi, K.; Takubo, T.; Takahashi, Y. Sen-i Kogyo Gakkaishi (J. Soc. Fiber Ind.) 1942, **8**, 119.
2. Schwenker Jr.,R.F.; Dusenbury,J.H. Text. Res. J. 1961, **30**, 800.
3. Ishikawa, H.; Hirabayashi, K.; Hayakawa, T. Sen-i Gakkaishi (J. Fiber Sci. Technol., Jpn.) 1969, **25**, 37.
4. Magoshi, J.; Magoshi, Y.; Nakamura, S.; Kasai, N.; Kakudo, M. J. Polym. Sci., Polym. Chem. Ed. 1977, **15**, 1675.
5. Magoshi, J.; Nakamura, S. J. Appl. Polym. Sci. 1975, **19**, 1013.

6. Magoshi, J. Kobunshi Ronbunshu 1974, **31**, 648.
7. Magoshi, J. Kobunshi Ronbunshu 1974, **31**, 765.
8. Magoshi, J. J. Soc. Rheo., Jpn. 1973, **1**, 22.
9. Nagura, M.; Goto, K; Ishikawa, H. Kobunshi Ronbunshu 1977, **34**, 389.
10. Magoshi, J. Zairyo 1973, **22**, 499.
11. Magoshi, J.; Nakamura, S. In Viscoelasticity of Biomaterials; Glasser, G.; Hatakeyama, H., Eds.; ACS Symp. Ser., No. 489; American Chemical Society: Washington DC, 1992; p.231.
12. Aoki, K. Nihon Sanshigaku Zasshi (J. Sericultural Sci., Jpn.) 1972, **41**, 429.
13. Aoki. K. Nihon Sanshigaku Zasshi (J. Sericultural Sci., Jpn) 1972, **41**, 429.
14. Aoki, K.; Takeuchi, T. Sen-i Gakkaishi (J. Fiber Sci. Technol., Jpn.) 1971, **27**, 486.
15. Magoshi, J.; Magoshi, Y.; Nakamura, S. J. Appl. Polym. Sci. 1975, **19**, 1013.
16. Nakamura, S.; Saegusa, Y.; Yamaguchi, Y.; Magoshi, J.; Kamiyama, S. J. Appl. Polym. Sci. 1986, **31**, 955.

RECEIVED July 7, 1993

Chapter 20

Mechanical and Chemical Properties of Certain Spider Silks

M. B. Hinman, S. L. Stauffer, and R. V. Lewis

Department of Molecular Biology, University of Wyoming, Laramie, WY 82071

Orb-web weaving spiders synthesize and use a variety of silks, each having different properties suited to their particular functions. These silks were gathered and subjected to mechanical testing and biochemical characterization. The dragline, minor ampullate, and cocoon silks of both *Nephila clavipes* and *Araneus gemmoides* were subjected to strain testing on an Instron universal test bed to compare their physical properties. The primary structures of two proteins composing the *N. clavipes* dragline silk are presented to initiate a comparative study of the structure/function relationships for proteins in these silk fibers.

Spiders have been specializing their use of silks over the course of millions of years of evolution, to the point that many are totally dependent on the silks for any evolutionary advantage. This is particularly true of the orb-web weaving spiders which spin highly complex webs from several different types of silk as well as use silk to protect their eggs and swathe their prey for storage. Orb-web spiders have six sets of silk-producing glands(*1*,*2*), each synthesizing a different protein fiber for use in a particular application. While having different mechanical properties, the silks do have several common features, including: i) being composed almost solely of protein; ii) a predominance of alanine, serine, and glycine in their amino acid compositions, but with the addition of varying quantities of glutamine, tyrosine, leucine, valine, and proline (unlike *Bombyx morii* silk); and iii) undergoing an essentially irreversible transformation from soluble proteins in their respective glands to insoluble fibers in their final state. The proliferation of silk proteins within one spider species and the ability to compare these silks to similar fibers used for

0097–6156/94/0544–0222$06.00/0

the same function from other species presents a unique opportunity to define structure/function relationships for the primary amino acid sequences of the proteins composing these fibers.

Dragline silk was one of the first of these protein fibers to be studied, due to its unusual combination of high mechanical strength coupled with the high elasticity, as shown in Table I.

Table I. Mechanical Properties of Dragline Silk

Material	Strength (N m^{-2})	Elasticity (%)
Dragline Silk	1×10^9	35
KEVLAR	4×10^9	5
Rubber	1×10^6	600
Tendon	1×10^9	5

Data derived from Gosline, et al.(*3*).

Dragline silks from many spiders have been extensively characterized both physically and chemically (*2,4-7*), since the silk is easily gathered by forced silking(*8*) and the gland producing it is large and readily accessible. The comparison of physical properties of dragline silks, minor ampullate silks, and cocoon silks from *Araneus gemmoides* to those of *Nephila clavipes* can be correlated to the primary, and higher order, structures of those silks as the protein sequences become available and higher order structures are defined by analytical means. However, the solvent- and enzyme-resistant nature (*6,7*) of silk fibers makes them very difficult to study biochemically, and the amino acid sequences of silk proteins have not previously been determined due to these problems. Recent studies in our laboratory (*9,10*) have resolved some of these difficulties, allowing us to present the partial amino acid sequences for two proteins we believe constitute the major subunits of *Nephila clavipes* dragline silk.

Comparison of Mechanical Properties of *N. clavipes* and *A. gemmoides* Silk Fibers

Dragline, minor ampullate, and cocoon silks were forcibly drawn (*9*) and fixed to lengths of black construction paper with 4 centimeter apertures. The ends of the sample were clamped into the heads of an Instron universal test bed and the sides of the aperture were cut, allowing the 4 centimeter length of spider silk to be suspended between the heads. The tension on the fiber was allowed to come to zero, then the heads were slowly separated. The force exerted as the heads separated was recorded by computer along with the separation distance between the heads as the experiment progressed. The strain (percentage elongation beyond the original fiber length) was determined from this measurement and since the diameter of the fiber was previously determined by microprojection onto a digital measuring pad, the stress (force per unit cross-sectional area) could be calculated. These data were graphed for 3 individual fibers, as shown in Figure 1 for the dragline silks from both *N. clavipes* and *A. gemmoides*.

The Nephila data (top) and the Araneus data (bottom) shows a biphasic response typical of many fibers, including *Bombyx morii* silk and manmade acetates. The change to a different stress/strain relationship occurs at about 5-7% elongation of the fiber for both species of spider. The final breaking point of the dragline silk fiber occurs at approximately 20-22% extension for the Nephila samples, with a final stress of 6.5 X 10^5 pounds per square inch (psi). The Araneus silk has a comparable final stress, averaging about 6.7 X 10^5 psi, but appears to break at approximately 16-17% extension. The overall mechanical characteristics of these two fibers are very similar as indicated by these two graphs. This observation can be contrasted with the amino acid compositions of the two silks, as Araneus dragline silk has three times the proline content of Nephila dragline silk(*11*). Proline is considered to have a major effect on protein secondary structure, particularly in the dragline silk, as will be discussed later.

Similar measurements and calculations were performed for the minor ampullate silk and cocoon silk from each species (graphs not shown). The minor ampullate silks for both species have a change in the stress/strain relationship at about 2-3% extension, but the Nephila silk breaks at around 22-25% extension while the Araneus silk breaks at 25-27% extension. The breaking stress of the minor ampullate silk for Araneus is 2 X 10^5 psi, higher than that for Nephila(1.4 X 10^5 psi). Cocoon silks from Nephila and Araneus have a change of behavior at 4-5% extension and both break at 18-20% final extension. However, Araneus cocoon silk is once again apparently stronger, breaking at 3.4 X 10^5 psi compared to 1.7 X 10^5 psi for the Nephila cocoon silk fiber.

A comparison of the fibers is facilitated by utilizing the concept of modulus, a measure of the stiffness (resistance to deformation under a load) of the fiber. The modulus is determined by taking the slopes of the lines found in the graphs in Figure 1, which will have two apparent moduli. Nephila and Araneus have comparable initial moduli at about 7 and 9 X 10^6 psi respectively. The secondary moduli are also comparable at 2 and 2.6 X 10^6 psi. As a comparison, the modulus for carbon steel is 30 X 10^6 psi and that of KEVLAR is 19 X 10^6 psi. The similar moduli for Nephila and Araneus dragline fibers once again indicate that the two fibers have comparable mechanical characteristics.

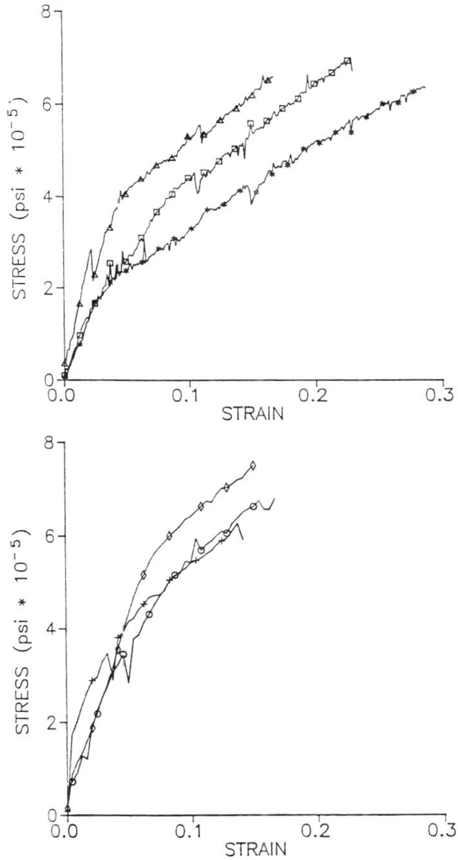

Figure 1. Mechanical Testing of Dragline Silks. Three separate samples of *Nephila clavipes* dragline silk (△,□,✳) shown in the top graph and three separate samples of *Araneus gemmoides* dragline silk (◇,○,+) shown in the bottom graph were tested for mechanical resistance as explained. The Instron cross-head speed was 5 mm per minute.

Biochemical Characterization of the Nephila Clavipes Dragline Silk

Work is progressing in our laboratory at obtaining the primary structure of *Araneus gemmoides* dragline silk, in order to facilitate comparison of structure/function relationships between this fiber and the *Nephila clavipes* dragline silk. We also are continuing studies aimed at obtaining sequence information for the minor ampullate and cocoon silks for both species. At this point, we have isolated partial cDNA clones for two proteins which we believe are the major subunits of *Nephila clavipes* dragline silk(*9,10*).

Peptides from partial acid digestion of *Nephila clavipes* dragline silk were isolated by high performance liquid chromatography. Twenty of these were subjected to amino acid sequencing yielding mostly small peptides, with an occasional penta- or hexapeptide. Synthetic, degenerate oligonucleotides were designed and produced (based on the most suitable, least degenerate of these sequences) to be used as probes of a cDNA library. The library was made from the mRNA of major ampullate glands from spiders which had been silked previously, which should have induced an increased synthesis of the appropriate mRNA for dragline silk. An initial screening with an oligonucleotide based on the pentapeptide GAGQG produced several positives, the largest of which was 2.3 kb in length. This clone was sequenced and the predicted amino acid sequence of the highly repetitive region of what we now call Spidroin 1 is shown in Figure 2, arranged to show the high degree of conservation and repetitiveness.

The peptide sequence of Spidroin 1 can be divided into three conserved regions; i) the core region, the first 15 amino acids comprised of GGX repeats, which is relatively unsubstituted, ii) the GGX variable region of 9 amino acids, in which there seem to be numerous deletions, and iii) the polyalanine region, which has 4 to 7 sequential alanines. One of the striking elements of the conservation of the GGX regions is that nearly all deletions are in sets of three amino acids, perhaps indicating the importance of this repeating structural element. Another feature worth noting is that the glutamines in the GGX regions are very highly conserved and are not ever substituted with asparagines, indicating that the glutamine amide side chain is probably critically involved with maintaining the secondary structure necessary for the function of dragline silk.

While there are apparently three regions of conservation, the basic peptide can be regarded as having only 2 structural domains; the domain having various lengths of GGX repeats, and the polyalanine domain. There are 63 amino acids not shown on Figure 2 at the carboxy-terminal end of Spidroin 1 in which the repetitive nature breaks down and finally disappears (see reference *9* for the full peptide sequence), but it has no apparent relation to the rest of the protein nor any striking clues to secondary structure. The structural significance for the two domains is discussed below.

```
-----------------------QGAGAAAAAA-
GGAGQGGYGGLGGQG-------------------
--AGQGGYGGLGGQG------AGQGAGAAAAAAA
GGAGQGGYGGLGSQGAGR---GGQGAGAAAAAA-
GGAGQGGYGGLGSQGAGRGGLGGQGAGAAAAAAA
GGAGQGGYGGLGNQGAGR---GGQ--GAAAAAA-
GGAGQGGYGGLGSQGAGRGGLGGQ-AGAAAAAA-
GGAGQGGYGGLGGQG-------------------
--AGQGGYGGLGSQGAGRGGLGGQGAGAAAAAAA
GGAGQ---GGLGGQG------AGQGAGASAAAA-
GGAGQGGYGGLGSQGAGR---GEGGAGAAAAAA-
GGAGQGGYGGLGGQG-------------------
--AGQGGYGGLGSQGAGRGGLGGQGAGAAAA---
GGAGQ---GGLGGQG------AGQGAGAAAAAA-
GGAGQGGYGGLGSQGAGRGGLGGQGAGVAAAAAA
GGAGQGGYGGLGSQGAGRGGLGGQGAGAAAAAA-
GGAGQRGYGGLGNQGAGRGGLGGQGAGAAAAAAA
GGAGQGGYGGLGNQGAGR---GGQ--GAAAAA--
GGAGQGGYGGLGSQGAGR---GGQGAGAAAAAA-
VGAGQEGIR---GQG-------------------
--AGQGGYGGLGSQGSGRGGLGGQGAGAAAAAA-
GGAGQ---GGLGGQG------AGQGAGAAAAAA-
GGVRQGGYGGLGSQGAGR---GGQGAGAAAAAA-
GGAGQGGYGGLGGAGVGRGGLGGQGAGAAAA---
GGAGQGGYGGV-GSG----------ASAASAAAA
SRLSS
```

Figure 2. Predicted Amino Acid Sequence of the Repetitive Region of Spidroin 1 from *Nephila clavipes* Dragline Silk.

A comparison of the sequence predicted by Spidroin 1 with the amino acid composition of intact dragline silk demonstrated the lack of appreciable proline in Spidroin 1, as well as a lower amount of tyrosine than expected. In addition, the sequence of Spidroin 1 did include many, but not all, of the peptides isolated by partial acid hydrolysis. The most notable exclusion was the peptide GYGPG, which includes both of the amino acids which were missing from the amino acid composition. Therefore, we rescreened the major ampullate gland library with an oligonucleotide probe based on this peptide and isolated another clone of 2 kb length which coded for a protein we named Spidroin 2 (*10*).

The repetitive peptide sequence of Spidroin 2 is shown in Figure 3, again arranged to show the maximum identity of the repetitive regions. There are distinct similarities and differences to the Spidroin 1 protein. Spidroin 2 can also be divided into three regions of conservation, with a subtle difference. The variable region of 20 amino acids, which contains the alternating pentapeptides GPGGY and GPGQQ is between the core region of 15 amino acids (containing the same pentapeptides) and the polyalanine region of 6 to 10 amino acids. In addition, there is another region of conservation which joins the pentapeptide repeats and the polyalanine, a much more distinct joining region than is evident in Spidroin 1. In the region of alternating pentapeptides, several items deserve consideration. First, all the apparent deletions are seen to be in sections of five amino acids, indicating the importance of this structural motif to the protein's structure. Second, GPGGY is never found next to itself, nor is GPGQQ, even in the section of the carboxy-terminal repetitive region where the repetitiveness and conservation are breaking down. Third, the apparent importance of the glutamine side chain is once again emphasized, as there are no asparagine for glutamine substitutions. And last, tyrosine and proline maintain a high degree of conservation, with no phenylalanine ever being substituted for a tyrosine.

The Spidroin 2 repetitive region can also be divided into two basic structural domains, the alternating pentapeptide domain which is rich in proline, and the polyalanine domain. It might also be argued that the highly conserved joining region is also a small structural domain. As with Spidroin 1, there is a carboxy-terminal extension of 95 amino acids which degenerates into non-repetitiveness and apparent structural heterogeneity (see reference *10* for the full sequence).

We believe that the Spidroin 1 and Spidroin 2 proteins constitute the only subunits of *Nephila clavipes* dragline silk. The sequences of Spidroin 1 and Spidroin 2 contain the sequences of the all of the peptides found from the partial acid hydrolysis of intact dragline silk. The amino acid content of intact dragline silk can be accounted for by a combination of the amino acid compositions of Spidroin 1 and Spidroin 2. It is very interesting to us that i)the proline content of this silk has been shown to vary two-fold along the length of the same fiber(*7*), ii) has been shown to be as low as 0.42% (*7*) and as high as 4.3% (*12*) proline, and that iii)the proline content from the dragline silk of a very similar species, *Nephila clavata*, has been shown to vary over the lifetime of the spider(*12*). These observations seem consistent with the presence of two proteins in the dragline silk but imply that the proteins are not always present in the same proportion, as the apparent contradictions are difficult to resolve any other way.

There are similarities between Spidroin 1 and Spidroin 2 at both the nucleotide and amino acid levels.There is a high degree of preference for the use of adenine or thymidine in the third or "wobble" position in the codons for almost all the amino acids. This preference is most apparent when examining the codon distribution for alanine, glycine, and glutamine in the repetitive regions of both Spidroins. In Spidroin 1 the preferences for using adenine or thymidine in the third position of the codons are 80.2%, 94.5%, and 98.5% respectively for these three amino acids. The codon preferences in Spidroin 2 for these three amino acids are very similar, being 74.2%, 88.6% and 97.3%, respectively. We believe the reason for this very large preference is that the secondary structure of the DNA and RNA would become very complex in a G/C rich gene such as the dragline silk, especially in the regions coding for the polyalanine stretches.

One of the most interesting similarities at the peptide level between Spidroin 1 and Spidroin 2 actually occurs outside the repetitive regions of the proteins. A 49 amino acid segment directly after the repetitive region of each protein shows an 80% identity between the two proteins. The area is shown below, with the 49 amino acids from Spidroin 1 over the same region from Spidroin 2, and the identities highlighted by underlining.

ASAAASRLSSPQASSRVSSAVSNLVASGPTNSAALSSTISNVVSQIGAS
ASAAASRLASPDSGARVASAVSNLVSSGPTSSAALSSVISNAVSQIGAS

The significance of this area in terms of any structural motif is obscure, but the presence of such a high degree of amino acid identity implies that the two Spidroins are derived from a common ancestral gene. This would be quite interesting as the differences between the GGX repetitive region of Spidroin 1 and the proline-rich repetitive region of Spidroin 2 are quite pronounced, and even the polyalanine regions of Spidroin 1 are consistently shorter than those of Spidroin 2.

Both Spidroin 1 and Spidroin 2 amino acid sequences were subjected to computer analysis of predicted secondary structure using MacVector, version 3.5 (International Biotechnologies, Inc.). This program uses both Chou-Fasman (CF) algorithms and Robson-Garnier (RG) algorithms to predict three types of secondary structure (α-helix, β-sheet, and β-turns) and then derives a consensus of the two (CfRg) to indicate where the two algorithms agree. To facilitate comparison, the secondary structure for a 200 amino acid section of each repetitive region (starting with the first polyalanine stretch in each protein) is shown in Figure 4, with the

```
----------------PGGYGPGQQGPGGYGPGQQGP--SGPGSAAAAAAAAAA
-----GPGGYGPGQQGPGGYGPGQQGPGRYGPGQQGP--SGPGSAAAAAA----
----------GSGQQGPGGYGPRQQGPGGYGQGQQGP--SGPGSAAAASAAASA
ESGQQGPGGYGPGQQGPGGYGPGQQGPGGYGPGQQGP--SGPGSAAAAAAAAS-
----------GPGQQGPGGYGPGQQGPGGYGPGQQGP--SGPGSAAAAAAAAS-
----------GPGQQGPGGYGPGQQGPGGYGPGQQGL--SGPGSAAAAAAA---
--------------------GPGQQGPGGYGPGQQGP--SGPGSAAAAAAAAA-
---------------GPGGYGPGQQGPGGYGPGQQGP--SGAGSAAAAAAA---
----------GPGQQGLGGYGPGQQGPGGYGPGQQGPGGYGPGSASAAAAAA--
--------------------GPGQQGPGGYGPGQQGP--SGPGSASAAAAAAAA
---------------GPGGYGPGQQGPGGYAPGQQGP--SGPGSASAAAAAAAA
---------------GPGGYGPGQQGPGGYAPGQQGP--SGPGSAAAAAAAAA-
-------------------------GPGGYGPAQQGP--SGPGIAASAASA---
---------------GPGGYGPAQQGPAGY----------GPGSAVAASAGA--
-------------------------GSAGY----------GPGSQASAAA----
SRLAS
```

Figure 3. Predicted Amino Acid Sequence of the Repetitive Region of Spidroin 2 from *Nephila clavipes* Dragline Silk.

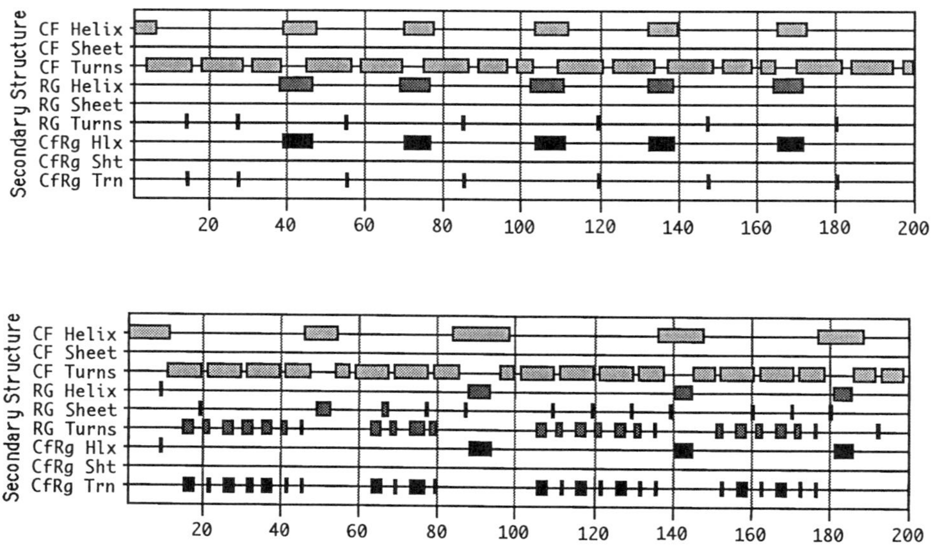

Figure 4. A Comparison of the Repetitive Secondary Structures Predicted for Spidroin 1 and Spidroin 2. The predictions were generated on a MacIntosh IIfx personal computer (Apple) using the program MacVector, version 3.5 (International Biotechnologies, Inc.)

predictions for Spidroin 1 as the top panel, and the Spidroin 2 predictions below. This figure emphasizes the repetitive nature of the secondary structures found in each of the Spidroins.

It is most useful to examine the consensus predictions for each protein. Many of the polyalanine regions of both Spidroins are predicted to be α-helices by both algorithms (the absence of such a prediction for the first polyalanine stretch is an artifact of the method, as this region is predicted to be a helix when the preceding amino acids are included). The GGX regions of Spidroin 1 are predicted to have a small probability for forming β-turns, while the proline-rich regions of Spidroin 2 are predicted to have a high probability for forming a number of β-turns. The low degree of probability for any of the structures to form a β-sheet is quite surprising, since low resolution X-ray diffraction and other studies indicate a high proportion (from 25-30%) of this structure is present in dragline silk (*3,4, 13-16*).

We originally believed that the predictions were correct in assigning the α-helix conformation to the polyalanine regions(*9*). The GGX region of Spidroin 1 was thought to be involved in β-sheet formation, since we expected from the X-ray diffraction studies that the unit cell size of *Nephila clavipes* would be similar to that for *Nephila senegalensis*, which has about 15.7 Å space for the side chains. This spacing would leave enough room for the bulky side chains of tyrosine, leucine, and glutamine to be incorporated into the sheets. However, the side-chain spacing for *Nephila clavipes* has been found to be 10.6 Å by studies in our lab and others (*17*). This corresponds to the side-chain spacing found in the spider *Antheraea mylitta*, known to be enriched in alanine. Additionally, we discovered in the literature that the GGX repeat is highly unlikely to form β-sheet regions (*18*). With the elucidation of the Spidroin 2 sequence, the probability of the proline-rich region being incorporated into a β-sheet became vanishingly small, leaving the polyalanine region to fill this role.

It was also known that while polyalanine polypeptides can form α-helices in solution, they were known to form β-sheets in crystals under low water conditions (*19*). Studies in our laboratory confirmed that under ambient conditions, the dragline silk from Nephila contained 6.8% water (Araneus dragline contained 5.3%). This corresponds to about one water molecule for every three amino acids, a decidely low water environment. As comparisons, we determined the water content of crystallized hemoglobin and lysozyme (49 and 73 % water, respectively), as well as that of two proteins we lyophilized before determining the water content, elastin and collagen (10 and 15% water). On the basis of this determination of low water content for the dragline silk fiber and the constraints imposed by the size of the unit cell, we assigned the β-sheet structure to the polyalanine regions of Spidroin 1 and Spidroin 2.

The β-turn structure is the most likely secondary structure for the proline-rich region of Spidroin 2 to form. The pentapeptide GPGQX is a predominant predictor sequence for β-turns (*20*), and the pentapeptide GPGGX is only slightly less likely to form the same structure. The β-turn conformation is predicted for such diverse proteins as gluten, synaptophysin, and elastin (*21,22*), all of which contain proline rich sequences in small peptide repeats. It is on the basis of these arguments that we believe that the Spidroin 2 proline-rich region is a series of linked β-turns, forming a ladder-like or even coiled structure. One consequence of this model, which we discovered on computer modeling of the 3-dimensional structure, is that the tyrosines of Spidroin 2 all appear on one side of the structure and the glutamines all appear on the other side. The significance of this observation is not clear. We also believe that the GGX region of Spidroin 1, already predicted to form β-turns by MacVector, has enough conformational flexibility to adopt the secondary structure of Spidroin 2, and may indeed be forced to by its close association with that protein in the fiber.

The β-turn model we are proposing for the Spidroins is reminiscent of a model proposed for elastin(*22*), in which the elastic force driving contraction of a stretched fiber is entropically driven. β-turns are known to have some degree of flexibility (*21,22*), so it is possible that Spidroin 2 forms a β-turn coil similar to that in elastin. The ultimate tensile strength in the fiber is probably due to the formation of β-sheet structures between the polyalanine segments, with those crystalline regions oriented along the axis of the fiber. It will be helpful to obtain similar sequence information for the dragline silks of other spiders. This model of a two protein fiber with alternating β-sheet and β-turn structures helps resolve many disparate studies, but we look forward to discussion and reformulation of the model as studies of dragline silk from this and other species more clearly define the roles of primary and secondary structure in the function of this unique fiber.

Literature Cited

1.Lucas,F. *Discovery* **1964**, *25*, 20-26

2.Koover,J. In *Ecophysiology of Spiders*, Nentwig, W., Ed.; Springer-Verlag, Berlin-Heidelberg, 1987; 160-186

3.Gosline, J.M., DeMont, M.E., Denny, M.W. *Endeavor* **1986,** *10*, 37-43

4.Gosline, J.M., Denny, M.W., DeMont, M.E. *Nature* **1986**, *309*, 551-552

5.Tillinghast,E.K., Townley,M. In *Ecophysiology of Spiders*, Nentwig, W., Ed.; Springer-Verlag, Berlin-Heidelburg, 1987; 203-210

6.Dong,Z., Lewis,R.V., Middaugh,C.R. *Arch. of Bioch. Biophys.* **1991**, *284*, 53-57

7.Work,R.W., Young,C.T. *J. Arachnol.* **1987**, *15*, 65-80

8.Work, R.W., Emerson, P.D. *J. Arachnol.* **1982**, *10*, 1-10
9.Xu,M., Lewis,R.V. *Proc. Natl. Acad. Sci. USA* **1990**, *87*, 7120-7124
10.Hinman, M.B., Lewis, R.V. *J. Biol. Chem.* **1992**, *267*, 19320-19324
11.Hinman, M., Dong,Z., Xu, M., Lewis, R.V. In *Biopolymers*; Case, S.T., Ed., Results and Problems in Cell Differentiation; Springer-Verlag: Berlin Heidelberg, 1992, Vol. 19; 227-254
12.Kaplan, D.L., Lombardi, S.J., Muller, W.S., Fossey, S.A. In *Biomaterials: Novel Materials from Biological Sources*; Byron, D., Ed.; Stockton Press: New York, NY, 1991; 3-53
13.Denny, M.W. *J. Exp. Biol.* **1976**, *65*, 483-505
14.Denny,M.W. In *Mechanical Properties of Biological Materials*; Vincent, L., Curry,M., Eds.; Cambridge University Press: Cambridge, 1987; 247-272
15.Work, R. *J. Exp. Biol.* **1985**, *118*, 379-404
16.Warwicker,J.O. *Faraday Soc. Trans.* **1960**, *52*, 554-557
17.Fraser,R.D.B., MacRae,T.P. In *Conformation in Fibrous Proteins*; Academic Press, New York, NY, 1973; 293-343
18.Lotz,B., Brack,A., Spach,G. *J. Mol. Biol.* **1974**, *87*, 193-203
19.Hempel,A., Camerman,N., Camerman,A. *Biopolymers* **1991**, *31*, 187-192
20.Wilmot,C.M., Thornton,J.M. *J. Mol. Biol.* **1988**, *203*, 221-232
21.Matsushita,N., Creutz,C.E., Kretsinger,R.H. *Proteins: Structure, Function, and Genetics* **1990**, *7*, 125-155
22.Wasserman,Z.R., Salemme,F.R. *Biopolymers* **1990**, *29*, 1613-1631

RECEIVED May 4, 1993

Chapter 21

Mechanical Properties of Major Ampulate Gland Silk Fibers Extracted from *Nephila clavipes* Spiders

Philip M. Cunniff, Stephen A. Fossey, Margaret A. Auerbach, and John W. Song

Natick Research, Development, and Engineering Center, Department of the Army, Natick, MA 01760–5019

The mechanical properties of major ampulate gland silk fibers extracted from Nephila clavipes (Araneae, Tetragnathidae) spiders are presented. Stress-strain response at conventional tensile testing rates (ca. 10%/s), and at very high strain rates (ca. 1,000,000%/s) are characterized. Fiber surface topography indicates evidence of some lobed (serrated) fiber surface. Fractured fibers indicate ductile failure at low strain rates, occasionally accompanied by a low degree of fibrillation; at high strain rates both brittle and ductile type fracture surfaces are observed. Dynamic mechanical properties and creep behavior of yarns are examined. Mechanical properties are thought to be sufficiently promising to warrant further research to produce genetically engineered man-made fibers. Ultimate axial tensile strength in excess of 2.9 GPa, ultimate axial tensile strain in excess of 11%, and modulus in excess of 60 GPa appear attainable from such man-made fibers. The very high strain rate response suggests applications in body armor systems.

The most important protein-based fibers, namely silkworm silk and wool, are used in the naturally occurring form. Although several attempts have been made to produce commercial fibers from natural proteins, none of the man-made fibers currently available are protein-based. However, protein-based fibers, in particular, spider silk fibers, offer a combination of unique characteristics that make them attractive. In principle, the fibers should have a precisely monodisperse molecular weight, should be stereoregular, should be spun from an environmentally benign solvent (water) and are processed under room temperature conditions.

Following the work of Lombardi and Kaplan [1-4], the prospect of mass-produced man-made fibers with properties and composition similar to those of natural Nephila clavipes (Araneae, Tetragnathidae) spider silk has renewed interest in the mechanical properties and morphology of spider silk fibers [5-7]. One possible application of silk fibers of interest to the U.S. Army is ballistic protective apparel (body armor).

The mechanical properties of natural (controlled silked) major ampulate gland N. clavipes spider silk are thought to provide some degree of insight into the mechanical properties of genetically engineered man-made fibers using the same protein sequence. The prospect of developing genetically engineered proteins offers the possibility of substantially improving fiber mechanical properties (particularly fiber crystallinity, and consequently fiber stiffness) through alteration of the molecular structure of the substituent proteins. Additionally, since fiber ultimate mechanical properties are most strongly influenced by macroscopic defect structure, the controlled processing conditions possible with man-made fibers offer the possibility of substantially improving the ultimate tensile strength and ultimate tensile strain of the fibers.

The morphology and deformation behavior of naturally occurring and controlled-silked spider silks has been investigated by a number of studies. Zemlin [8] examined the mechanical properties of several orb-web spinning spiders, including N. clavipes. Modulus values as high as 29.8 GPa (250 gpd[1]) with tensile strength values of 1.87 GPa (15.25 gpd)[2] and ultimate tensile strain of 17.4% were reported; however, errors in the determination of fiber density and fiber diameter of at least 8% and as much as 64% were suggested. Warwicker [9] estimated that 40-45% of the total protein in silkworm silk fibers was crystalline. Gosline et al. [10] reported that these crystalline regions, estimated to occupy 30% of the volume of the fiber, are connected by cross-linked amorphous rubbery regions that account for the majority of the mechanical properties of spider silk fibers. Additionally, Gosline [11] showed that supercontracted spider silks are best described as a rubber; fiber elasticity arises almost entirely from entropy.

Experimental Techniques

Controlled Silking: Controlled silking was conducted in a manner similar to that described by Work and Emerson [12]. Silking rates between 1.5 and 12.2 cm/s were controlled using a variable speed motor. The extrusion rate was selected to be greater than the minimum rate reported by Magoshi [13] who identified a critical extrusion rate of 0.83 cm/s before β-pleated sheet formation occurs in Bombyx mori fibers. The silking rate was also lower than the rate used by Zemlin [8] who noted that at silking speeds of between 10.2 and 61.0 cm/s (22 and 120 ft/min.) there was slight evidence to suggest that higher strength fibers were obtained at lower silking rates, while modulus and elongation properties were essentially unaffected. Larger fiber diameters were

---

1. For convenience, the commonly used textile unit of gram-force per denier (gpd) will be used in conjunction with SI units in this paper. Note that:

Kg force = 9.806 N
9000 denier = g/m
gpd = 11.33/ρ GPa; where ρ = fiber density (g/cm^3)

2. These properties compare favorably to high performance fibers. For example compared to Kevlar 29® fibers (Modulus 62.0 GPa; Strength 2.76 GPa, Ultimate tensile strain 4%) the strength of silk is 70% of Kevlar's, silk's modulus is approximately half of Kevlar's, but silk's ultimate tensile strain is over 4 times as great as Kevlar's.

found to be produced at increasingly higher silking velocities, and considerable damage to the fibers was noted at these rates.

Tensile Tests (Instron Tests): Tensile specimens were prepared by attaching single fibers to paper tabs using 3m Co. Scotch-Weld® 1838 B/A epoxy adhesive; the fiber gauge length used was 5.08 cm. Prior to tensile testing, each sample was inspected at 320X magnification in a light microscope to determine the presence of more than one fiber. The epoxy was cured for at least 24 hours; samples were mounted between pneumatic fiber grips in a model 4201 Instron testing machine, and the edges of the paper tabs were cut away. The strain rate used in tensile testing was 10%/s. Approximately 30 samples were tested for each fiber type. Data was collected on an IBM PC type computer and later transferred to a Sun Microsystems SPARCstation computer for data analysis. Fiber strain was determined from testing rate and elapsed time. All tests were carried out at approximately 70 °F and 50% relative humidity.

Microscopic Examination: Samples were mounted on aluminum sample stubs using a double faced adhesive tape. Samples were coated with approximately 30 nm of AuPd at a 72 mm working distance and 15 mA current in a Blazers SCD040 sputter coater and viewed in an Amray 1000A scanning electron microscope (SEM). Fiber diameters were measured using the image acquisition and processing program on a Noran TN-5500 X-ray analyzer. SEM photomicrographs were taken at working distances ranging from 12-15 mm using an electron beam acceleration potential of 10 kV.

Dynamic Mechanical Properties: Determination of dynamic mechanical properties was conducted using a Seiko DMS210 dynamic mechanical analyzer on 82.27 denier yarns. The sample dimensions were 7 mm in length and approximately 0.007 mm^2 in cross sectional area. The initial axial load applied was 200 grams. A fixed frequency of 1 Hz was used throughout the testing program; a temperature range of -150 °C to 220 °C with a heating rate of 4 °C/min was used.

Creep: Creep behavior was observed using a TA Instrument Thermal Mechanical Analyzer (TMA) on 82.27 denier yarns. Gauge lengths of the samples was approximately one inch (25mm). A constant stress of one Newton (102g force) was applied during the test. The experiments were conducted at four different temperatures (-40 °C, 25 °C, 100 °C and 150 °C).

Dynamic (Ballistic Impact Rate) Stress-Strain Response: Single yarn impact tests were conducted using approximately 75 denier yarns. The yarns were made up from the silk of between 5 and 8 spiders that were simultaneously silked. Twelve-inch gage length yarns were twisted to approximately 1 turn/cm, and clamped between rigid supports in the ballistic impact test range, as illustrated in Figure 1. The yarns were impacted transversely with the impact point at the center of the yarn by a 1.1 gram steel projectile propelled by a compressed helium charge through a 0.22 calibre (0.558 cm) smooth bore gun. A multiple microflash photograph of the impact event was obtained using Polaroid Type 57 (ASA 20,000) film, and an EG&G Model 502A microflash unit. Each photograph contained 15 exposures of the projectile and yarn at different times after impact. The time between subsequent microflashes was set to approximately

35 μs. Projectile velocity was measured before (striking velocity) and after (residual velocity) the impact event using light screens and counter-timers. A projectile striking velocity of about 300 m/s was used in each of the four impact tests conducted. Figure 2(a) is a typical multiple microflash photograph of a spider silk yarn under impact at 300 m/s; Figure 2(b) and 2(c) are microflash photographs of high tenacity nylon and Kevlar 29 yarns under similar impact conditions. Photographs were digitized using an Apunix digitizer and a Sun Microsystems SPARCstation computer for analysis. The projectile line of flight, yarn initial position, and yarn deformed position were reduced from the digital images, and used in conjunction with a numerical model for the determination of the response of transverse ballistic impact of yarns to iteratively approximate yarn constitutive properties.

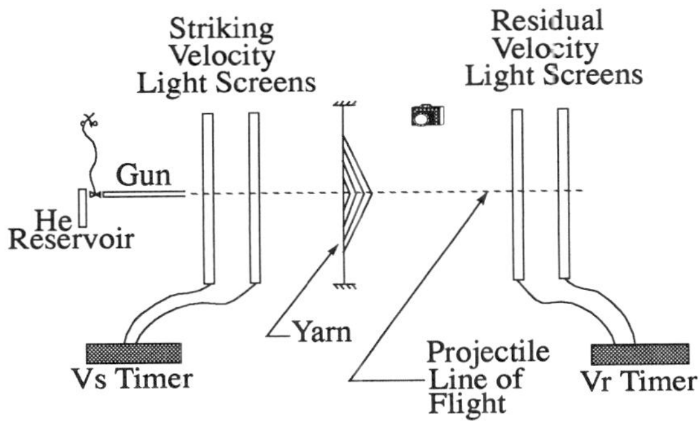

Figure 1. Single Yarn Impact Test Range.

Numerical Techniques

Impact Response of a Constrained Yarn: Under certain circumstances, a rate independent theory of impact may be used to predict the impact behavior of a yarn under transverse impact [14-16]. The more general condition of transverse impact is amenable only to computer simulation. The method used in this work, referred to as "direct analysis", was developed by Davis [17], and applied to transverse impact of yarns by Lynch [18]; the method was extensively modified by Roylance [19].

When a long yarn is impacted transversely by a projectile, a series of strain wavelets propagate away from the impact point in both directions. The velocity (c) of each of the wavelets of the series may be expressed in a lagrangian coordinate system, attached to and moving with points on the yarn, by:

$$c = \sqrt{\frac{1}{\rho}\frac{d\sigma}{d\varepsilon}}$$

where: σ Fiber Stress; $\frac{d\sigma}{d\varepsilon}$ Slope of the dynamic stress-strain curve

ε Fiber Strain; ρ Fiber linear density

In the region between the two longitudinal wave fronts, a V shaped transverse deflection is formed from the inwardly flowing material with the projectile at the vertex of the V. At the edge of the transverse wave, the inward flow of material ceases abruptly, and is replaced by a transverse particle velocity equal to the projectile velocity; the strain and tension in the yarn are unchanged across this wave front.

The direct analysis method provides transverse and longitudinal deflections, transverse and longitudinal velocities, strain and tension in each element at times corresponding to the time steps of the procedure, and at positions corresponding to the laboratory coordinates of each element at the respective time. Transverse deflection data from the experimental portion of the work was available at times corresponding to the microflash times, and at positions corresponding to approximately one mm distances along the length of the yarn. In practice, a constitutive model is postulated for a material, and the model is executed iteratively until the error between the model predictions and experimental measurements is minimized.

Results and Discussion

Surface Topography And Fiber Geometry: In general, the spider silk fibers examined in the scanning electron microscope had a regular unfaceted surface; no apparent gross defects were noted on the surface of the majority of fibers. Figure 3 is a scanning electron micrograph of silk fibers drawn during a single silking of a spider. The mean diameter of the sample illustrated in Figure 3 was 3.00 μm with a standard deviation of 0.62 μm

The surface topography of some of the spider silk fibers showed irregular asymmetrical lobes, as illustrated in Figure 4. Figure 4 is a compilation of five separate scanning electron micrographs of a single 1 cm section of a silk fiber drawn at the same rate as the fibers illustrated in Figure 3. The range of variation seen in the fibers is typical of the range that may be expected from spiders at the silking rates examined in this work. The increase in the serrated appearance of the fiber surface appears to be related to the fiber diameter, but insufficient data was available to quantify the relationship.

The lobes that occasionally occur in the fiber are somewhat similar to the structure of cellulose acetate or viscose rayon. Viscose rayon yarns have a distinct difference in morphology between the skin and core of the fiber; the skin of these fibers develops immediately after extrusion, and then collapses after removal of solvent [20]. The extent of development of the internal morphology of spider silk fibers upon extrusion from the spider is unclear; a skin/core morphology may exist in these fibers.

Since the spider's spinneret is living tissue, some irregularity in the surface geometry of the fiber may be attributable to variations in the geometry of the spinneret during fiber production. Additionally, since the variations do not occur in the majority of the fibers, it is perhaps reasonable to attribute them to momentary fluctuations in the spider's spinneret geometry. Similarly, while the fibers were drawn at a constant draw speed, variations in the spider's ability to produce silk may have varied with time in a manner sufficient to account for the differences in observed fiber diameter. These variations in the production rate of silk, coupled with variations in the geometry of the spider's spinneret, are likely causes for the observed poor fiber quality.

Fiber diameters were measured from SEM images. Five to eighteen measurements were taken along the length of a single fiber. The mean, maximum, and

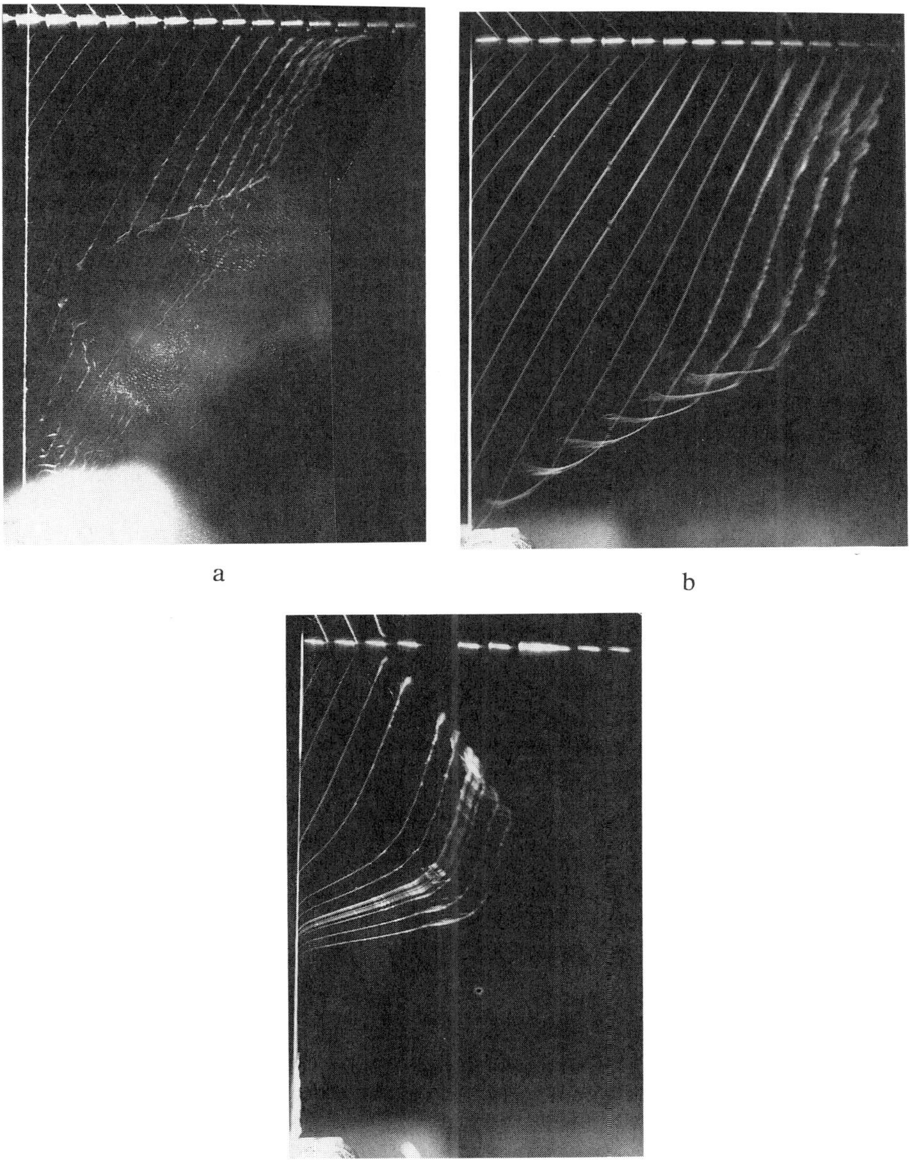

a b

c

Figure 2. (a) Photograph of a single yarn of silk under impact at approximately 300 m/s. (b) Photograph of a single nylon yarn under similar impact circumstances, (c) Photograph of a single Kevlar yarn under similar impact circumstances.

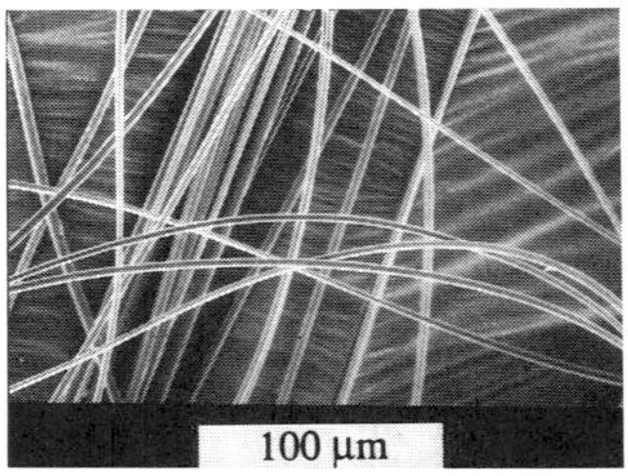

Figure 3. Scanning electron micrograph of a number of typical spider silk fibers.

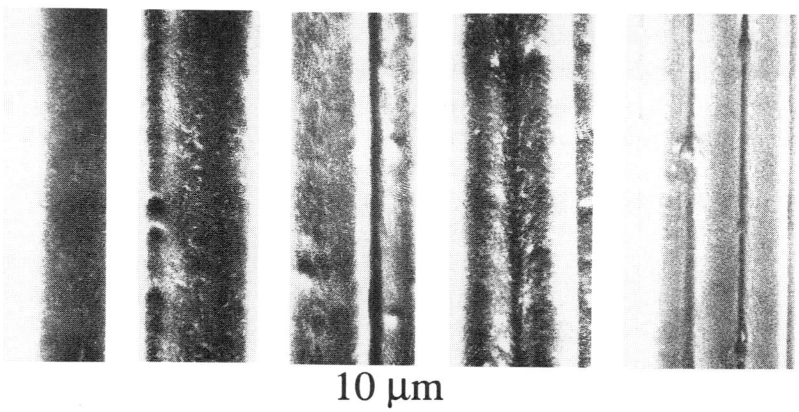

Figure 4. Surface topography and variation in fiber diameters.

minimum fiber diameter measurements for a number of fibers investigated in this study are tabulated below. The mean diameter and variations in fiber diameter for several different fibers drawn at the same draw speed (6.1 cm/s) appears to be representative of all of the draw speeds considered here. Circular cross sections were assumed for the calculation of fiber cross sectional area, as suggested by Zemlin's work [8] (Table I).

The errors introduced by the use of a numerical mean fiber diameter in the determination of fiber cross sectional area are mitigated by averaging the calculated stress over a fairly large number of test specimens.

Table I: Variation in Fiber Diameter With Draw Speed

Draw Speed (cm/s)	Mean Diameter (μm)	Minimum Diameter (μm)	Maximum Diameter (μm)
1.5	2.45	2.19	3.01
3.1	4.49	3.97	4.85
6.1	2.85	2.23	3.91
6.1	3.43	3.22	3.62
6.1	4.37	2.91	5.36
6.1	3.74	3.61	3.86
12.2	4.31	2.95	6.29

Uniformity Of Stress-Strain Response: Work [21] noted that the variability of the stress-strain response of fibers from controlled-silked spiders was greater than the variability from fibers obtained from orb-webs. Inspection of Zemlin's controlled-silked fiber data [8] indicates that the average tensile strength of all the N. clavipes silk fibers tested was approximately 0.95 GPa (8 gpd) with a standard deviation of 381 MPa (3.2 gpd), the mean modulus was 12 GPa (101 gpd) with a standard deviation of 5.2 GPa (43.4 gpd), the mean elongation at break was 16.9% with a standard deviation of 5.2%. The variability in the stress-strain data collected in this work is no less pronounced, as indicated in Figure 5.

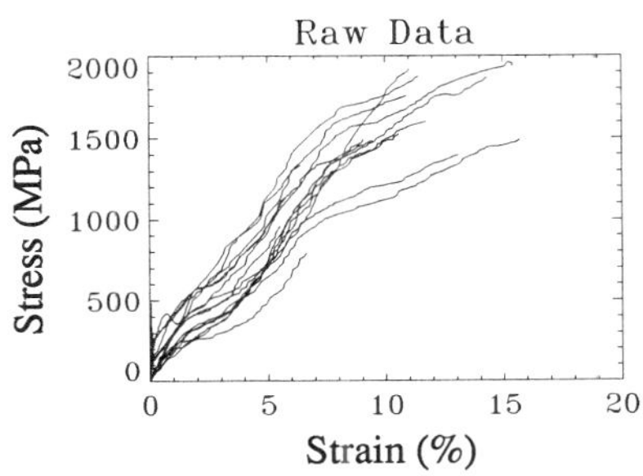

Figure 5. Typical variability in the response of single fibers after outriders have been removed.

Although a single fiber diameter has been assumed in preparing Figure 5, it was recognized that there was a distribution of fiber diameters among the different fibers, and also some variability in fiber diameter along a single sample length. It was assumed, however, that the stress-strain behavior of each fiber drawn from a single spider during a single silking session would be identical. Given this assumption, errors in the calculation of stress caused by assuming a single fiber diameter cancel one another when the response is averaged (provided the distribution about the mean fiber diameter is uniform).

Figure 6(a) illustrates the variability in the stress-strain response from spider to spider (curves labeled a, b and c in the figure), and from the same spider on different days (curves labeled b in the figure). Each sample was drawn from spiders at 6.1 cm/s draw rate; only the averaged data sets are illustrated. Apparently, the difference in stress-strain response observed within one lot of fiber is representative of the variability from lot to lot within the same draw speed.

Figure 6(b) illustrates the response of four different lots of fibers, each drawn at different rates. The variability from lot to lot at different draw speeds is only slightly more pronounced than the variability within a lot. The data of Figure 6(b) indicates an average initial modulus of about 22 GPa (190 gpd), an average secant modulus of 13 GPa (110 gpd), an average ultimate tensile strain of about 9%, and an average tensile

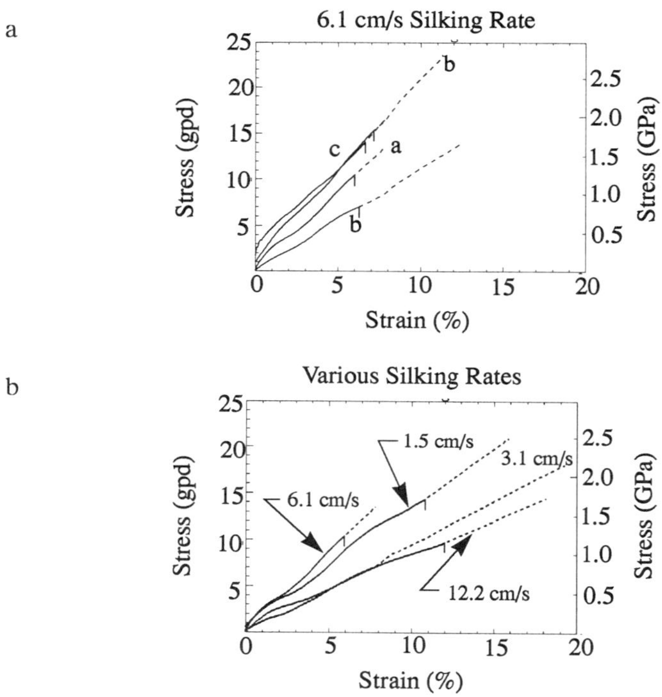

Figure 6(a). Typical variability in the response of single fibers drawn at the same rate. (b) Variability in the response of silk drawn at the different rates.

strength of 1.1 GPa (10 gpd). These average properties do not indicate the silk fibers have exceptional potential as a high strength fibers. However, two observations imply that the potential exists to significantly improve properties in a synthetically produced fiber. First, failure strength of the fiber is dominated by the distribution of gross defects in the fiber. Provided that the controlled-silked fiber is considered poor relative to a synthetically spun fiber, one would expect the properties of a recombinantely produced synthetically spun fiber to be similar to the best naturally produced fiber rather than similar to the mean. Secondly, the fiber diameter used to calculate stress can have a significant effect on the reported strength. Fibers were observed to contain variations in diameter along short distances in the fiber (relative to the gauge length used in tensile tests). The strength of such a variable cross section fiber is dominated by the minimum cross section present in the test sample; implying that the properties may be more nearly like those calculated based on a minimum fiber diameter.

Given the qualification that recombinantely engineered spider silk fibers should be at least as good as spider silk, and with the large variability in the mechanical behavior of controlled-silked spiders and the errors associated with determination of fiber cross sectional area notwithstanding, it can be tentatively stated that genetically engineered man-made fibers should have ultimate tensile strength of at least 2.9 GPa (24 gpd), modulus of at least 60 GPa (500), and ultimate tensile strain of at least 11% (i.e. at least as good as the properties of the curve illustrated in Figure 6(a), curve b).

Fracture Topography: The fibers in Figure 7(a), (b) and (c) were broken at a strain rate of 10%/s. The fracture surfaces are rough, which is thought to be indicative of a ductile fracture mode. No gross defects were observed in the immediate region of the fracture area, which is thought to be indicative that the fibers were uniformly strained to nearly the failure point at the time of failure.

The fibers in Figure 8(a), (b), and (c) were strained to break in less than 200 μs. Due to the dynamic wave propagation characteristics typical during impact, a determination of the actual strain rate is complicated. Nevertheless, it can be stated with certainty that the strain rate was considerably in excess of 50,000%/s. Many of the fibers fractured under these conditions showed the same type of fracture surfaces as those fractures under quasi-static conditions, as illustrated in Figure 8(a). Some of the fibers showed a convoluted lateral deformation as illustrated in Figure 8(b), and smooth fracture surfaces thought to be indicative of brittle fracture surfaces, as illustrated in Figure 8(c). The fiber illustrated in Figure 8(c) (a view of the fracture surface) shows evidence of the high lateral compressive stresses associated with transverse impact.

Some of the fibers showed a low degree of fibrillation as illustrated in Figure 9. It is believed that fibrils, which appeared relatively infrequently, originated from axial fractures along the lobes in fibers such as those illustrated in Figure 4. It is unclear when the convoluted surfaces illustrated in Figure 8(b) and the fibrillation illustrated in Figure 9 occurred. Post failure damage of Kevlar and poly(para phenylene benzobisoxazole) (PBO) fibers has been shown to be related to the limited compressive strengths of these fibers [22].

High Strain Rate Response: Quasi-static stress-strain tests are conducted on a time scale where appreciable creep or stress relaxation may occur. While a viscoelastic response of polymers is still possible even at very high strain rates, time effects are

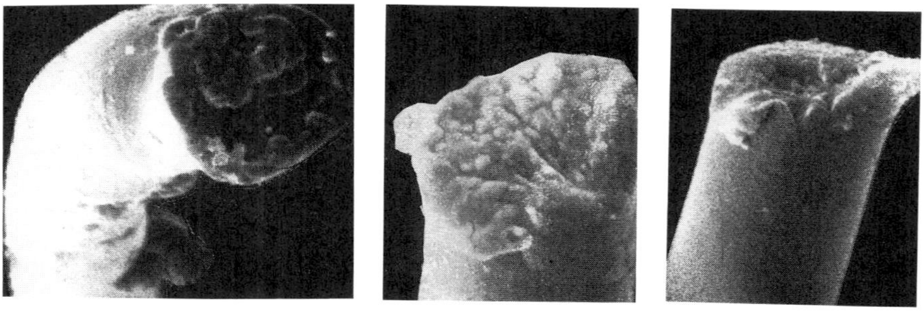

5 μm   5 μm   5 μm

a   b   c

Figure 7. Fracture topography (static fracture rate).

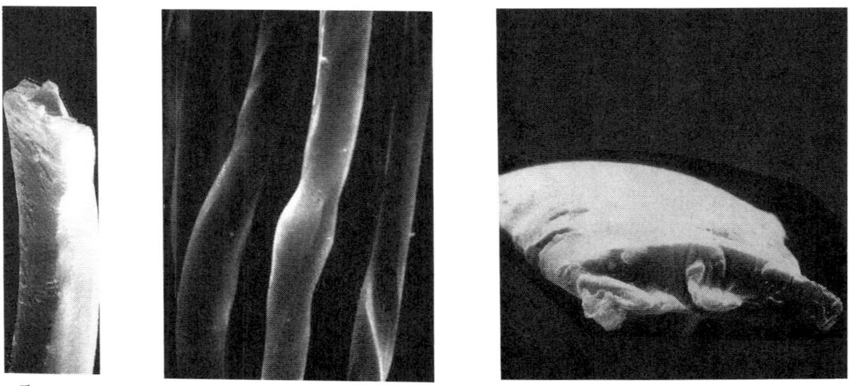

5 μm   10 μm   5 μm

a   b   c

Figure 8. Fracture topography (ballistic fracture rate).

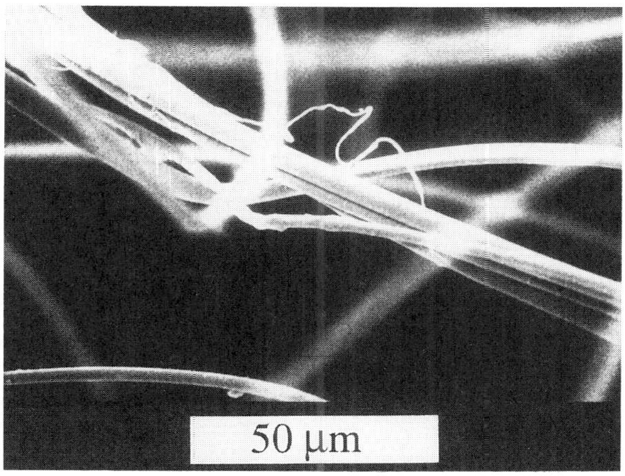

Figure 9. Fracture topography (ballistic fracture rate).

much less pronounced at these rates [21]. At very high strain rates, the macromolecular motions responsible for creep are less likely to occur.

The stress-strain response of silk fibers is non-linear at quasi-static strain rates, as illustrated in Figures 5 and 6; initially, the fiber is relatively stiff, at higher strain levels, the fiber becomes more compliant. The overall quantitative relationship between stress and strain in the fiber at quasi-static conditions could be approximated by, say, a polynomial type equation. A quantitative definition of the mechanical behavior at high strain rates could be determined within that framework by determining a set of coefficients for this polynomial equation using the single yarn impact method. However, such an approach would complicate the numerical procedure, and would add little to the present work. The immediate goal of the high strain rate work reported here is to determine if the high strain rate behavior of natural silk fibers is indicative of the behavior of fibers that would be good body armor materials. A constitutive equation that provides a good fit to the experimental data is therefore sufficient for the present study.

For simplicity, a standard linear solid constitutive model was selected for use with the single yarn impact model. The standard linear solid model allows for an increase in fiber compliance with increasing time after the application of load. Consequently, when used in the single yarn impact procedure, the standard linear solid model captures the essential feature of increased fiber compliance with increasing loads. Good agreement of model to experiment was obtained using this procedure, as illustrated in Figure 10.

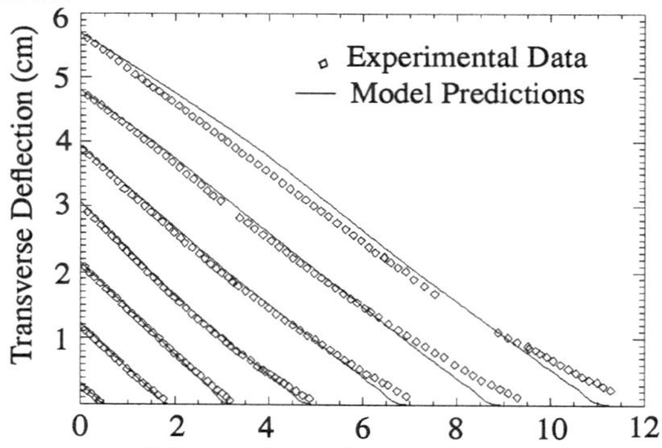

Figure 10. Comparison of results from numerical model with experimental yarn impact data (see Figure 2a).

The stress-strain response of spider silk yarns tested at high strain rates is quantitatively similar to the quasi-static response of silk fibers. Initially the high strain rate yarn modulus of 20.3 GPa is similar to the average initial quasi-static modulus of 22 GPa; the fully relaxed modulus of 10.1 GPa (corresponding to longer times after impact, and consequently higher strain levels) is similar to the quasi-static secant modulus of 13 GPa.

Dynamic Mechanical Analysis: Figure 11 shows typical dynamic mechanical properties of an 82 denier yarn sample. In Figure 11, curves a and a' are the initial run storage modulus and tan δ, respectively for the yarn. Temperature was held at 180 °C for 30 minutes, and the scan was re-run. Curves b and b' are the storage modulus and tan δ for the second run on the same yarn. Two transitions are apparent. The transition near -70 °C is considered to be associated with local motion within the amorphous regions and the transition observed near 210 °C can be explained as segmental motion of the main chain of the molecule. Young's modulus (E') at room temperature region is approximately 20 GPa and is maintained until about 170 °C followed by an abrupt drop. A slight increase in E' between 0 °C to 100 °C is caused by shrinking of the yarn due to the moisture evaporation process during the first run. AS illustrated in Figure 11, Young's modulus increased slightly and the magnitude of tan δ decreased in the heat treated sample. This result indicates that the macroscopic structural change from a rubber-like initial structure to a more perfectly aligned

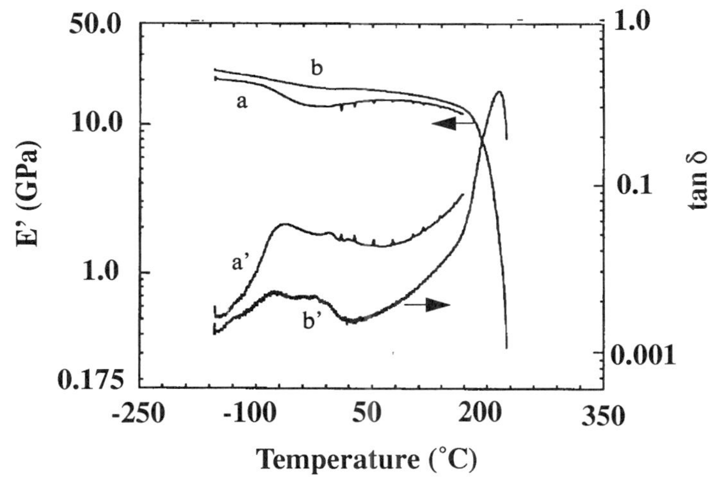

Figure 11. Typical scans of the dynamic mechanical properties of spider silk yarns.

structure through moisture evaporation under tension and heat. Microscopic changes through improved crystal perfection and/or increased crystallinity might be responsible for the increased stiffness upon heat treatment; alternately, the role of bound water may have infleunced the stiffness of the heat treated sample.

From differential scanning calorimetry and dynamic mechanical property studies of amorphous regenerated films cast from degummed Bombyx mori silk fibers Magoshi et al. [23,24] reported the glass transition temperature (Tg) of 175 °C when restricting their analysis to the temperature range between 20 and 220 °C. However, from the dielectric loss tangent measurement between -180 and 220 °C, they observed transition temperatures at -40 °C and at 175 °C (1 Hz test frequency). Furthermore, they concluded that the transition observed around 175 °C seems to be due to the segmental motion of main chains and crystallization of silk fibroin. This result suggests that the transition that appeared around 175 °C may be associated with motion within the crystals. From their studies on proteins, Morozov and Gevorkian [29] observed a transition temperature near -70 °C and identified it as a mechanical glass transition temperature. Based on our preliminary evaluations, thermal and dynamic mechanical behavior of our samples are somewhat different than previously reported similar materials. Detailed analyses are currently underway.

Thermal Mechanical Analysis: Figure 12 is a plot of creep compliance under the constant force of one newton as a function of time and temperature. Initial creep is strongly dependent on temperature as illustrated in Figure 13. After initial elongation, creep behavior does not seem to be significantly affected by test temperatures. This observation suggests that rubberlike cross-linked fiber structures become taut under initial load leaving small room for creep. Another possibility for the weak dependance of creep on testing temperature is due to the large amount of hydrogen bonding that is expected in the fiber; hydrogen bonds may act as cross-links, limiting long term creep.

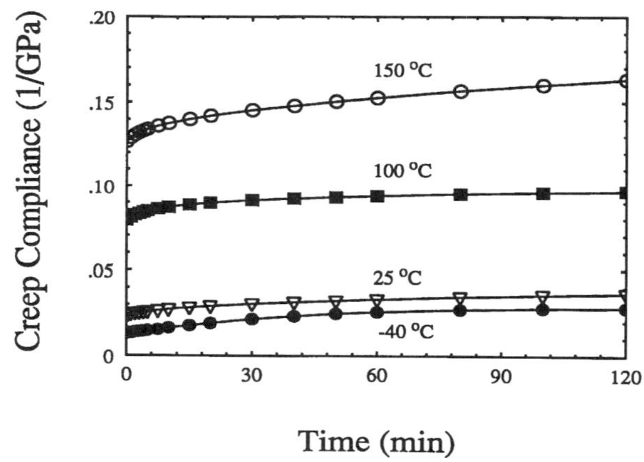

Figure 12. Creep compliance under constant load as a function of time and temperature.

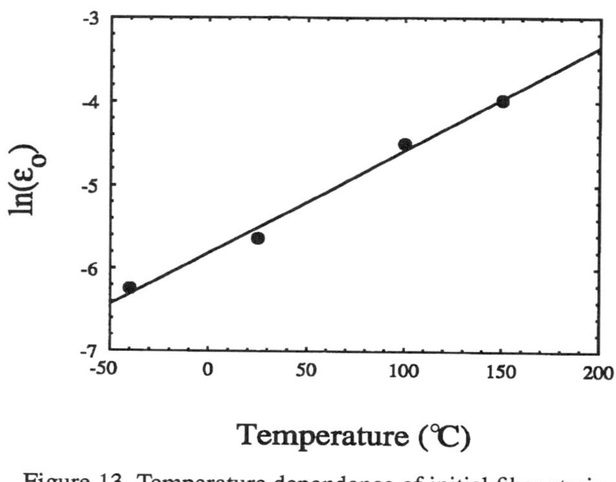

Figure 13. Temperature dependance of initial fiber strain.

Summary

SEM analysis of spider silk fibers indicates variations in the fiber's diameter, possibly due to variations in the production rate of silk coupled with variations in the geometry of the spider's spinneret. SEM analysis of fractured fibers indicates a ductile failure mode at low strain rates, occasionally accompanied by a low degree of fibrillation. SEM analysis also indicates ductile fracture modes with occasional brittle fracture modes in spider silk fibers strained to break under high strain rates.

The average tensile properties of spider silk fibers observed in this study are roughly comparable to other commercially available textile fibers. The average tensile properties of spider silk yarns are tabulated in Table II along with the properties of several comparable textile fibers and several high strength fibers.

Table II: Summary of Fiber Tensile Properties

Fiber	Modulus (GPa) / (gpd)	Strength (GPa) / (gpd)	Elongation at Break (%)
Major Ampulate Gland N. clavipes Spider Silk (Average)	22 / 190	1.1 / 10	9
high tenacity nylon	5 / 50	0.9 / 8.7	18
industrial polyester	15 / 125	0.9 / 7.5	13
Kevlar 29	62 / 520	2.8 / 22	3.5
Kevlar 49	124 / 1040	2.8 / 22	2.5
Spectra 1000	171 / 2000	3.0 / 35	2.7
High Modulus Graphite	393 / 2394	2.6 / 15.8	0.6

The high strain rate tensile of the spider silk determined from ballistic testing are similar to the average properties at quasi-static strain rates, as indicated in Table III. The model used for the determination of high strain rate properties predicted a relaxed modulus which is roughly comparable to the secant modulus of the quasi-static tests.

Table III: Comparison of High Strain Rate Properties to Quasi-Static Properties

Property	Quasi-Static	Ballistic
Initial Modulus (GPa) / (gpd)	22 / 190	20.3 / 175
Secant Modulus (GPa) / (gpd)	13 / 109	N/A[a]
Relaxed Modulus (GPa) / (gpd)	N/A[a]	10.1 / 85
Elongation at Break (%)	9	10

a. Not Applicable.

The best fiber properties that were observed in this study exhibited high elongation at break, high initial tensile modulus, and high strength. The problems of moisture absorption notwithstanding, if genetically engineered fibers could be produced with these properties, the unique combination of these properties should provide materials with outstanding performance characteristics in applications such as body armor. It is recognized, however, that the solution of the problems associated with moisture absorption may be ultimately resolved by producing increasingly more crystalline materials. These more crystalline materials are likely to be much more stiff and less extensible than the fibers observed in this study. Figure 14 illustrates a relationship between initial tensile modulus and elongation at break for high strength fibers [30]. The performance of these fibers will depend strongly on the changes in mechanical properties associated with the processing required to make the fiber more crystalline.

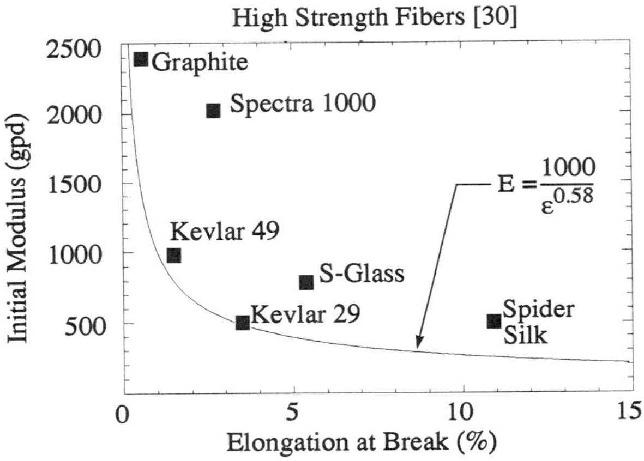

Figure 14. Relationship between initial tensile modulus and elongation at break for high strength fibers [30].

Spider silk fibers have been found to have a low thermal transition temperature (-70 °C), which is thought to be indicative of materials with good impact performance. Preliminary examination indicates that some degree of crystalline melting may have occurred at 210 °C, followed by the thermal decomposition of the fiber. The decomposition temperature of the fiber is low relative to commercially available high temperature fibers (see, for example [30]), but is roughly comparable to other fibers that are thought to have a high potential in impact applications. Compared, for example, to the melting temperature of polyvinylalcohol fibers of 257 °C, or to the melting temperature of ultra high molecular weight polyethylene of 147 °C, spider silk fibers have a reasonable temperature performance.

Literature Cited

1. Lombardi S.J.; Kaplan D.L., Poly Preprints, Div. Poly. Chem., Amer. Chem. Soc. **1990**, *31(1)*, 195-196
2. Lombardi S.J.; Kaplan D.L., J. Arachnol. **1990**, *18(3)*, 297-306

3. Lombardi S.J.; Kaplan D.L., Acta. Zool. Fennica **1990**, *190*
4. Lombardi S.J.; Fossey S.; Kaplan D.L., In Proc. Amer. Soc. Composites 5th Tech. Conf., Technomic Pub. Co., Lancaster PA. **1990**, 184-187
5. Kaplan D.L.; Lombardi S.J.; Muller W.S.; Fossey S. A., Macmillian Press **1991**
6. Fossey S.A.; Nemethy G.; Gibson K.D.; Scheraga H.A., Conformation of Fibrous Protiens, Academic Press, New York, **1990**
7. Kerkam K.; Viney C.; Kaplan D.; Lombardi S., Nature **1991**, *349(14)*, 596-598
8. Zemlin J.C., U.S. Army Natick Laboratories, Natick MA. Technical Report TR 69-29-CM (AD 684333) **1968**
9. Warwicker S.O., J. Molec. Biol., **1960**, *3*, 350
10. Gosline J.M.; DeMont M.E.,; Denny M.W., Endeavour, **1986**, *10*, 37-43
11. Gosline J.M.; Denny M.W.; and DeMont M.E., Nature, **1984**, *309(7)*, 551-552
12. Work R.W.; Emerson P.D., J. Arachnol., **1982**, *10*, 1-10
13. Magoshi J.; Magoshi Y.; Nakamura S., Polymer Comm., **1985**, *26(10)*, 309-311
14. Fenstermaker C.A.; Smith J.C., Appl. Poly. Symp. **1965**, *1*, 125.
15. Crout P.D.; Pilsworth M.N.; Hoge H.J., U.S. Army Natick Laboratories, Technical Report NATICK/69-18-PR. **1968**
16. Smith J.C.; McCracken F.L.; and Schiefer H.F., Text. Res. J., **1958**, *4*, 288.
17. Koeing H.A.; Davis N., Int J Solids Structures, **1968**, *4*, 643.
18. Lynch F.S., Army Materials and Mechanics Research Center, Watertown MA, Technical Report AMMRC TR70-16, **1970**
19. Roylance D.K., J. Appl. Mech., **1973**, 143-148
20. Encyclopedia of Polymer Science., J. Wiley and Sons, Inc., N.Y.,**1986**, *6*, 699
21. Work R.W., Text. Res. J., **1976**, *7*, 485-492
22. Allen S.R., J. Matr. Sci., **1987**, *22*, 853
23. Work R.W., Text. Res. J., **1977**, *10*, 650-662
24. Work R.W., J. Exper. Bio., **1985**, *118*, 379-404
25. Vollrath F.; Edmonds D.T., Letters to Nature, **1989**, *340*, 305-307
26. Work R.W.; Morosoff N., Text. Res. J., **1982**, *5*, 349-355
27. Magoshi J.; Magoshi Y.; Nakamura, S.; Kasai, N.; Kakudo, M., J. of Poly. Sci., Polymer Physics ed., **1977**, *15*, 1675
28. Magoshi,J.; Magoshi, Y., J. Poly. Sci. Polymer Physics Ed., **1975**, *13*, 1347
29. Morozov V. N.; Gevorkian, S. G., Biopolymers, **1985**, *24*, 1785
30. Yang H.H., Aromatic High Strength Fibers, J. Wiley and Sons, N.Y., **1989**

RECEIVED July 15, 1993

Chapter 22

Initial Degradative Changes Found in *Bombyx mori* Silk Fibroin

Mary A. Becker[1,3] and Noreen Tuross[2]

[1]Materials Science and Engineering Department, Johns Hopkins University, Baltimore, MD 21218
[2]Conservation Analytical Laboratory, Smithsonian Institution, Washington, DC 20560

Color evaluation, surface morphology, amino acid composition, solubility and molecular weight distributions are used to describe the initial degradative changes in *Bombyx mori* silk fibroin resulting primarily from irradiation with UV and visible light. The biochemical techniques are more sensitive detectors of the initial chemical alterations of the fibroin molecule than either color evaluation or morphology. The data suggest selective destruction of amino acids accompanied by peptide bond cleavage. The peptide bond cleavage mechanism is initially selective and becomes random with increased irradiation.

Silk fibers are predisposed to environmental degradation, making the deterioration mechanism of interest to both museums and protein biochemists. A comprehensive knowledge of the physico-chemical properties of the fibrous protein, fibroin, is necessary for the elucidation and confirmation of deterioration mechanism(s). Understanding the basic chemical reactions of the deterioration process is imperative before intervention to preserve and maintain historic and artistic collections is undertaken. The degradation of a large number of fragile silk objects in museum collections is the impetus for the physico-chemical characterization of the protein silk fibroin. The purpose of this study was threefold: (1) to describe the initial degradative chemical alterations found in silk fibroin caused by light exposure; (2) to investigate the chemical mechanism(s) involved in silk fibroin degradation in the solid state; and (3) to determine if biochemical techniques are more sensitive monitors of these alterations than the more commonly used methods of breaking strength, color difference, and morphology.

[3]Current address: Maurice Morton Institute of Polymer Science, The University of Akron, Akron, OH 44325–3909

0097–6156/94/0544–0252$06.00/0

Biochemical analysis of the fibroin requires separation of the β-pleated protein from the sericin coating and the insolubility of the extruded, or as-spun, protein hampers use of the many traditional protein purification techniques. In characterizing as-spun silk, it is important to accurately distinguish the fibroin from the sericin, and use the shortest extraction time to minimize any chemical attack on the fibroin itself. A characteristic feature of the fibroins in general is the high proportion of the smaller side group amino acids, glycine, alanine and serine, while sericin is approximately one third serine.

The average molecular weight of the fibroin molecule in the silk gland was $3.65 \times 10^5 \pm 0.20 \times 10^5$ in either 6 M guanidine·HCl or 8 M urea solution at pH 7.0 (*1*). The fibroin molecule from the silk gland consists of two subunits having molecular weights of 350 kd and 25 kd connected by a disulfide bond in a 1:1 ratio (*2*). These two fractions are associated with a heavy (H-) and a light (L-) chain DNA. A small portion of the fibroin H-chain DNA has been sequenced (*3, 4*), and the H-chain fibroin gene has been shown to consist of alternating crystalline and amorphous coding sequences (*5, 6*). The amino acid sequence derived from the fibroin L-chain gene has been recently determined and contains many of the amino acids with bulkier side chains (*7*).

The effect of light on silk has long been considered the primary cause of its degradation and the most noticeable alteration of aged silk fabric is yellowing. As deterioration of many organic materials progresses, chemical reactions leading to increased absorption characteristics result in yellowing of the material. Reflectance spectra of UV-irradiated silk fabrics have been used to qualitatively describe the color change associated with the photochemical degradation (*8*). Often the decrease in breaking strength, a macroscopic manifestation of degradation, is the criterion for assessing deterioration (*9-13*). Maximum degradation by oxidation, photolysis and photo-oxidation will initially occur at the exposed surface of a material (*14*). For example. irradiated stained nylon filaments examined by microscopy confirmed that the surface structure is altered first (*15*). Textural and physical damage associated with UV-irradiated poly(ethylene terephthalate) films was not found, however surface cracking appeared on the exposed side after flexural testing (*16*). Alterations of irradiated silk's structure have been observed using an electron microscope after various wet treatments (*17*). The increase in ammonia nitrogen has also been used to monitor the deterioration and has provided evidence for oxidative changes (*9*).

In a recent study, stress-strain measurements, thermal analysis, x-ray diffraction patterns and amino acid compositions were used to assess the effect of different monochromatic wavelengths of radiation on silk (*13*). The shorter wavelength radiation, 254 nm, was determined to be more detrimental to the silk than the 365 nm radiation (*13*). Unfortunately, only the amino acid compositions for the 254 nm irradiation were reported, but an overall weight loss as a function of irradiation time was noted (*13*). As irradiation time increased, a total weight lost was the result of the release of ammonia gas produced by photochemical reactions, although the specific method of calculating this weight loss was not reported (*13*). These results are used to support the previous finding of random chain scission in the amorphous regions which eventually extend into the crystalline domains.

The effects of polypeptide chain breakage on the building block chemistry in silk has not been well described. Studies aimed at the investigation of decomposition reactions in fossils and soils have assessed the thermal decomposition products of amino acids and peptides, including racemization. The primary pathways of singular amino acid decomposition were determined to be decarboxylation and deamination based on pyrolysis studies (*18*). Glycine and alanine are among the more thermally stable amino acids.

Extrapolating singular amino acid pyrolytic data to the decomposition rate of proteins however neglects the effects from interactions facilitated by the protein structure, reactions with absorbed molecules, mechanical action, oxidation and photochemical reactions (*18*). Since all reactions took place in dilute solutions, the possibility of recombination or polymerization was remote (*18*). The likelihood of a reaction being kinetically favored due to diffusion of volatile degradation products from the solid state or, in the case of fossil materials, loss by solution processes, was also not discussed. The most significant outcome of Vallentyne's decomposition studies was the proposed "selective" destruction of amino acids (*18*).

Deamination, decarboxylation and oxidation result from the interaction of radiant energy with amino acids and proteins. Photochemical reactions are difficult to quantify since the location of absorption within the molecule is often uncertain. Reaction rates are also affected by the absorption characteristics of side chains and the peptide bond as well as the diffusion rates of gases through the polymer. The absorption characteristics of the protein have a similar effect on the reaction rates of photolysis as steric hinderance does on hydrolysis reactions. Photolysis is the only known means of hydrolyzing amino acids to their respective hydroxy acids and ammonia, as this reaction has not been observed under physiological conditions where deamination is oxidative (*19*).

The absorption change in the near-UV and visible region can result from modifications of the side groups, molecular aggregation, radiation induced fragmentation and the formation of absorbing degradation products (*20*). Tyrosine, tryptophan and phenylalanine, identified by the phenol, indole and phenyl chromophores, are the principle absorbing groups in proteins and are therefore associated with color formation. DOPA is formed from tyrosine upon exposure to UV radiation (*21*) and also absorbs strongly at 280 nm (*24*). Studies of the effect of light on wool have shown that while the chromophoric groups in proteins are partially destroyed by light exposure, there was no explicit correlation between the amino acids destroyed and color change (*23*). These color results support the conclusion that although spectral reflectance data does quantify color change, it is insufficient to infer specificity in the amino acids residues involved.

In summary, color evaluation and morphology indicate that a chemical reaction has occurred, but do not quantify the extent of deterioration. The biochemical techniques offered the possibility of obtaining detailed quantitative chemical information on the state of preservation of a silk object with minimal sampling.

Materials

New, commercially degummed silk habutae (Testfabrics, Inc., style #609), a plain weave fabric having 126 ends/inch (37.6 denier) and 117 picks/inch (37.6 denier), was irradiated using specific wavelengths of light and exposure times. Removal of the sericin by commercial degumming was confirmed by the amino acids detected in material extracted (0.5 % of fabric weight) after boiling fabric in deionized water for one half-hour (Table I). Fabric swatches in sample holders with exposed areas of 42 mm x 50 mm were irradiated in an Atlas Ci-65 Weather-ometer for varying lengths of time up to a total dose of 1000 kilojoules/m^2 (kJ/m^2). The silk fabric was irradiated using a xenon arc light

Table I. Amino Acid Composition of *Bombyx mori* Silk Proteins (mol %)

Amino Acids	New Silk[a]	Hot Water Extract[a]	Cocoon Fibroin[b]	Posterior silk gland fibroin[d] large subunit	small subunit	Urea Extract 1000 kJ/m^2 Filter A	Filter B	Filter C
Glycine	43.7	36.7	43.7	49.4	10.0	24.5	33.8	45.4
Alanine	29.3	15.9	28.8	29.8	16.9	18.6	25.8	34.4
Serine	10.7	11.9	11.9	11.3	7.9	10.9	11.2	9.5
Tyrosine	5.5	1.5	5.1	4.6	3.4	2.7	1.8	-
Valine	2.4	4.2	2.2	2.0	7.4	5.2	3.7	5.8
Aspartic Acid*	1.8	10.9	1.3	0.65	15.4	14.6	8.5	3.4
Glutamic Acid*	1.2	5.9	1.0	0.70	8.4	10.5	6.8	1.5
Threonine	0.9	3.5	0.90	0.45	2.8	4.3	2.7	-
Phenylalanine	0.9	2.8	0.61	0.39	2.7	2.1	1.2	-
Isoleucine	0.7	3.6	0.71	0.14	7.3	4.9	2.8	-
Leucine	0.6	2.7	0.51	0.09	7.2	1.8	1.5	-
Arginine	0.5	n.d.	0.46	0.18	3.8	n.d.	n.d.	n.d.
Lysine	0.3	n.d.	0.31	0.06	1.5	n.d.	n.d.	n.d.

* Both Asparagine and Glutamine are converted to their respective acid form upon hydrolysis.
[a] Silk habutae, Testfabrics, Inc. #609 This study. The unexposed silk had a 0.5 % weight loss after the hot water extraction.
[b] Lucas, F., Shaw, J.T.B., Smith, S.G., *Biochem. J.*, **1960,** *83*, 164.
[c] Komatsu, *Bull. seric. Exp. Stn Japan*, **1975,** *26,* 170.
[d] Shimura et al. *J. seric. Sci. Tokyo* **1982,** *51,* 20.
n.d.= not determined

source with three different combinations of glass filters; borosilicate-soda lime, borosilicate-borosilicate, and quartz-quartz with lower wavelength cut-offs at 300, 285, and 230 nm respectively. These filter combinations are referred to as Filter A (300 nm), Filter B (285 nm), and Filter C (230 nm) and correspond to exposures of electromagnetic radiation spectra simulating sunlight-through-glass, sunlight and nearly direct output from the xenon arc light source, respectively (*24*).

The total energy, dose, in kilojoules per square meter (kJ/m^2), is dependent on the power, or irradiance (Watt/m^2) of the incident radiation and the wavelength of light at which the energy is measured. The radiant energy was monitored at 420 nm for Filter A at an irradiance of 0.76 W/m^2. Radiant energy was monitored at 340 nm and an irradiance of 0.40 W/m^2 for both Filter B and Filter C. The irradiance of the sun at sea level is 151 W/m^2 at 340 nm and 733 W/m^2 at 420 nm with an overall irradiation of 740 W/m^2 when taking into account variations in energy intensity, atmospheric scattering and absorption (*25*). Thus, the Atlas Weather-ometer 1000 kJ/m^2 Filter B exposure monitored at 340 nm is equivalent to 1.8 hours of sunlight at sea level based on the irradiation data of the sun at sea level at 340 nm. A dose of 1000 kJ/m^2 at an illumination level of 5 footcandles is approximately equal to 494 days using Filter A, or 471 days for Filter B, or 383 days for Filter C (*26*).

Alternating light and dark exposures were used in order to simulate the cyclical radiative environment that occurs during the lifetime of museum objects. The light/dark cycling apparatus allowed the fluctuation of relative humidity and ambient air temperature within the chamber. The average temperature experienced by the test silks for all specimens was 21 °C and the relative humidity varied between 55-63 %.

Methods

Color Evaluation. Color was assessed using spectral reflectance curves obtained on a Hunterlab Ultrascan spectrophotometer and compared to a white background tile used as a standard. Illuminant D_{65}, recommended by the Commission International de ℓ'Éclairage (CIE), with a spectral distribution similar to natural daylight over the range 300-830 nm was used for all measurements (*27*). Each spectral reflectance curve represents the average of five reflectance readings at intervals of 5 nm from 375 nm to 750 nm.

The silk samples were analyzed with the 10° 1964 CIE Supplementary Standard Observer (large field of vision); the specular reflectance was not measured (*27*). All measurements were taken with a sample thickness of 5 layers of fabric to avoid background effects (*28*).

Morphology. A Jeol JXA-840A scanning electron microscope (SEM) was used to examine the surface topography of silk fabric samples at high magnification. Samples were mounted onto aluminum stubs using double-sided tape and coated with a gold-palladium (Ag-Pd) thin film before SEM examination. Only minimal charging of the specimen occurred when an accelerating voltage of 15 KeV was used for the photomicrography of the silk textiles. Backscattered electron images were recorded as micrographs using Polaroid 55 P/N type film. Micrographs were made of the silk samples, before and after a denaturing treatment.

Amino Acid Analysis. Microgram samples were hydrolyzed in excess 6N HCl under standard conditions, 110°C for 22 hours. After hydrolysis, samples were dried in a vacuum evaporator at 50 °C and stored dry at -20 °C until just prior to chromatographic separation. For amino acid analysis, samples were diluted with pH 2 buffer. All amino acid compositions are based on daily calibrations to a standard solution (Sigma AA-S-18) containing 100 or 125 picomoles of each amino acid reported.

Amino acid compositions and ammonia content of the irradiated silk fabrics were determined by ion-exchange chromatography (*29, 30*). Primary amino acids were separated by ion-exchange high pressure liquid chromatography (IE-HPLC) and detected by post-column derivatization with *o*-phthaldialdehyde/2-mercaptoethanol to form fluorescent substituted isoindole products.

Two runs were required in order to adequately separate ammonia and the basic amino acids from the interfering buffer peaks (*31*). Histidine was not resolved in either analysis and is not reported. Asparagine and glutamine residues are deamidated to aspartic and glutamic acid upon acid hydrolysis (*32*) and therefore reported as the sum of both the acid and amide (Asx and Glx). Proline, a secondary amine, does not react with OPA, the fluorescing derivatization reagent. Tryptophan is known to be essentially completely destroyed during hydrolysis, serine, threonine and tyrosine can also show losses to some extent (*33*). No corrections were made for possible losses resulting from hydrolysis.

Solubility. Silk samples ranging in weight from 65 to 74 mg were soaked in 1 mL of 7 M urea, 0.05 Tris (pH 8) solution. Urea was chosen as the solvent for this study because of its known denaturing properties (*34*) which allow protein molecules to assume a random-coil conformation. The samples were extracted for one week with constant rocking and occasional vortex mixing to facilitate dispersion of the solution throughout the fabric. After one week, the urea-Tris solution was removed, and the remaining fabric was rinsed a minimum of five times with distilled water. The fabric samples were lyophilized to dryness. Percent weight loss was calculated using the comparison of pre-extraction to post-extraction dry weights.

Molecular Weight Distribution Profiles. Protein fragments extracted from the silk fabric samples under denaturing conditions (7 M Urea, 0.05 M Tris, pH~8), as described above in the solubility section, were analyzed by gel permeation chromatography to determine their molecular weight range. A Perkin-Elmer fast pressure liquid chromatography (FPLC) system equipped with a Superose 6 column (5,000-$5X10^6$ MW) was used. A solution of 7 M urea, 0.05 M Tris (pH 8) was used as the eluting mobile phase. Eluting protein fragments were detected using absorption intensity at 280 nm.

Results and Discussion

The spectral reflectance curves of the white silk fabric were altered by successive exposures to light for increasing amounts of incident ultra-violet radiation although the fine trends in the reflectance curves varied with each filter. All exposure conditions resulted in an increase in the absorbance of wavelengths in the lower region of the visible spectrum (Figure 1). The increase in the absorbance of the spectral reflectance curves for the Filter A exposure condition indicated a gradual, energy-dependent, color-formation mechanism. The color formation observed in the Filter B fabrics appeared to require a specific amount of energy and thereafter was unaffected by increasing exposure. The fact that exposures to Filters A and B produced the same final color of the silk samples indicated that the color-formation mechanism was initially energy dependent, but the reaction went to apparent

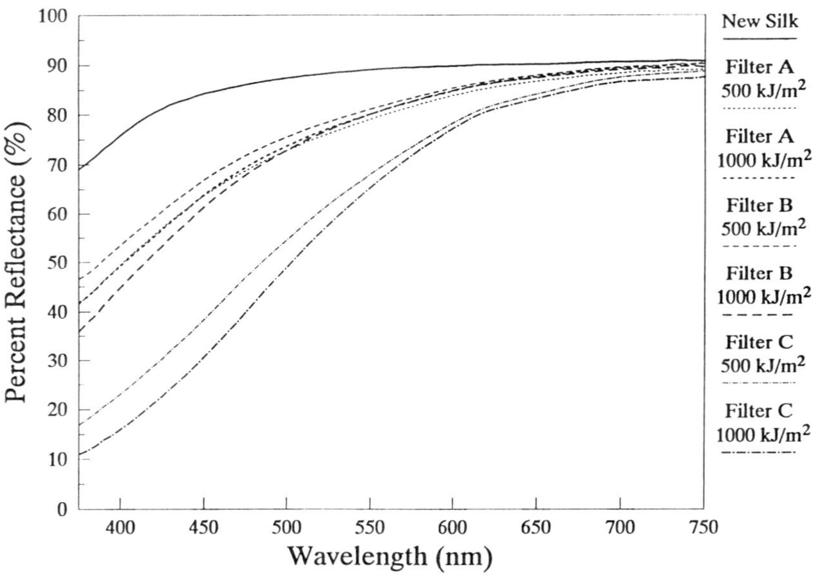

Figure 1. Spectral reflectance curves of new silk fabric irradiated for doses of 500 and 1000 kJ/m^2 for Filters A, B and C (Adapted from ref. 37).

completion under these experimental conditions. The spectral reflectance curves for the Filter C exposure provided evidence of additional color-forming mechanism(s) that appear to be strongly energy dependent.

The pattern of increasing absorbance in the 375-550 nm region suggest a threshold effect, an energy barrier which must be exceeded, in order for yellowing to continue (Figure 1). The two possibilities for a threshold effect in the yellowing reaction(s) are a wavelength-dependent reaction and an overall energy dependence reaction. In a totally wavelength dependent effect the formation of energy absorbing species with Filter C exposures could be a entirely different color-forming reaction precipitated by the increased energy and quantum efficiency of incident UV radiation compared to the milder exposures. Alternatively, the apparent UV-dependence of color forming species could indicate a total dose-dependent induction period which is accentuated by the high energy content of incident UV radiation of Filter C, but is not necessarily wavelength-dependent. The first explanation seems more realistic since the quantum yields of UV radiation are higher than for lower energy wavelengths and energy dissipating mechanisms, such as fluorescence, phosphorescence and non-radiative transitions, would decrease with the rate of total dose-dependent reactions. This study was not designed to allow explicit discrimination between wavelength-dependent mechanisms and dose-dependent reactions, but provides a basis for future studies.

Silk fibers are characteristically smooth and have a uniform triangular-shaped cross-section. Occasionally fibrils emerge from the smooth surface of new silk habutae, possibly caused by loosening of ordered regions during the commercial degumming process (Figure 2a). There was no visual indication of any material loss from the surface of new, Filter A and B silk samples after extraction of the soluble fibroin degradation products for one week using 7 M urea, 0.05 M Tris (pH 8) as the denaturant.

Obvious alterations in the morphological appearance occur in the 1000 kJ/m^2 Filter C silk sample (Figure 2b). The fibers appear to fuse together and thin in the regions contiguous with the cross-over-points of the weave and considerable material appears to have eroded away after denaturing treatment (Figure 2c). This "melted" appearance does not occur in the Filter A or Filter B samples which were exposed to less UV radiation, even for the same dose of incident energy.

In general, there appears to be an induction period before morphological changes become apparent. Only the Filter C samples exposed to UV-containing radiation show this "melted" appearance, these reactions which might not otherwise be present upon exposure to lower energy radiation are induced by the high energy content of the incident UV-radiation.

The amino acid composition (mol %) of the unexposed silk habutae used in this study was nearly identical to reported *Bombyx mori* fibroin composition (Table I). An overall loss of hydrolyzable amino acids per 1 μg of dry weight of fibroin was calculated for the hydrolysates of irradiated silks as compared to unexposed silk (Figure 3). There was essentially no change in recoverable amino acids for the 1000 kJ/m^2 Filter A sample, the percentage of amino acids recovered was within the range of experimental error. A 76 % loss of recoverable amino acids was determined for the 1000 kJ/m^2 Filter B sample and a substantial loss of 66 % was determined for the 1000 kJ/m^2 Filter C sample (Figure 3).

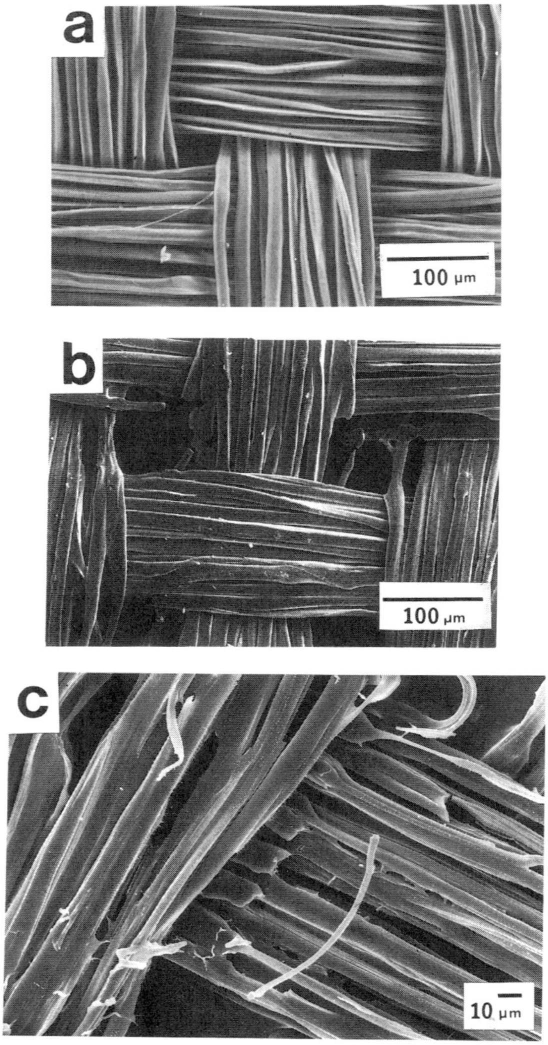

Figure 2. New silk fabric before artificial aging treatment (a), after exposure to 1000 kJ/m^2 using Filter C (b) and a detail of the 1000 kJ/m^2 Filter C sample (c).

There was a notable decrease in the amount of tyrosine recovered for both increasing UV content and dose (Figure 4a). While silk exposed using Filters A and B exhibited linear decreases in recoverable tyrosine, Filter C samples showed a logarithmic loss of tyrosine with increased exposure. This apparent change in the rate of loss supports evidence for threshold effects and participation of other mechanisms with both increasing UV content and increasing exposure. A pronounced increase in the amount of hydrolyzable ammonia was also found in the irradiated silk fabrics as both dose and UV content were increased (Figure 4b).

Linear regression analysis was used to determine the dose dependent rate of loss of individual amino acids for Filter A, B, and C silk samples (Table II). Data used in these determinations were the percent recovered, i.e. the picomoles recovered normalized to unexposed silk. Overall, the rate of loss of each amino acid was slowest for Filter A and fastest for Filter C. Glycine and alanine, together accounted for approximately 70 % of the overall amino acid composition and each amino acid exhibited a rate of loss based on percent recovery ranging from $5x10^{-3}$ mol %-m^2/kJ for Filter A to $20.5x10^{-3}$ mol %-m^2/kJ for Filter C. The largest percentage loss was observed for tyrosine, the rates of loss range from $3x10^{-2}$ mol %-m^2/kJ for Filter A to $6x10^{-2}$ mol %-m^2/kJ for Filter C, and this corresponded to losses ranging from 188-448 nmol-m^2/kJ per 1 μg dry weight of silk.

Table II. The Rate of Loss of each Amino Acid as a Function of Exposure (x 10^{-2} mol%-m^2/kJ)

Amino Acid	Filter A	Filter B	Filter C
Aspartic Acid	1.4 ± 1.1	3.8 ± 0.5	3.2 ± 0.4
Threonine	1.5 ± 0.9	3.9 ± 1.1	3.7 ± 0.9
Serine	0.9 ± 1.0	3.0 ± 0.6	3.3 ± 0.5
Glutamic Acid	0.8 ± 1.3	2.8 ± 0.3	2.6 ± 0.3
Glycine	0.6 ± 1.1	2.0 ± 0.6	2.2 ± 0.5
Alanine	0.5 ± 1.0	2.0 ± 0.7	1.9 ± 0.6
Valine[b]	0.9 ± 1.0	2.8 ± 1.6	1.9 ± 1.3
Isoleucine	0.2 ± 1.5	2.0 ± 1.5	1.9 ± 1.0
Leucine[b]	+1.1 ± 1.8	0.6 ± 2.0	+0.2 ± 1.6
Tyrosine[a]	2.6 ± 0.8	6.2 ± 0.5	6.3 ± 0.5
Phenylalanine[b]	1.1 ± 1.5	0.7 ± 4.1	1.6 ± 3.4
Lysine[a]	1.6 ± 0.4	2.7 ± 0.4	4.6 ± 0.3
Arginine	+0.3 ± 0.6	2.3 ± 0.3	3.9 ± 0.3
Ammonia[a]	+10.9 ± 1.2	+27.7 ± 1.1	+38.7 ± 0.8

[a] show significant loss or gain ($r^2 > 0.50$) for all artificial aging conditions
[b] no significant loss ($r^2 < 0.50$) for all artificial aging conditions
Adapted from ref. 37.

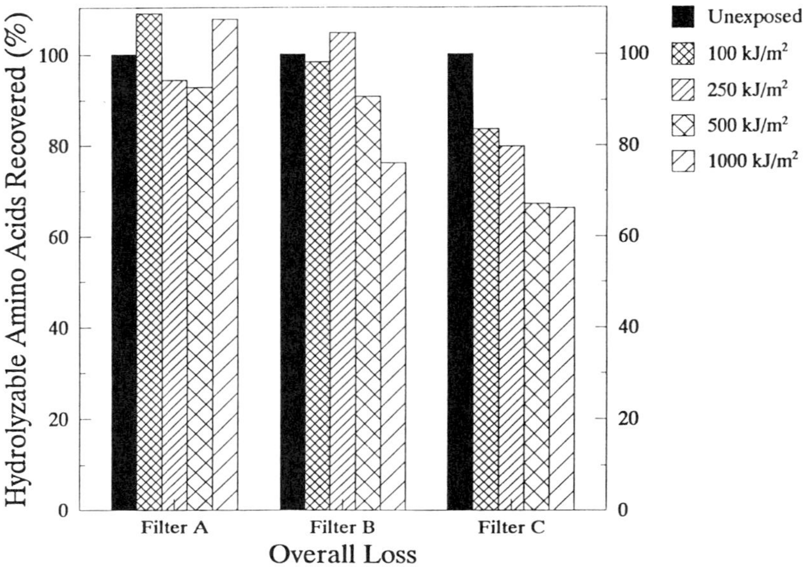

Figure 3. The percentage of hydrolyzable amino acids recovered from 1 µg dry weight of artificially aged silk normalized to unexposed silk as a function of dose for the three filters (Adapted from ref. 37).

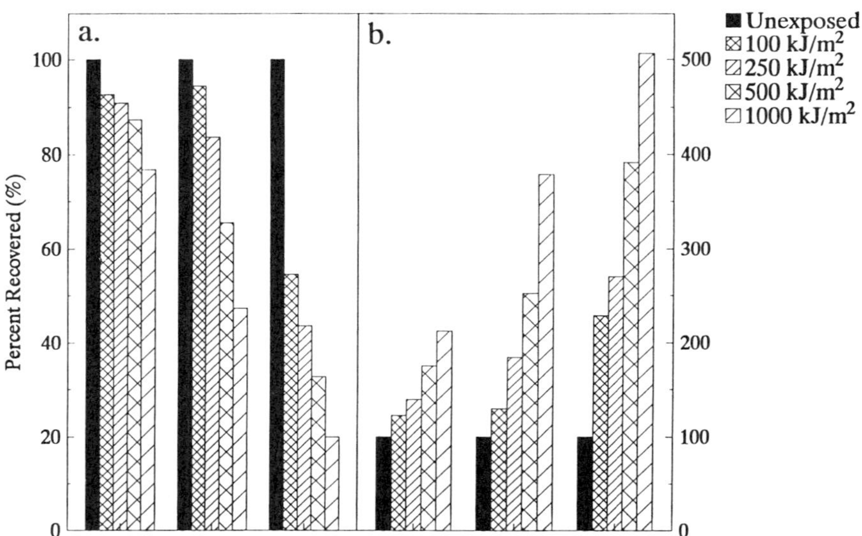

Figure 4. The percentage of tyrosine (a) and hydrolyzable ammonia (b) recovered for the 500 and 1000 kJ/m² exposures to all three filters normalized to unexposed silk (Adapted from ref. 37).

Tyrosine disappeared most rapidly under all three filter conditions. For the more aggressive Filter B and C exposures, the amino acids exhibited statistically different rates of change. For Filter C, the rate of loss of tyrosine was 6.3×10^{-2} mol%-m^2/kJ and appeared to take on logarithmic characteristics (Figure 5). Although the amino acids appeared in a slightly different order for Filter C compared to Filter B, the acidic and hydroxylated amino acids were lost at faster rates than the aliphatic amino acids. Glycine and alanine remained the more stable amino acids in Filter C exposures with rates of loss of 2.2×10^{-2} mol %-m^2/kJ and 1.9×10^{-2} mol %-m^2/kJ, respectively. Lysine, arginine, threonine, serine, aspartic acid, and glutamic acid have rates of loss intermediate to tyrosine and glycine in both Filter B and Filter C silk samples (Figure 5). The rate of loss between Filters A, B, and C was statistically equivalent for three amino acids; isoleucine, leucine and phenylalanine. That is, the rate of loss was independent of the UV content of incident radiation, and proportional to incident energy. All the other amino acids evaluated showed a dependence on the UV content of the incident radiation in their rate of degradation.

In summary the aliphatic amino acids; valine, leucine, isoleucine, glycine and alanine were relatively more stable under the treatment conditions, although all these amino acids showed substantial loss with the most extreme treatments. Valine, leucine and isoleucine are known to be more stable to hydrolysis due to steric considerations *(35)*. In addition to the substantial loss in tyrosine content, lysine, arginine, threonine, aspartic acid, serine and glutamic acid were also lost at appreciable rates. Thus the amino acids with hydroxylated and ionizable side chains were degraded at a faster rate than the aliphatic amino acids.

The solubility of fibroin was defined as the percent weight loss after extraction under denaturing conditions (7 M Urea, 0.05 M Tris, pH 8). The irradiated silk samples showed an increase in solubility as the dose increased and as more UV light was included in the incident radiation (Figure 6). New unexposed silk habutae had a 3 % weight lost after extraction under denaturing conditions (7 M Urea, 0.05 M Tris, pH 8). New silk habutae boiled in deionized water for one half-hour only lost 0.5 % of its initial dry weight and the hot water extract had an amino acid composition similar to the L-chain region, not sericin, (Table I). The Filter A samples ranged in solubility from 1 % for the 100 kJ/m^2 dose to 5 % for the 1000 kJ/m^2 dose, Figure 6. The Filter B fabric lost 18 % of its initial dry weight for the 1000 kJ/m^2 dose and 38 % under the more harsh Filter C conditions, Figure 6.

The amino acid compositions of the urea-extracts containing the fibroin degradation products were determined by post-column OPA IE-HPLC (see Amino Acid Section) after removing the salt solution chromatographically. The compositions of the extracted degradation products were compared to those of the L-chain (associated with the amorphous region) and fibroin (the L- and H-chain) (Figure 7). The amino acid composition of the L-chain *(7)*, has a higher percentage of aspartic acid, glutamic acid, valine, isoleucine and leucine than the large subunit protein from the silk gland *(36)*.

Overall, the amino acid composition of the extracted degradation products was heavily weighted with the amino acids found in L-chain segment and the amorphous regions associated with the H-chain of the fibroin molecule *(5, 6)* in the less severe treatment. The 1000 kJ/m^2 Filter B sample showed increasing amounts of glycine and alanine and decreasing percentages of aspartic acid, threonine and glutamic acid

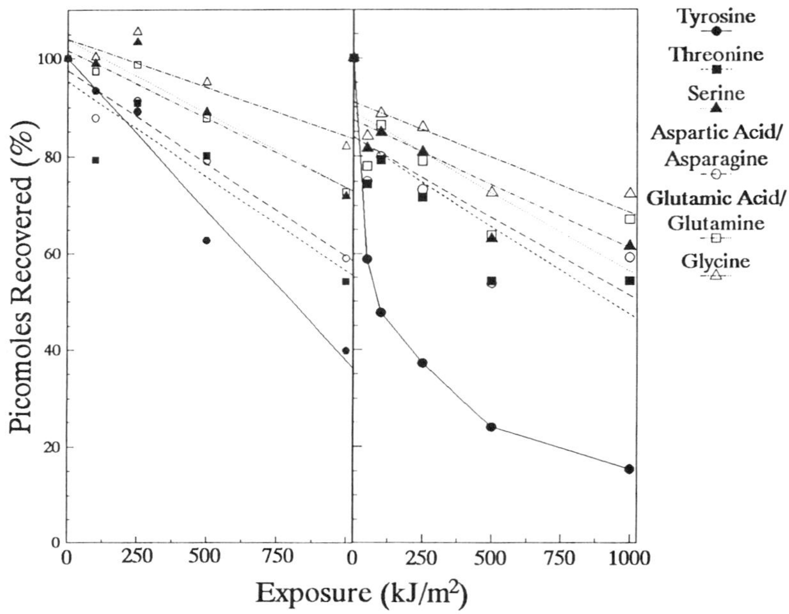

Figure 5. The rate of loss of six amino acids with increasing exposure for Filter B and Filter C (Adapted for ref. 37).

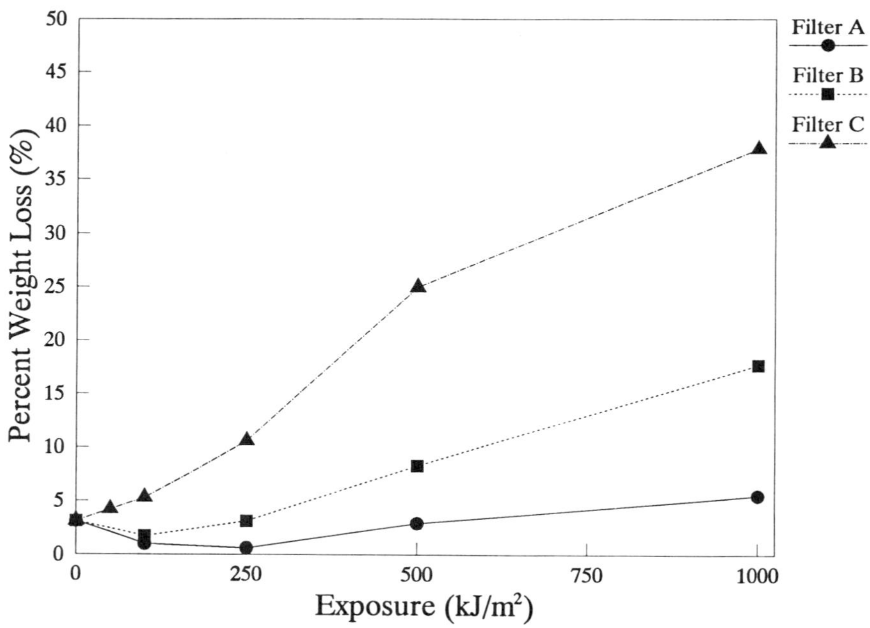

Figure 6. Solubility profiles as a function of increasing dose of Filters A, B and C (Adapted from ref. 37).

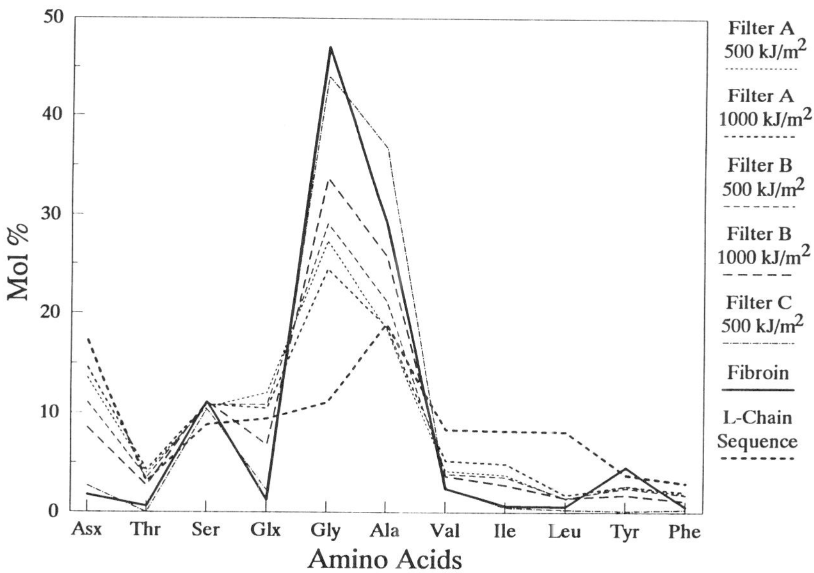

Figure 7. The amino acid composition of the extracted degradation products from artificially aged silk fabric for Filters A, B and C compared to fibroin and the L-chain sequence (Adapted from ref. 37).

in the compositions as the UV content increased (Figure 7). The amino acid composition of the material extracted from 1000 kJ/m^2 Filter B sample was intermediate to the L-chain (or amorphous) and fibroin, with more of the amino acids representative of fibroin than found in the milder Filter A exposure. The 1000 kJ/m^2 Filter C sample approached the composition profile nearly identical to fibroin (Figure 7). While the percentage of glycine and alanine extracted under denaturing conditions increased with both increasing UV in the incident radiation and increasing dose, the percentage of serine remained nearly constant for all exposures. Thus, for the three most abundant amino acids in fibroin, glycine, alanine and serine, only serine appeared evenly distributed throughout both the crystalline and amorphous regions.

In general, under mild artificial aging conditions, the amorphous regions were degraded first and then as the exposure and dose increase, the whole fibroin matrix, crystalline and amorphous regions were affected. Thus providing further evidence for threshold effects in the deterioration process, in which a selective deterioration mechanism is followed by more random peptide chain scission with increasing UV irradiation.

Gel permeation chromatography was used to characterize the molecular weight distribution profiles of the proteinaceous material extracted from fabrics under denaturing conditions. Molecular weight distribution profiles, acquired for the silk samples irradiated from 100 to 700 kJ/m^2 using Filter A, showed a noticeable change in the distribution pattern with increasing exposure (Figure 8). The degradation products included higher molecular weights as the dose increased. The succession of molecular weight products recovered implied the degradation process was energy dependent.

Conclusions

The initial biochemical degradative changes in *Bombyx mori* silk fibroin are (1) the appreciable loss of tyrosine, (2) the loss of acidic and ionizable amino acids, (3) the increase in the amount of hydrolyzable ammonia recovered in the acid hydrolyzate, and (4) cleavage of the peptide bond as indicated by increased solubility and broader molecular weight distribution profiles of extractable proteinaceous material for the artificially-aged conditions studied.

The overall pattern of degradation elucidated by both amino acid analysis and extracted degradation products of irradiated silk is a pattern of destruction of amino acids and cleavage along the length of the silk molecule. The deterioration of silk fibroin is dominated by photolytic degradation of the ionizable amino acid constituents and peptide bond cleavage, with supporting evidence for acidic cleavage similar to the mechanisms proposed for thermal processes (*18*). Thus, an initially selective deterioration process becomes a more random chain breakage with both increasing light exposure and increasing UV content.

Amino acid compositions and solubility are more sensitive indicators of chemical alterations taking place in the fibroin than the more traditional textile testing methods. They detect immediate changes where there is a macroscopic induction time before these changes are detected by color evaluation and morphology.

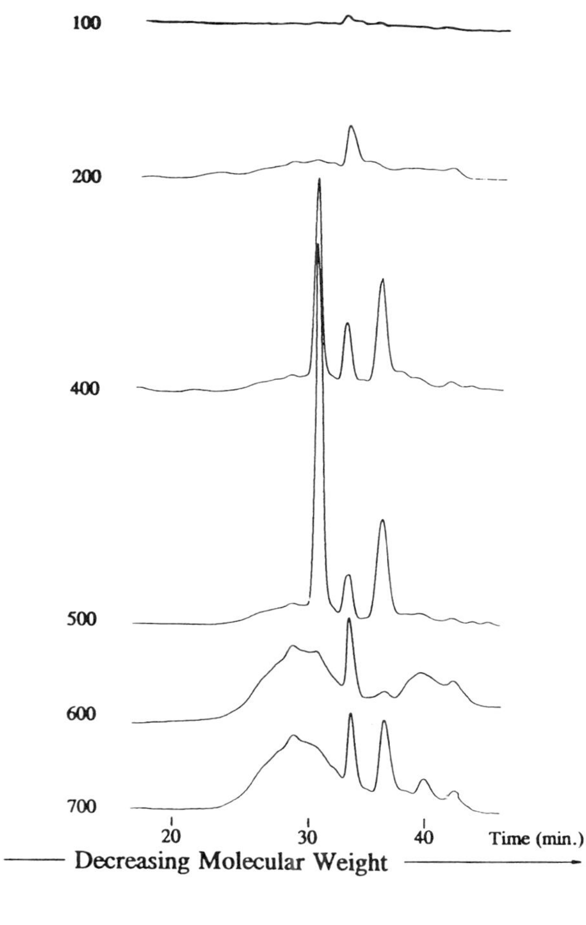

Figure 8. A series of consecutive molecular weight distribution profiles of the degradation products extracted under denaturing conditions for silk artificially aged by exposure to simulated sunlight through glass (Adapted from ref. 37).

Acknowledgments

Funding for this work was provided by the Conservation Analytical Laboratory, Smithsonian Institution, Washington DC.

Literature Cited

1. Sasaki, T,; Noda, H. *Biochim. Biophys. Acta* **1973**, *310*, 76-90.
2. Sasaki, T,; Noda, H. Biochim. Biophys. Acta **1973**, *310*, 91-103.
3. Tsujimoto, Y.; Suzuki, Y. *Cell,* **1979**, *16,* 425-436.
4. Tsujimoto, Y.; Suzuki, Y. *Cell,* **1979**, *18,* 591-600.
5. Gage, L.P.; Manning, R.F. *J. Biol. Chem.*, **1980**, *255(19),* 9444-9450.
6. Manning, R.F.; Gage, L.P. *J. Biol. Chem.*, **1980**, *255(19),* 9451-9457.
7. Yamaguchi, K.; Kikuchi, Y.; Takagi, T; Kikuchi, A.; Oyama, F.; Shimura, K.; Mizuno, S. *J. Mol. Biol. 1989,* 210, 127-139.
8. Watanabe, A.; Tagawa, M.; Osawa, R. *Kaseigaku zasshi (J. Home Econ.* Japan) **1979**, *30(8)*, pp 56-60.
9. Harris, M. *American Dyestuff Reporter* **1934**, *23*, 403-405.
10. Rutherford, H.A.; Harris, M. *American Dyestuff Reporter* **1941**, *30*, 345-346, 363-364.
11. Hirabayashi, K. *Kobunkazai no Kagaku (Scientific Papers on Japanese Antiques and Art Crafts)* **1981**, *no. 26*, 24-34.
12. Hirabayashi, K.; Takei, M.; Kondo, M. *Archaeology and Natural Science* **1985/84**, *no. 17*, 61-71 (in Japanese).
13. Hirabayashi, K.; Yanagi, Y.; Kawakami, S.; Okuyama, K.; Hu, W. *J. Seric. Sci. Jpn.* **1987**, *56(1)*, 18-22.
14. Jellinek, H.H.G. In *Physicochemical Aspects of Polymer Surfaces, Vol. 1*, Mittal, K.L., Ed.; Plenum Press: New York, 1983; pp 255-281.
15. Kato, K. *Text. Res. J.*, **1962**, *32*, 181-184.
16. Blais, P.; Day, M.; Wiles, D.M. *J. Appl. Poly. Sci.* **1973**, *17*, 1895-1907.
17. Kuwahara, A. *Sen-i Gakkaishi* **1968**, *24(7)*, 330-335.
18. Vallentyne, J.R. *Geochim. et Cosmochim. Acta*, **1964**, *28,* 157-188.
19. Weizmann, C.; Bergmann, E.; Hirshberg, Y. *J. Amer. Chem. Soc.* **1936,** *58* 1675-1678.
20. McLaren, A.D.; Shugar, D. *Photochemistry of Proteins and Nucleic Acids*; Pergamon Press: New York, 1964.
21. Arnow, L.E. *J. Biol. Chem.* **1937**, *120*, 151-153.
22. *Merck Index, The, 9th Ed.;* Merck & Co.: New Jersey, 1976; no. 5310.
23. Inglis, A.S.; Lennox, F.G. *Text. Res. J.*, **1963**, *33*, pp 431-435.
24. Atlas Ci-65 Weather-ometer® Instruction Manual; Atlas Electric Devices: Chicago, IL; 1986; p I-17.
25. Moon, P. *J. Franklin Inst.* **1940,** *230,* 583-617.
26. Thomson, G. *The Museum Environment, 2nd Ed.* Butterworths: London, 1986; 293 pp.
27. Billmeyer, Jr., F.W.; Saltzman, M. *Principles of Color Technology;* John Wiley & Sons: New York, 1981; pp 37-44.

28. *AATCC Technical Manual, Vol 61*; American Association of Textile Chemists and Colorists Research Triangle Park, NC; 1986; p 261-266.
29. Benson J.R.; Hare P.E. *Proc. Nat. Acad. Sci. USA* **1975**, *72(2)*, 619-622.
30. Stathoplos, L. Ph.D. Dissertation, University of Rhode Island, Kingston, RI, 1989.
31. Hare, P.E. *Methods in Enzymology, Vol. XLVII, Part E*; Hirs, C.H.M.; Timasheff, S.N., Eds.; Academic Press: New York, 1977; pp 3-18.
32. Robinson, A.B.; Rudd, C.J. *Current Topics in Cell Reg.* **1974,** *8,* 247-294.
33. Hunt, S. *Chemistry and Biochemistry of the Amino Acids*; Barrett, G.C., Ed.; Chapman and Hall: New York, 1985; Chapter 12.
34. Stryer, L. *Biochemistry* W.H. Freeman and Company: New York, 1988; pp 220-221.
35. Hill, R.L. *Adv. Prot. Chem.* **1965,** *20,* 37-107.
36. Shimura, K. *Experientia,* **1983,** *39(5),* 455-461.
37. Becker, M.A. Ph.D. Dissertation, Johns Hopkins University, 1993.

RECEIVED June 29, 1993

Chapter 23

Molecular Modeling Studies on Silk Peptides

Stephen A. Fossey and David Kaplan

Biotechnology Division, Natick Research, Development, and Engineering Center, U.S. Army, Natick, MA 01760–5020

A model for the meta-stable state, silk I, of *B. mori* silk fibroin based on conformational energy calculations on representative silk peptides is discussed(*1*). In addition the model and other models for the silk I form are compared with the available experimental data from X-ray and electron diffraction and NMR and IR spectroscopy.

Silk-like proteins are of interest for a number of applications. Among these are high performance fibers (*2*), enzyme immobilizing substrates (*3*) and polymers with highly controlled crystal morphologies (*4*). The last area capitalizes on the ability of genetic means of production to precisely control the primary sequence of the polymer and the self-assembly properties of proteins.

Silk fibroin is a block copolypeptide WITH crystalline domains characterized by a Gly-X repeat where X is Ala or Ser, and of less crystalline domains in which the Gly-X repeat is common but which contain a higher fraction of residues with large side chains(*5*). The crystalline domains of *Bombyx mori* can exist in one of two different morphologies. The more stable one is known as silk II (*6,7*). A detailed structural model for silk II was first proposed by Marsh *et al.* (*8*) and refined by Fraser *et al.* (*9,10*) Its basic feature, a packing of antiparallel, β-pleated sheets, is generally accepted. More recently Takahashi *et al.* (*11*), based on the intensity of X-ray diffraction reflections have proposed that most of the β-sheets in silk fibroin are anti-polar rather than polar as proposed by Marsh *et al.* (see Methods for a description of polar and anti-polar).

Despite a long history of interest in the less stable silk I form it has remained poorly understood. Attempts to induce orientation of the silk I form for studies by X-ray or electron diffraction or solid state NMR cause the silk I form to convert to the more stable silk II form. The silk I form can be obtained by letting the contents of the silk gland dry undisturbed (*9*). If the contents of the gland are mechanically sheared or treated in a number of other ways, the silk II form is obtained (*9,12*).

The sequence of the repeating unit of the crystalline fraction of *B. mori* silk fibroin is (*13*)

$$\text{G-A-G-A-G-S-G-A-A-G[S-G-(A-G)}_n\text{]}_8\text{Y}$$

where n is usually 2 and averages to 2. Poly(L-Ala Gly) has often been used as a model for the crystalline fraction of *B. mori* silk fibroin since the two crystalline forms of poly(L-Ala-Gly) have been shown to be isomorphous with the two crystalline forms of silk.

Experimental studies of silk I have consisted primarily of powder X-ray diffraction (*5,7,14,15*), electron diffraction, (*14*) and solid state ^{13}C-NMR spectroscopy (*12,14-18*). From these studies, it has not been possible to define a unique structure for silk I. Attempts to determine a structure were based on model building and comparison of the predictions of these models with experimental data. A number of models have been proposed, including the "crankshaft" model (*14*) and one based on a loose fourfold helix (*19*).

Conformational energy calculations on representative model polypeptides have been applied to the structure of collagen(*20,21*), poly(glycine) (*22,23*), poly(alanine)(*24*), and copolymers of alanine and glycine (*24*). These efforts have proceeded from model building to comparison with experimental data, but have had some success in elucidating more structural detail than available from experimental results.

Methods

Fossey et al. (*1*) have used conformational energy calculations to investigate the structure of silk-like peptides. The calculations were carried out by using the ECEPP/2 (Empirical Conformational Energy Program for Peptides) algorithm(*25-27*). The details of the calculations have been presented elsewhere, therefore only an overview of the approach is given here.

Starting Conformations. In all cases, β-sheets (regions E and C of Figure 1.) consisting of five strands of six residues, each of alternating alanine and glycine, were studied. The work of Chou *et al.* (*28*) shows that there was little difference in the dihedral angles of sheets composed of either four or eight residues or of two or three strands. The primary effect of increased chain length observed was a slight decrease in the twist generally associated with β-sheets in globular proteins. A six residue peptide was chosen to anticipate the inclusion of serine, which in *B. mori* silk crystalline domains appears once in each six residues. Because of the two fold symmetry of β-strands the sidechains of consecutive residues along the strands project from opposite sides of the β-sheet (*29*). Therefore, in a poly (L-Ala-Gly) strand, all Ala side chains point to the same side. There are then two ways in which adjacent strands can be related. The methyl side chains of the alanine residues in these strands can point to either the same side or to opposite sides of the β-sheet. The sheets in which methyl groups of adjacent strands point to opposite sides of the sheet are referred to as "out-of-register". Those in which all methyl groups point to the same side of the sheet are referred to as "in-register." Takahashi et al. (*11*) have used the nomenclature polar (Figure 2a) and anti-polar (Figure 2b) to refer to in-register and out-of-register respectively, this more common usage will be adopted for the remainder of the paper. Both parallel and antiparallel as well as polar and anti-polar β-sheets were considered. In addition, various packing arrangements for the sheets were considered in the calculations. This approach constitutes a search of conformational space for sheet structures.

Determination of the Unit Cell. To be consistent with prior studies on silk the *b* axis is taken along the strand, the *a* axis is perpendicular to the strand in the plane of the sheet, and the *c* axis is in the direction between sheets. The dimensions of a unit cell correspond to the distance between repeating units in each direction, two residues (Ala-

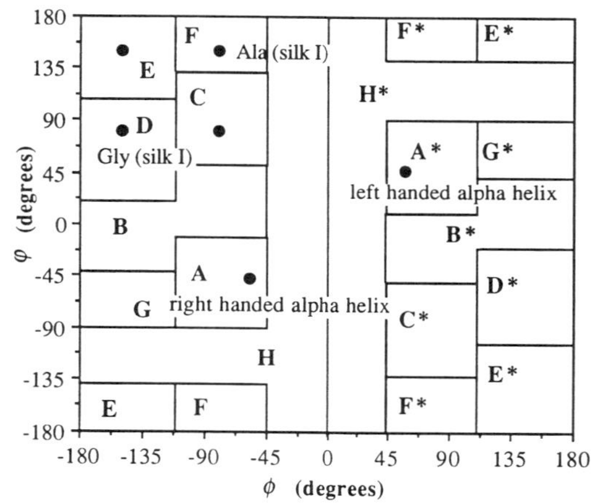

Figure 1. φ−φ **Map (adapted from ref. 31)**

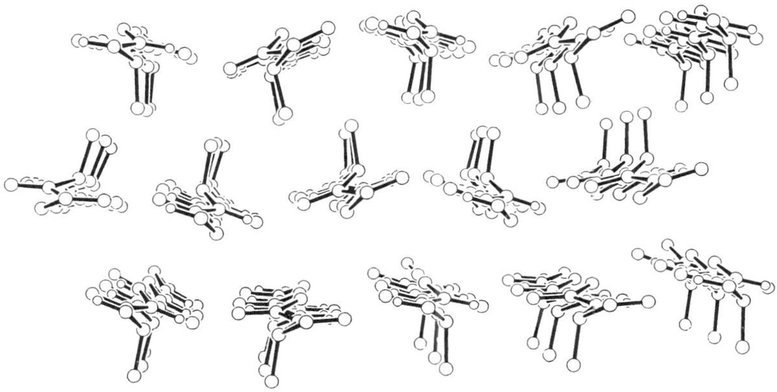

(a)

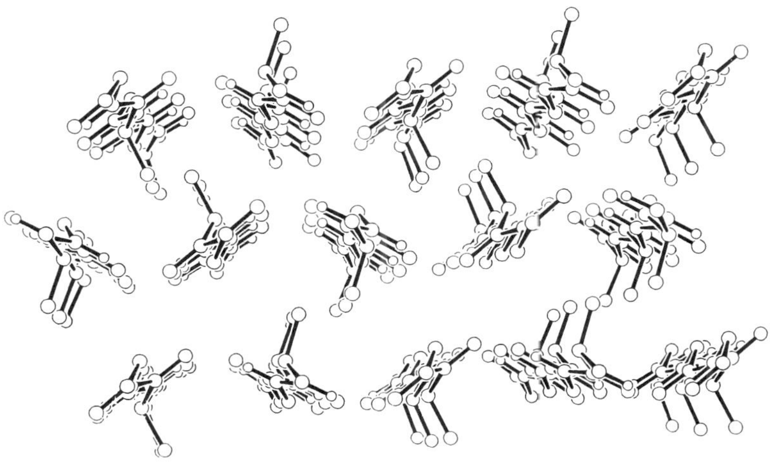

(b)

Figure 2. Silk II (a) polar, (b) anti-polar.

Gly) along the *b* axis and two adjacent chains in the sheet along the *a* axis. In the direction of sheet packing (*c* axis), the repeating unit contains two adjacent sheets for most structures discussed here, with one exception: in the structure referred to later as monoclinic, the repeating unit along the *c* axis contains only one sheet instead of two, because strands in adjacent sheets are identical.

Results

Conformational Energies. All energy-minimized single-sheet structures had a right-handed twist, as found previously (*28,30*). For each of the four types of sheets, energy minima were found in the C and E regions (*31*) of the ϕ–φ map. Figure 1 shows the regions of the ϕ–φ map as well as location of the dihedral angles for the FD conformation. All of those minima that were located in the C region had dihedral angles of approximately (ϕ, φ)=(-80°, 80°) for every residue and all minima located in the E region occurred approximately at (ϕ, φ)=(-150°, 150°) for every residue. In addition to these two minima, the anti-polar arrangement for an antiparallel sheet gave a third minimum with a different conformation, in which most of the alanyl residues adopted dihedral angles near (ϕ, φ)=(-80°, 150°) and most of the glycyl residues adopted dihedral angles near (ϕ, φ)=(-150°, 80°) . The components of the conformational energy and the characteristic repeats of each single sheet minimum are shown in Table I.

Table I.
Conformational Energies (kcal/mol) of Single Sheets

Descriptor[a-c]	Total	Inter-Strand	Intra-Strand	Repeat Distance Along Chain Direction (nm)	Repeat Distance Across Chain Direction (nm)
E, A, Polar	-73.4	-121.3	47.9	0.713	0.480
C, A, Polar	-91.1	-108.0	16.9	0.566	0.459
E, P, Polar	-43.1	-90.9	47.8	0.715	0.460
C, P, Polar	-91.9	-107.7	15.7	0.563	0.456
E, A, A-Polar	-76.8	-128.4	51.7	0.708	0.459
C, A,A-Polar	-104.8	-115.5	10.6	0.566	0.460
FD, A, A-Polar	-79.9	-133.6	53.8	0.645	0.449
E, P, A-Polar	-72.1	-125.8	53.7	0.716	0.448
C, P, A-Polar	-90.2	-105.4	15.2	0.567	0.483

SOURCE: Adapted from ref. 1.
a The letter code of Zimmerman et al.(*31*) is used. FD indicates conformations near (ϕ, φ) = (-80°,150°) for Ala, (ϕ, φ) = (-150°,80°) for Gly. For the other structures, the letter code pertains to the conformational states of both Ala and Gly (see Figure 1).
b A: antiparallel sheet; P: parallel sheet.
c Polar (in-register) sheet; A-Polar: anti-polar (out-of-register) sheet.

For both the polar and anti-polar single sheets, the minima in the C region are clearly of lower energy. From the intra- and interchain components of the energy, it can be seen that the minima in the C region are favored because of a much lower intrachain energy.

This lower intrachain energy is the result of the formation of a bent intrachain hydrogen bond in the C conformation, while interstrand hydrogen bonds are also retained (*32,33*). The minima in the C region show poor agreement with the crystal lattice repeats found experimentally for the two crystalline forms of poly(L-Ala-Gly). The disagreement disappears, however, when the energy of three stacked sheets is considered, as discussed below.

Conformational Energies of Three Stacked Sheets. Table II lists the conformational energy contributions for stacked β-sheets. The C-region minima are not listed because their total energy is much higher by at least 300 kcal/mol (*i.e.* at least about 3 kcal/ mol of residues). Although the isolated sheets with C conformations contain short interstrand hydrogen bonds, in addition to the bent intrachain hydrogen bonds as mentioned above (*32,33*), many of these interstrand hydrogen bonds are weakened or broken when the energies of the stacked sheets are minimized.

Table II. Conformational Energies (kcal/mol) of Three Stacked Sheets

Number	Description[a-c]	Total	intra-strand	inter-strand[d]	inter-sheet
1e	E, A, Polar	-481	164	-297	-348
2f	E, A, Polar	-437	152	-338	-251
3	E, P, Polar	-420	153	-390	-183
4	E, A, A-Polar	-299	152	-334	-117
5g	FD, A, A-Polar	-394	177	-387	-184
6h	FD, A, A-Polar	-388	191	-189	-390
7	E, P, A-Polar	-260	155	-226	-189

SOURCE: Adapted from ref. 1.
a-c See corresponding footnotes of Table I. d Sum of interstrand energies between strands of the same sheet. e Figure 2a. f Unlike the other polar structures, these sheets have Gly faces stacked against Ala faces of adjacent sheets. g Orthorhombic structure, Figure 3a. h Monoclinic structure, Figure 3b.

The computed lowest energy structure is formed by antiparallel β-sheets with polar chains and residues in the E conformational state (line 1 in Table II). As discussed below, this structure is essentially identical with the model that had been proposed by Marsh *et al.* (*8*) and its unit cell parameters agree with the observed values for silk II and the silk II like structure of poly(L-Ala-Gly). The lowest energy anti-polar structure is composed of antiparallel sheets in which the residues adopt the dihedral angles alternating near (ϕ, φ)=(-80°,150°) for Ala and near (ϕ, φ)=(-150°, 80°) for Gly (Figure 1). These dihedral angles correspond to right-handed and left-handed twisting of sheets (*34*), respectively, with approximately equal magnitudes of the twist. Their alternation results in a twofold axial repeat, with no net twist of the sheets. This structure can exist as two variations with slightly different packing arrangements that differ in total energy by only 6 kcal/mol. or <0.1 kcal/mol of residues (lines 5 and 6 in Table II).

The higher energy modification of the anti-polar structure (line 6 of Table II) has a monoclinic unit cell (Figure 3b). The dihedral angles of the orthorhombic and

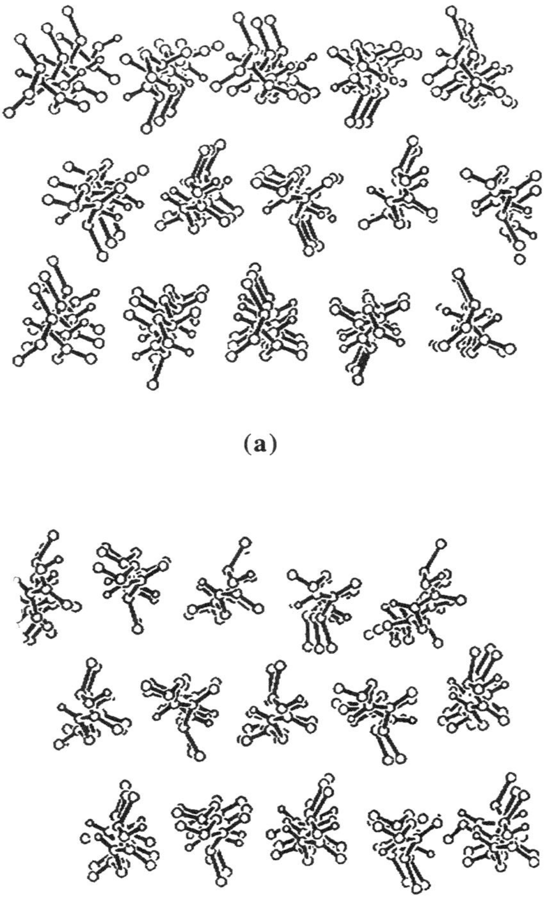

(a)

(b)

Figure 3. Silk I (a) orthorhombic, (b) monoclinic (adapted from ref 1).

monoclinic modifications are very close to each other, differing on the average by less than 2°. The unit cell dimensions in the plane of the sheets (axes *a* and *b*) are within 0.01 nm of each other. Because of the change in the intersheet packing, however, the lengths of the *c* axes differ in the two structures. The length of the c axis in the monoclinic structure is 0.595 nm, inclined at the unit cell angle β= 106.6°, *i.e.* the distance between second nearest-neighbor sheets is 1.190 nm, measured along the *c* axis, in contrast to the corresponding *c* axis length of 1.126 nm in the orthorhombic structure. The unit cell geometry in the monoclinic structure, on the other hand, results in a perpendicular separation of 1.14 nm between second-neighbor sheets, which is only slightly higher than the 1.126 nm distance in the orthorhombic structure. The c = 1.126 nm unit cell dimension of the orthorhombic structure agrees better with observed X-ray data, as shown below. For these reasons, and because of the difference in energies, we propose the orthorhombic modification as the model for silk I, as discussed below. However, given that the two structures are quite close in energy it seems reasonable that both variations would be observed as a disordered packing of sheets.

The energy of the polar silk II structure (line 1 of Table II) is 87 kcal/mol lower than that of the proposed anti-polar silk I structure (line 5 of Table II), *i.e.* it is lower by almost 1 kcal/mol of residues. This suggests why silk II is more stable than silk I, as observed experimentally. On the other hand, the energy difference is sufficiently small, so that silk I can exist as a less stable form.

Comparing Tables I and II, it can be seen that the order of the energy minima is different for single and for stacked sheets. Two differences are of particular note. First, if one considers only conformational states E and FD, the differences in the total energy are dominated in both cases by the interstrand component, but the intersheet energy may also make a large contribution in some structures. In the single sheet, anti-polar chains are favored over polar ones in the E conformation by the interstrand energy, because the strands can be packed more closely, allowing an interchain spacing of 0.46 nm (as compared to 0.48 nm for in-register antiparallel sheets). On the other hand, for stacked sheets, an antiparallel polar sheet with an E conformation (line 1 of Table II) has the lowest energy. It is favored over anti-polar sheets, because the low intersheet energy, arising from interactions of two Gly faces in contact with each other and of two Ala faces in contact with each other, more than compensates for the less favorable intrasheet chain-chain interactions. This is seen by a comparison of lines 1 and 2 of Table II, where it is indicated that the stacking of like faces in contact (line 1) is highly favored by over 100 kcal/mol in intersheet energy over the arrangement in which the Gly faces pack against Ala faces (line 2). Second, conformational states FD and E have nearly the same total energy in antiparallel anti-polar single sheets (Table I), but the FD state allows better stacking, as indicated by the much more favorable intersheet energies (*cf.* line 4 with lines 5 and 6 of Table II) .

Discussion

The comparison of the low energy structures with the available experimental data is discussed below. The low energy polar antiparallel form corresponds to a β-pleated sheet (with a structure similar to silk II) (*8*). Fossey *et al.* (*1*) proposed, based on a comparison with X-ray and electron diffraction data as well as the calculated density, that the anti-polar antiparallel sheet with dihedral angles which differ from those of a beta sheet should serve as a model for silk I.

Poly(L-Alanine-Glycine), Silk I Form. The structure proposed by Fossey *et al.* (*1*), shown in Figure 3a, is a hydrogen-bonded antiparallel sheet in which the alanyl residues adopt dihedral angles near (ϕ, φ)=(-80°, 150°) and the glycyl residues adopt dihedral angles near (ϕ, φ)=(-150°, 80°) (Table III). Since the proposed structure is composed of anti-polar chains, the methyl side chains of alanyl residues in adjacent strands point to opposite sides of the sheet. The side-chain distribution in the anti-polar sheet is thus symmetrical, in contrast to the polar silk II form, which has all alanyl residues on one side and all glycyl residues on the other. The sheets in both forms are stacked such that the strands of adjacent sheets are offset laterally (*i.e.* horizontally in Figure 3) by one half the interstrand distance (*9*).

Table III.
Observed and Calculated Unit Cell Dimensions for Poly(L-Ala-Gly) Silk II Form

Unit Cell Axes (A)[a]	Observed[b]	This Work[c]	Colonna-Cesari *et al.*[d]
a	9.44	9.48	(9.44)
b	6.94	7.06	(6.94)
c	8.96	8.66	8.72
Intersheet[e] spacings (A)			
Ala-*c*	5.17	5.02[f]	5.02
Gly-*c*	3.79	3.72[f]	3.70

SOURCE: Adapted from ref. 1.
a The labeling of the unit cell axes, as used in various publications, refers to different directions for the two forms of silk, in relation to the sheet structure. The notation used here follows the notntion for silk II, as described in Methods.
b Reference 4, p. 312, and Ref. 5.
c Structure 1 of Table II, shown in Figure 2a.
d Reference 24. In this calculation, the *a* and *b* axes were held fixed at their observed values (shown in parentheses), and only the c axis was allowed to vary. For the parameter *a*, an interstrand distance of 4.72 A was used in Ref. 24. The doubled value, correspondlng to the *a* axis of the unit cell, is listed here.
e Ala-*c* is the spacing in the *c* direction across Ala-Ala faces of stacked sheets; Gly-*c* is the spacing across Gly-Gly faces of stacked sheets.

f The intersheet spacing is calculated as an average distance of the C^{α}'s of the top and bottom sheets from a least-squares plane that was fit to the C^{α} atoms of the middle sheet.

There are two crystalline forms of the proposed silk I sheets, one in which the unit cell is orthorhombic with $a = 0.894$ nm, $b = 0.646$ nm, $c = 1.126$ nm. The *a* dimension is twice the interstrand distance, the *b* dimension is the repeat for two residues along the chain, and the c dimension is twice the intersheet distance. The other is a monoclinic unit cell with $a = 0.894$ nm, $b = 0.646$ nm and $c = 1.213$ nm and $\beta = 106.6°$. The observed spacings reported by Lotz and Keith (*14*) are consistent with both unit cells. Table IV shows the corresponding spacings for the orthorhombic unit cell.

Table IV.
Observed and Calculated Spacings for the Silk I Form of Poly(L-Ala-Gly) and Silk I

Poly(L-Ala-Gly) in the Silk I Form[a] spacing(nm)	intensity[c]	Silk I[b] spacing (nm)	intensity[c]	calculated structure spacing (nm)
0.780	w-d			
0.721	s	0.729	ms	0.700
0.560	w-d	0.561	w	0.563
0.486	w-d	—	—	
0.449	vs	0.450	s	0.447
0.420	w-d	—	—	
0.394	ms	0.402	w	0.375
0.360	ms	0.362	ms	
0.316	ms-vd	0.315	ms	0.323
0.272	w-d	0.273	w	0.282
0.240	ms-vd	0.244	mw	
0.225	vw-d	0.224	vw	0.225
0.207	vw-d	—	—	

SOURCE: Adapted from ref. 1.
a Reference 14.
b Reference 15.
c Abbreviations used: w, weak; d, diffuse; s, strong; vs, very strong; ms, medium-strong; vw, very weak; vd, very diffuse.

The hydrogen bonds in the proposed structures are of normal length, with an average H • • • O distance of 0.195 nm. The average N—H • • • O angle of 154° is close to those in observed antiparallel beta-sheets of globular proteins(*35*), 160° ± 10°, and it is the same as the value of 154° reported(*36*) for β-poly(L-Ala). The average C=O • •H angle is 134°, which is less than the 158° reported for β-poly(L-Ala). The out-of-plane component of the C = O • • • H angle is 39°, which is considerably larger than the median of the absolute value of the out-of-plane angle, 15°, in antiparallel, β-sheets (*35*). However, in small molecules there is considerable variation in the C = O • • • H angle, with large out of-plane angles being common. In fact, the preferred C = O • • • H angle is 135° in many crystals of small molecules, which are free to adopt the most advantageous geometry (*35*). The hydrogen bonds formed in the proposed structure are thus somewhat different from those observed in a β-pleated sheet, but they appear to be reasonable, in comparison with hydrogen bonds observed in organic crystals (*35*).

The X-ray diffraction results of Asakura *et al.* (*15*) for silk I and Lotz and Keith (*14*) for the silk I form of poly(L-Ala-Gly) are shown in Table IV, in comparison with the characteristic repeat distances from the proposed structure. From X-ray or electron diffraction, two lattice dimensions are characteristic of the silk I form (*14*); in a third, the repeat is less certain. We assign the strong observed reflection at 0.45 nm to the interstrand distance. The average interstrand spacing in our model is 0.447 nm. The computed fiber axis repeat is 0.646 nm for two residues or 0.323 nm per residue, close to the observed 0.316 nm reflection. A spacing of 0.700 nm in our model corresponds to a strong reflection indexed as 101. This spacing is closest to the observed 0.721 nm spacing for poly (L-Ala-Gly). Given that the ECEPP/2 potential produces a shorter intersheet distance for the silk II form than the observed spacing this seems to be a reasonable match. Thus, most of the observed reflections, especially the strong ones, can be indexed in a consistent manner using the model proposed here (Table IV). The calculated density of the overall structure is 1.30 g/cm^3 while Lotz and Keith (*14*) report that the density of wet crystals of poly(L-Ala-Gly) in the silk I form is 1.33 g/cm^3, and when corrected for water content, 1.25 g/cm^3.

A number of models for silk I have been proposed earlier. In the "crankshaft model" of Lotz and Keith (*14*), the alanyl residues are in a β-sheet conformation and the glycyl residues in a left-handed α-helical conformation. The resulting unit cell has the dimensions a = 0.472 nm, b = 0.96 nm, and c = 1.44 nm.A fiber axis repeat of an integer multiple of 0.32 nm has been proposed by Lotz and Keith (*14*). Using the crankshaft model, they assigned to the length of the unit cell the value of 0.96 nm, *i.e.* four residues per chain axis repeat. In the model proposed here, b = 0.646 nm, i.e., an average axial rise of 0.323 nm per residue accounts well for this reflection. When potential disorder, which eliminates or reduces certain reflections, is taken into account the crankshaft model provides a good fit to the X-ray diffraction spacings and intensities.

We have also calculated the conformational energy of three sheets of five strands of the antiparallel crankshaft structure proposed by Lotz and Colonna Cesari (*37*). The total energy was -365 kcal/mol., which is considerably higher than the values computed for our model (Table II). The structure proposed by Lotz and Keith's(*14*) and the one presented here are alternative models for silk I, because both are consistent with the fiber X-ray diffraction data.

There is also structural information from infrared and NMR spectroscopy available. Brack and Spach (*38*) observed IR spectra which corresponded to neither an α nor a β form. While the crankshaft model has alanine residues in a β-sheet conformation and glycine residues in an α-helical conformation. Additionally, the NH stretching mode is split into two frequencies (3280 cm^{-1} and 3315 cm^{-1}) which Lotz and Keith (*14*) suggest could be attributed to two hydrogen bond geometries. Two hydrogen bond geometries require that both parallel and antiparallel sheet arrangements be present. The silk I model of Fossey *et al.* (*1*) has two hydrogen bond geometries based on the relation of the hydrogen bond to the peptide plane of the residues, but it has not been shown that this would account for the splitting observed.

A number of workers have published NMR data which is relevant to the crystal structure of silk. Using ^{13}C NMR Saito *et al.* (*19*) conclude that the alanine residues of silk I cannot be in the β-sheet conformation as in the crankshaft model. Asakura *et al.* (*39*) using [1-^{13}C]Gly labeled *B. mori* fibroin in solution measured the long range coupling constant $^3J_{C'\text{-}N\text{-}C^{\alpha}\text{-}H}$. These coupling constants were then converted by the Karplus-like relation of Bystrov (*40*) to obtain the dihedral angle ϕ for alanine and serine residues. The values of ϕ allowed are -143° to -147°, -93° to -97°, -4° to -6°, and 124° to 126°. Since the silk I form is obtained by drying silk solutions with out disturbance the silk I conformation is probably quite close to solution conformation. The ϕ value for alanine in the crankshaft model is -124° and -104° while in the model of Fossey *et al.*. (*1*) it is -80°. Recently, Nicholson *et al.* (*41*) have reported solid state NMR measurements for the silk II form. Unfortunately the oriented samples needed cannot be obtained for silk I. Although these data are not definitive they do tend to support the model of Fossey *et al.* (*1*) over the crankshaft model.

Several other models have been proposed in which the chains adopt various helical conformations (*42-45*). It has been shown for several models (*14,37,42,44*) by conformational energy computations that their strand conformations correspond to relatively low-energy conformations of an isolated poly(L-Ala-Gly) chain. Most of these models lead to less consistent assignments of the available experimental diffraction data with regard to the translational repeat than those of the present model.

Silk I to Silk II Transition. The proposed silk I model has a low intrachain energy barrier for the conversion to the silk II form. To form the polar model of silk II requires the breaking and reforming of all hydrogen bonds and rotation of every other chain by 180° around the chain axis. Interestingly, to form the anti-polar silk II form doesnot require breaking any hydrogen bonds or chain rotations. This lower energy barrier could account for the increased population of antipolar silk II, as proposed by Takahashi (*11*), despite the fact that the polar form has a lower conformational energy.

Summary

A model for silk I and the silk I form of poly(L-alanine-glycine) has been proposed by Fossey et al (*1*), based on conformational energy calculations. The computed silk model provides good agreement with the available experimental data, because it can account for most spacings in the observed fiber X-ray diffraction pattern of silk I and of the corresponding form of poly(L-Ala-Gly). The alanyl residues adopt a right handed sheet-like conformation and the glycyl residues a left-handed sheet-like conformation in the proposed structure. The dimensions of its orthorhombic unit cell are $a = 0.894$ nm, $b = 0.646$ nm, and $c = 1.126$ nm. Silk fibroin, when in the gland, is a concentrated solution, and when it is allowed to dry undisturbed, is found to be in the silk I form (*5*), therefore, it is likely that the liquid crystalline structures also consist of silk I. The proposed silk I model may also represent the structure of the liquid crystal phase recently described by Kerkam *et al.* (*46*), who showed that concentrated silk fibroin solutions formed nematic liquid crystals.

LITERATURE CITED

1. Fossey, S. A.; Nemethy, G.; Gibson, K. D.; Scheraga, H. A. *Biopolymers* **1991,** *31*,1529.
2. Lombardi, S. J.; Fossey, S. A.; Kaplan, D. K. Proceedings of the American Society for Composites - Fifth Technical Conference. **1991**
3. Demura, M.; Asakura, T. *Biotech. and Bioeng.* **1989,** *33*, 598.
4. McGrath, K. P.; Fournier, M. J.; Mason, T. L.; Tirrel, D. A. *J. Am. Chem. Soc.* **1992,** *114*, 727.
5. Fraser, R. D. B.; MacRae, T. P. *Conformation in Fibrous Proteins and Related Synthetic Polypeptides,* Academic Press: New York, NY, 1973; Chapt. 13.
6. Kratky, O.; Schauenstein, E.; Sekora, A. *Nature* **1950**, *165*, 319.
7. Kratky, O. *Trans. Faraday Soc.* ^**1956,** *52*, 558.
8. Marsh, R. E.; Corey, R. B.; Pauling, L. *Biochem. Biophys. Acta* **1955**, *16*, 1.
9. Fraser, R. D. B.; MacRae, T. P.; Stewart, F. H. C.; Suzuki, E. *J. Mol. Biol.* **1965**,*11*, 706.
10. Fraser, R. D. B.; MacRae, T. P.; Stewart, F. H. C. *J. Mol. Biol* **1966,** *19*, 580.
11. Takahashi, Y.; Gehoh, M.; Yuzuriha, K. *J. Poly. Sci. Pt. B: Poly. Phys.* **1991,** *29*, 889.
12. Ishida, M.; Asakura, T.; Yokoi, M.; Saito, H. *Macromolecules* **1990**, *23*, 88.
13. Strydom, D. J.; Haylett, T.; Stead, R. H. *Biochem. Biophys. Res. Comm.* **1977**, *79*, 932.

14. Lotz, B.; Keith, H. D. *J. Mol. Biol.* **1971**, *61*, 201.
15. Asakura, T.; Kuzuhara, A.; Tabeta, R.; Saito, H. *Macromolecules* **1985**, *18*, 1841.
16. Kricheldorf, H. R.; Muller, D.; Ziegler, K. I *Polymer Bull.* **1983**, *9*, 284.
17. Saito, H.; Tabeta, R.; Kuzuhara, A.; Asakura, T. *Bull. Chem. Soc. Japan* **1986**, *59*, 3383.
18. Asakura, T.; Yoshimizu, H.; Yoshizawa, F. *Macromolecules* **1988**, 21, 2038.
19. Saito, H.; Tabeta, R.; Asakura, T.; Iwanaga, Y.; Shoji, A.; Ozaki, T.; Ando, I. *Macromolecules* **1984**, *17*, 1405.
20. Scheraga, H. A.; Nemethy, G. In *Molecules in Natural Biosciences—Encomium for Linus Pauling*, **1991**.
21. Nemethy, G. & Scheraga, H. A. *Bull. Inst. Chem. Res. Univ. Kyoto* **1989**, *66*, 398-408.
22. Lotz, B. *J. Mol. Biol.* **1974**, *87*, 169.
23. Colonna-Cesari, F.; Premilat, S.; Lotz, B. *J. Mol. Biol.* **1974**, *87*, 181.
24. Colonna-Cesari, F.; Premilat, S.; Lotz, B. *J. Mol. Biol.* **1975**, *95*, 71.
25. Momany, F. A.; McGuire, R. F.; Burgess, A. W.; Scheraga, H. A. *J. Phys. Chem.* **1975**, *79*, 2361.
26. Nemethy, G.; Pottle, M. S.; Scheraga, H. A. *J. Phys. Chem.* **1983**, *87*, 1883.
27. Sippl, M. J.; Nemethy, G.; Scheraga, H. A. *J. Phys. Chem.* **1984**, *88*, 6231.
28. Chou, K.-C.; Pottle, M.; Nemethy, G.; Ueda, Y.; Scheraga, H. A. *J. Mol. Biol.* **1982**, *162*, 89.
29. Pauling, L.; Corey, R. B. *Proc. Natl. Acad. Sci., U.S.A.* **1953**, *39*, 253.
30. Chou, K.-C.; Nemethy, G.; Rumsey, S.; Tuttle, R. W.; Scheraga, H. A. *J. Mol. Biol* **1986**, *188*, 641.
31. Zimmerman, S. S.; Pottle, M. S.; Nemethy, G.; Scheraga, H. A. *Macromolecules* **1977**, *10*, 1.
32. Chou, K.-C.; Scheraga, H. A. *Proc. Natl. Acad. Sci., U.S.A.* **1982**, *79*, 7047.
33. Chou, K.-C.; Nemethy, G.; Scheraga, H. A. *J. Mol. Biol.* **1983**, *168*, 389.
34. Chou, K. C.; Nemethy, G.; Scheraga, H. A. *Accts. Chem. Res.* **1990**, *23*, 134.
35. Baker, E. N.; Hubbard, R. E. *Prog. Biophys. Mol. Biol.* **1984**, *44*, 97.
36. Arnott, S.; Dover, S. D.; Elliot, A. *J. Mol. Biol.* **1967**, *30*, 201.
37. Lotz, B.; Colonna-Cesari, F. *Biochimie* **1979**,*61*, 205.
38. Brack, A.; Spach, G. *Biopolymers* **1972**, *11*, 563.
39. Asakura, T.; Wanatabe, Y.; Itoh, T. *Macromolecules* **1984**, *17*, 2421.
40. Bystrov, V. F. *Prog. NMR Spectrosc.* **1976**, *10*, 41.
41. Nicholson, L. K.; Asakura, T.; Demura, M.; Cross, T. A. *Biopolymers* **1993**, *33*, 847.
42. Konishi, T.; Kurokawa, M. *Sen-i Gakkaishi* **1984**, *24*, 550.
43. Asakura, T.; Yamaguchi, T. *J. Seric. Sci. Jpn.* **1987**, *56*, 300.
44. Ichimura. S.; Okuyama, K. *Polym. Prep. Jpn.* **1989**, *38*, E48.
45. Oka, M.; Baba, Y.; Kagemoto, A.; Nakajima, A. *Polym. J. (Jpn.)* **1990**, *22*, 416.
46. Kerkam, K.; Viney, C.; Kaplan, D.; Lombardi, S. *Nature* **1991**, *349*, 596.

RECEIVED July 15, 1993

Chapter 24

Approaches to Modeling and Property Prediction of Model Peptides

Ruth Pachter, Robert L. Crane, and W. Wade Adams

Materials Directorate, Wright Laboratory, WL/MLPJ, Wright-Patterson AFB, OH 45433–7702

Molecular simulations that predict a 'spring-like' molecular mechanical response of alpha-helical biopolymers with a reinforcing *intra*-molecular hydrogen bonding network are presented in this study. Mechanical properties of extended biopolymer strands based on naturally occurring amino acids, *e.g.*, poly(L-Ala), and for comparison poly(L-Glu) and poly(Ala-Gly) *versus* synthetic PPTA containing an amide bond, are compared to those assuming alpha-helical structures. Thus, the pivotal role of such motifs in biological systems utilizing superior compressive mechanical properties, especially of silk, can be inferred.

Research that is currently underway in our laboratory in the design of opto-electronic materials with unique nonlinear optical and mechanical properties, particularly biopolymers that may be based on silk-type model systems, is found to be advanced by molecular modeling and properties prediction. We are investigating the use of biopolymers in pursuing the objective of developing new materials *(1)* to overcome existing mechanical disadvantages, at least at the molecular level. For example, although high-performance 'rigid-rod' polymer fibers *(2,3,4)* demonstrate high tensile properties *(5,6)* that may enable their incorporation into aerospace structures, these materials show low compressive strength, and their failure mode has been confirmed *(7,8)*. On the other hand, flexible arrangements of alpha-helical biopolymer motifs found in Nature have important mechanical functions, as shown by the structures found in keratin in hair, myosin and tropomyosin in muscle, epidermin in skin, and fibrin in blood clots, but mostly by the elasticity of spider silk, thus lending support to the notion of designing polypeptide structures that mimic Nature. In particular, the *coiled coil* super-structure of the myosin tail leads to a hierarchy of *intra-* and *inter-* reinforcing interactions to accomplish a highly ordered array of a thick filament that provides the means for mechanical function in muscle contraction *(9)*. Similarly, membrane skeletons that consist of spectrin networks which enable erythrocytes to resist strong shearing forces in blood flow, are comprised of flexible triple-stranded alpha-helical *coiled coils* providing the underlying mechanical stability and resilience of the erythrocyte membrane *(10)*, while other membrane skeletons in erythrocytes are also important in altering and stabilizing the shapes of various types of cells *(11)*. Moreover, it is supposed that alpha-helical strands are acting as a rubbery reinforcement matrix to provide elasticity to spider silk *(12,13)*, validated to some extent by our experimental study of these fibers in compression *(14)* that indicate no 'kinking' *(7,15)* failure. Thus, the design of new biopolymers based on silk-type model systems is of special interest since spider silk arrangements were found to be governed by the same principles that apply to synthetic materials.

Furthermore, since the synthesis and characterization of new polymeric nonlinear optical materials are of increasing interest *(16,17)*, especially photochromic polypeptides *(18)*, we have been involved in a continuing research effort to study the conformational flexibility in these systems *(19,20)*. In addition to designing these materials to have superior mechanical properties by mimicking spider-silk structures, it is important to understand the underlying effects of the chromophore on the resulting polypeptide response to enable the design of photoresponsive polypeptides with controlled behavior. For example, in the photomodulation of the helix/coil equilibrium of poly(spiro-L-glutamate), the light-adapted polymer exists in an alpha-helical conformation with the photochromic side-chain in its spiropyran form, while during dark adaptation a slow helix-to-coil dark reaction occurs, as the chromophores convert to the merocyanine form. Molecular simulations were used to study the effects of the nonlinear optical moieties, *e.g.*, spiropyran and merocyanine species, on the alpha-helical core of poly(spiropyran-L-glutamate) *(19)*. To further discern these effects, the influence of the chromophore conformation on a model of poly(spiropyran-L-tyrosine) was evaluated *(21)*, and also of amphiphatic *coiled coils* *(22)*. Although methods for modeling conformation and dynamics of polypeptides of defined secondary structure, *e.g.*, silk model peptides *(23)* has been reviewed *(24)*, and the motions of an alpha-helical polypeptide were examined by molecular and harmonic dynamics (25), none were carried out on the conformational flexibility of photochromic polypeptides. In this study we describe briefly the information from molecular dynamics simulations of β-sheet type strands to mimic the spider dragline sequence, for an initial insight into the possible stability of such sequences.

Computational chemistry has indeed become a valuable tool for analyzing the dynamics and structure of macromolecular systems, and also for the theoretical prediction of properties, *e.g.*, the ultimate mechanical properties *(26)*. Attempts to calculate moduli of polymers have been reported *(27,28,29)*, but these methods rely on molecular mechanics empirical potential energy functions, while no molecular orbital calculations were reported. A series of such calculations, namely of 'rigid-rod' polymers (poly(*para*-phenylene benzobisoxazole), poly(*para*-phenylene benzobisthiazole), poly(*para*-phenylene benzobisimidazole) *(30,31)*), polyethylene *(32,33)*, and of highly conjugated structures (poly(*para*-phenylene) *(34)* and C_{60} *(35)*), have offered insight into deformations at the molecular level and the strain-dependent stiffness behavior. It is therefore of particular interest to apply this approach for the calculation of the mechanical response of biopolymer chains that model silk-type peptides. In this study we predict theoretically the 'spring-like' microscopic mechanical response of alpha-helical biopolymer models (poly(Ala) and poly(Glu) that are helix 'makers') having a reinforcing *intra*-molecular hydrogen bonding network, as compared to a silk-type sequence of poly(Ala-Gly) that is expected to assume a β-sheet geometrical configuration, and also of a synthetic polymer containing an amide bond, *i.e.* poly(*para*-phenylene terephthalamide) (PPTA) [KEVLAR™]. This theoretical verification of the absence of compressive buckling in alpha-helical biopolymers rationalizes the molecular elasticity and resistance to 'kinking' of those strands, manifested by the prevalence in Nature for *coiled coils*. This new insight of the role of the alpha-helix in these systems as a requirement for superior compressive mechanical properties may enable new guidance for the synthesis of motifs consistent with the silk-type molecular frameworks as optimized by Nature.

Results and Discussion.

Method. In the calculations of mechanical properties we employ a rigorous method recently developed and applied *(30)*. Within this approach *(36)* we apply the cluster method *(37)* that characterizes a polymer by a translation vector Tv *(38)*. The moduli calculations for a polymer model are performed by incrementally increasing or decreasing Tv of the equilibrium geometry of the cluster, representing molecular tensile or compressive strain, and the geometry under this constraint is optimized. The dependence of the heat of formation on molecular strain is established and the stiffness of the polymer chain given by the second derivative of the heat of formation *vs.* strain relationship. For each of these strained geometries a normal mode analysis may also be carried out in order to determine the strain-dependence of the system's characteristic frequencies *(32)*.

The elastic modulus of a polymer is given by the ratio of the stress σ (force/unit area), and strain ε (fractional change in cluster length $\Delta L/L_{eq}$):

$$E = \frac{\sigma}{\varepsilon} = \left(F/A_{eq} \right) / \left(\Delta L/L_{eq} \right) = KL_{eq}/A_{eq}$$

where K is the force constant (N/m), L_{eq} the equilibrium length, and A_{eq} the cross-sectional area of the polymer. K is determined by fitting ΔH_{ff} (change in heat of formation as a function of strain) to a third-order polynomial *(38)*:

$$\Delta H_{ff} = a_1(\Delta L)^3 + a_2(\Delta L)^2 + a_3(\Delta L) + a_4 .$$

A_{eq} (Å^2) is derived from X-ray diffraction data or estimated from the polymer density d (g/cm^3):

$$d = m/V = m/A_{eq}L_{eq} = nM_w/A_{eq}L_{eq} .$$

Thus, $A_{eq}=M_w/dN_AL_{eq}$ where M_w is the molecular weight of the cluster (g/mol), n number of moles, V volume, and N_A Avogadro's number.

Comparison of Extended Biopolymer Chains and PPTA. First we compare biopolymer strands of poly(L-Ala) and poly(L-Glu) that assume an extended configuration but not a beta-sheet type structure, to the synthetic PPTA polymer model *(39)*. The cluster length requisite for this calculation is determined by the electron delocalization length within the polymer *(40)*. Since we used six repeat units for poly(L-Ala) and five repeat units for poly(L-Glu) in the unit cells (maximum number that is computationally feasible) no non-zero off-diagonal elements in the density matrix resulted. The geometries calculated by using the amide bond correction to the Hamiltonian were compared to those optimized with a classical force-field *(41)*, confirming that the alpha-helical configuration is conserved within our approach. Figure 1 outlines the mechanical response of extended poly(L-Ala) and poly(L-Glu), where for poly(L-Ala) we obtain a modulus of 160 GPa (A_{eq}~20 Å^2, L_{eq}=21.7 Å, and K=145 N/m), and for poly(L-Glu) a modulus of 230 GPa (A_{eq}~20 Å^2, L_{eq}=14.3 Å, and K=320 N/m). High tensile strengths are clearly predicted for the extended biopolymers, while only a small change in energy is observed in compression since the molecular structure enables more conformational variation resulting from the additional rotatable bond, as compared to similar calculations on rigid-rod polymers (*30*).

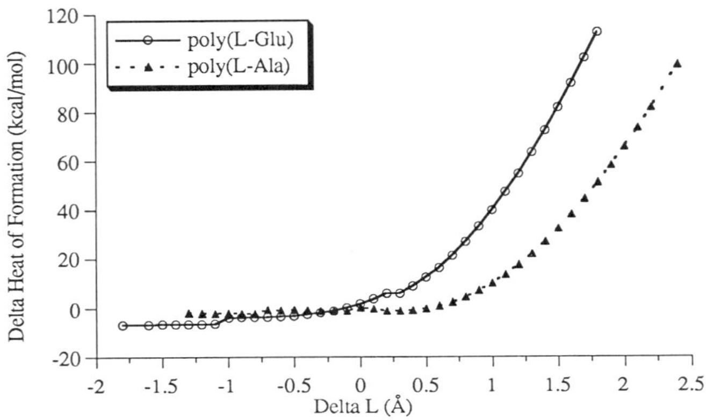

Figure 1: ΔH_{ff} *vs.* ΔL dependence for the extended biopolymer chains (Adapted from ref. (*42*)).

The semi-empirical AM1 (Austin Texas 1) Hamiltonian used in this method (*30*) has been developed to be able to reproduce hydrogen bonding in studies of biologically interesting systems. The assignment of a number of spherical Gaussians to each atom in the AM1 approach mimics the correlation effects.

The ΔH_{ff} *vs.* ΔL dependence shown in Figure 2 for PPTA results in a

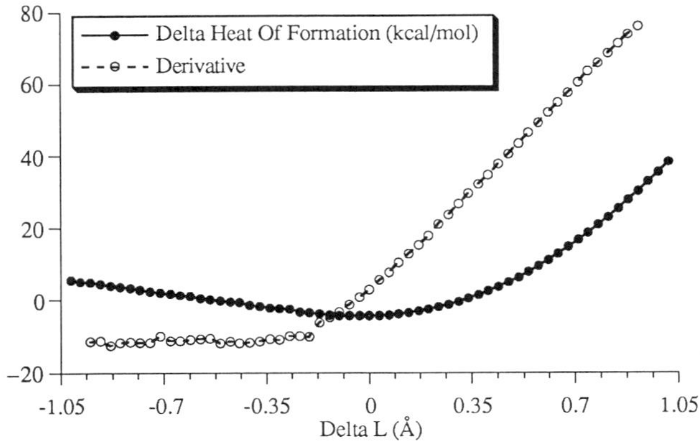

Figure 2: ΔH_{ff} *vs.* ΔL dependence for PPTA (Adapted from ref. (*42*)).

modulus of *ca.* 330 GPa, where d=1.5 g/cm^3 *(43)*, L_{eq}=13.1 Å, M_w=238.25 g/mol, A_{eq}=20 Å^2, and K=50.6 N/m., overestimating the experimental range of 120-200 GPa (*44,45*), that can be partially corrected by biasing the Hessian with experimental vibrational frequencies *(46)* to 290 GPa. This is comparable to results derived from atomistic simulations of PPTA *(47)*. The dashed line in Figure 2 is the force curve (10^{-10} N) obtained by a numerical derivative of the heat of formation, showing that 'kinking' occurs at approximately 2.5% compressive strain (the force curve is the molecular equivalent of a strain-stress curve). The overall response is similar to that of other 'rigid-rod polymers' *(30)*, with the calculated molecular deformations supporting in part the supposition of a conformational change in PPTA upon compression *(48,49)*.

Alpha-Helical Biopolymer Strands. A distinctly different response of an alpha-helical chain to the application of strain when compared to that of other polymers, is shown in Figure 3. No buckling is shown in compression, with a modulus of 60 GPa (A_{eq}~ 20 Å^2, L_{eq}=10.2 Å, and K=107 N/m). The force (10^{-10} N) curve (squares) shows that in distinct contrast to PPTA (Figure 2), there is no evidence for 'kinking'. Small energy variations are calculated in tension, with a similar response revealed for poly(L-Glu). Note that even at *ca.* 40% tensile strain there was incomplete extension of the chain with breaking of the hydrogen bonds, and thus no large increase in the heat of formation. At higher extensions a structure approaching more β-sheet type may result. Although overestimated, the predicted tensile modulus for poly(L-Glu) is in qualitative agreement with experiment *(50)*,. The calculated ΔH_{ff} *vs.* ΔL dependence results in a modulus of 60 GPa (A_{eq} ~ 20 Å^2, L_{eq}=5.2 Å, and K=240 N/m).

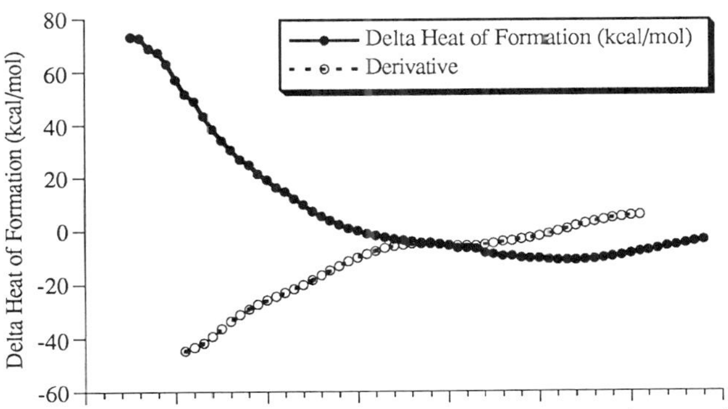

Figure 3: ΔH_{ff} *vs.* ΔL dependence for the alpha-helical model of poly(L-Ala).

A comparison of the alpha helical models of poly(Ala) and poly(Glu) chains to preliminary results calculated for a silk-type repetitive sequence of poly(Ala-Gly) (Figure 4) that is expected to assume more of a β-sheet geometrical arrangement at its lowest energy configuration, reveals some interesting results. Since straining the alpha-helical structure to an extended architecture is more favorable in this case than for poly(Ala) or poly (Glu), the energy decreases markedly on tension, and not even a local energy minimum is observed. Moreover, since there would be less repulsive interactions in this case in compression due to the Glycine side-chains with no substitution, the energy changes less in compression, as compared to poly(Ala) and poly(Glu). The modulus is 28 GPa, where $A_{eq} \sim 20$ Å^2, L_{eq}=13.1 Å, and K=43 N/m.

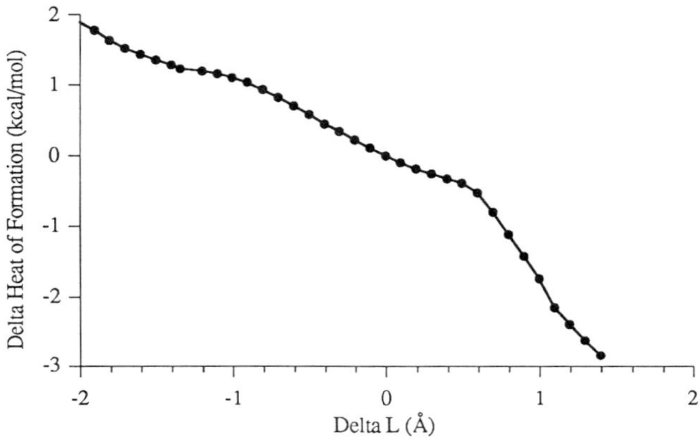

Figure 4: ΔH_{ff} *vs.* ΔL dependence for the alpha-helical model of a repetitive sequence of poly(Ala-Gly).

These results demonstrate a 'spring-like' behavior of alpha-helical biopolymer chains in contrast to the extended systems. Interestingly, it has been shown experimentally that synthetic high performance polymers fail by 'kinking', *(7,8)* whereas spider silk fibers that contain alpha-helical structures on the surface do not 'kink' *(8)* consistent with our calculations. Our analysis helps explain the molecular elasticity of biopolymers that consist of alpha-helical strands and hence the apparent prevalence in Nature for *coiled coils*. These data suggest new design possibilities for the 'molecular engineering' of polymers with enhanced compressional strength.

Conclusion.

This study reveals the potential of molecular simulations to predict the macroscopic mechanical response of alpha-helical biopolymers with a reinforcing *intra*-molecular hydrogen bonding network, *viz.*, a 'spring-like' behavior. This calculation explains the pivotal role of such motifs in biological systems requiring superior compressive mechanical properties. Our foreknowledge of the absence of 'kinking' and the understanding of the structure-to-function relationships in biopolymers is significant in that it may enable the synthesis of structural motifs consistent with molecular frameworks optimized by Nature, *e.g.*, spider silk-type polypeptides, while molecular dynamics simulations will provide an understanding of the conformational flexibility predicted for those structures *(51)*. Moreover, direct comparisons of our predictive capability with experimental studies of the mechanical properties of biomolecules, as have been recently initiated, for example, for DNA strands *(52)*, will further enhance the ability to design new materials.

Literature Cited.

1. Allcock, H. R.; *Science* **1992**, *255*, 1106.
2. Kovar, R. F.; Arnold, F. E.; *J. Poly. Sci., Poly. Chem. Ed.* **1976**, *14*, 2807.
3. Helminiak, T. E.; Arnold, F. E.; Benner, C. L.; 19*Polymer Preprints, ACS Poly. Div.* **1975**, *16*, 659.
4. Wolfe, J. F.; Loo, B. H.; Arnold, F. E.; *Polymer Preprints, ACS Poly. Div.* **1978**, *19*, 1.
5. Kumar, S.; *International Encyclopedia of Composites* **1991**, *4*, 51.
6. Wolfe, J. F.; *Encyclopedia of Polymer Science and Engineering* **1988**, *11*, 601.
7. McGarry, F. J.; Moalli, J. E.; *Polymer* **1991**, *32*, 1816, and references therein.
8. Vezie, D. L.; *Ph.D. Dissertation*, Massachusetts Institute of Technology, **1993**.
9. Warrick, H. M.; Spudich, J. A.; *Ann. Rev. Cell Biol.* **1987**, *3*, 379.
10. Byers, T. J.; Branton, D.; *Proc. Nat. Acad. Sci.* **1985**, *82*, 6153.
11. Marchesi, T; *Ann. Rev. Cell Biol.* **1985**, *1*, 531.
12. Kaplan, D. L.; Lombardi, S. J.; Muller, W. S.; Fossey, S. A.; "Silks: Chemistry, Properties, and Genetics-Biomaterials: Novel Materials from Biological Sources" **1991**, Macmillan Press.
13. Vollrath, F.; *Scientific American* **1992**, *70*, and references therein.
14. Mahoney, D.; Vezie, D.L.; Eby, R. K.; Adams, W. W.; manuscript in preparation.
15. LeTeresa, S. J.; Porter, R. S.; Farris, R. J. J.; *Mater. Sci.* **1988**, *23* 1886.
16. Yitzchaik, S.; Berkovic, G.; Krongauz , V.; *Macromolecules* **1990**, *23*, 3539.
17. Cooper, T. M.; Obermeier, K. A.; Natarajan, L. V.; Crane, R. L.; *Photochem. Photobiol.* **1992**, *55* 1.
18. Ishii, T.; Wada, T.; Garito, A. F.; Sasabe, H.; Yamada, A.; *Mat. Res. Soc. Symp. Proc.* **1990**, 129.
19. Pachter, R.; T.M. Cooper, T. M.; Natarajan, L. V.; Crane, R. L.; Adams, W. W.; *Biopolymers* **1992**, *32*, 1129.
20. Cooper, T. M.; Pachter, R.; Stone, M. L.; Crane, R. L.; *Biopolymers* submitted.
21. Pachter, R.; T.M. Cooper, T. M.; Crane, R. L.; Adams, W. W.; manuscript in preparation.
22. Sennett, B. W.; Pachter, R.; Cooper, T. M.; Crane, R. L.; Adams, W. W.; *ACS Polymer Preprint*, **1993**.
23. Fossey, S. A.; Nemethy, G.; Gibson, K.D.; Scheraga, H. A.; "Conformation of Fibrous Proteins", Academic Press, New York, **1990**.
24. Scheraga, H. A.; *Chemica Scripta* **1989**, *29A*, 3.
25. Perahia, D.; Levy, R. M.; Karplus, M.; *Biopolymers* **1990**, *29*, 645.
26. Roe, R. J.; "Computer Simulation of Polymers" **1991**.
27. Rutledge, G. C.; Suter, U. W.; *Polymer* **1991**, *32*, 179.
28. Sorensen, R. A.; Liau, W. B.; Boyd, R. H.; *Macromolecules* **1988**, *21*, 194.
29. Tashiro, K.; Kobayashi, M.; Tadokoro, H.; *Macromolecules* **1978**, *11*, 908.

30. Wierschke, S. G.; Shoemaker, J. R.; Pachter, R.; Haaland, P. D.; Adams, W. W.; *Polymer* **1992**, *33*, 357.
31. Shoemaker, J. R.; *M.Sc. Thesis*, Air Force Institute of Technology, **1991**.
32. Shoemaker, J. R.; Horn, T. R.; Haaland, P. D.; Pachter, R.; Adams, W. W.; *Polymer* **1992**, *33*, 351.
33. Horn, T. R.; Haaland, P. D.; Pachter, R.; Adams, W. W.; *Polymer*, in press.
34. Haaland, P. D.; Pachter, R.; Adams, W. W.; *Polymer,* in press.
35. Haaland, P. D.; Pachter, R.; Pachter, M.; Adams, W. W.; *Chemical Phys. Lett.* **1992**, *199*, 379.
36. Stewart, J. J. P.; MOPAC (Version 5.0), Quantum Chemistry Program Exchange #455.
37. Similar to the Hückel treatment of cyclic bonding in simple cases developed by Stewart, J. J. P.; *New Polym. Materials* **1987**, *1*, 53.
38. Klei, H. E.; Stewart, J. J. P.; *Int. J. Quant. Chem., Quant. Chem. Symp.* **1986**, *20*, 529.
39. Kwolek, S.; U. S. Patent No. 3 600 356 1971.
40. Perkins, P. G.; Stewart, J. J. P.; *J. Chem. Soc. Faraday Trans. II* **1980**, *76*, 520 .
41. Brooks, B. R.; Bruccoleri, R. E.; Olafson, B. D.; States, D. J.; Swaminathan, S.; Karplus, M.; *J. Comp. Chem* **1983**, *4*, 187.
42. Pachter, R.; Haaland, P. D.; Crane, R. L.; Adams, W. ;W; *Materials Research Society, Biomimetics Symposium Proceedings* **1993**, in press.
43. Adams, W. W.; Eby, R. K.; *MRS Bulletin* **1987**, 23.
44. Allen, S. R.; *Polymer* **1988**, *29*, 1091.
45. Tashiro, K.; Kobayashi, M.; Tadokoro, H.; *Macromolecules* **1977**, *10*, 413.
46. Karasawa, N.; Dasgupta, S.; Goddard, W. A.;199 *J. Phys. Chem.* **1989**, *95*, 2260.
47. Rutledge, G. C.; Suter, U. W.; Papaspyrides, C. D.; *Macromolecules* **1991**, *24*, 1934.
48. Tanner, D.; Fitzgerald, J. A.; Phillips, B. R.; *Adv. Mater.,* No. 5 **1989**, 151.
49. Klunzinger, P.K.; Eby, R. K.; manuscript in preparation.
50. G. Wegner, private communication, estimates the modulus as ca. 30 GPa.
51. Preliminary calculations (Pachter, R.; Anderson, G.; Crane, R.L.; Adams, W. W.; unpublished results) of polypeptide models such as those identified in spider silk (*12*) were performed. Molecular dynamics (MD) calculations (for 100 ps) using the CHARMm force-field were carried out after appropriate heating and equilibration periods to establish the relative stability of three such sequences. A superposition of the average structure during the MD simulation on the initially minimized one shows that of the three sequences modeled, Spidroin 1 that includes Ala and only one Arg residue results in a lower overall destabilization of the structure (RMS deviation of *ca.* 3Å). Changes in the peptide backbone ($\{\phi, \psi\}$ dihedral angles) during the dynamics simulation show the range of values of the center residues to assume a β-strand structure.
52. Bustamante, C.; Vesenka, J.; *Polymer Preprints, ACS Poly. Div.* **1992**, *33*, 743.

RECEIVED July 15, 1993

Processing and Applications

Chapter 25

Mechanism of Fiber Formation of Silkworm

Jun Magoshi[1], Yoshiko Magoshi[2], and Shigeo Nakamura[3]

[1]National Institute of Agrobiological Resources, Tsukuba, Ibaraki 305, Japan
[2]National Institute of Sericultural and Entomological Science, Tsukuba, Ibaraki 305, Japan
[3]Department of Applied Chemistry, Faculty of Engineering, Kanagawa University, Kanagawa-ku, Yokohama 221, Japan

Larvae of the silkworm, *Bombyx mori*, control the molecular orientation of their cocoon fiber with several sophisticated spinning techniques. This chapter reviews the interactions of behavior, morphology, and biochemistry in the production of silkworm silk. Numerous physical and chemical variables can induce transitions among the four known phases of this silk: random-coil, α, β, and well-oriented β form conformation. Fibroin solution becomes fibrous as it moves anteriorly through the silk gland. Cations mediate a gel-to-sol transition. Viscosity falls and strength increases as pH drops to about 4.9 and the incipient fibers assume a liquid crystal character. Strain-rate effects on yield stress are profound in lab-drawn fibers, the larva pulls fibers by segments in a figure-eight pattern and cannot spin if its movements are restrained. The resulting fiber is superdrawn.

New textures of synthetic fiber under development include silk-like(1), spun-like(2), and leather-like(3) materials. These are intended to serve as imitations of natural fibers. Much of the development of synthetic-fiber technology progressed by imitation of the spinning of silkworms.

Attractive features of silk fiber include its pearl-like gloss and its light velvety touch. Silk cloth can be easily dyed. It is comfortable to wear. Silk is fashionable fiber.

0097–6156/94/0544–0292$06.00/0

In this chapter we examine the details of crystallization and fiber formation during cocoon spinning.

EXPERIMENTAL

Materials

For experiments on the mechanism of fiber formation, liquid silk was obtained from the posterior part of the middle division of the silk gland in full-grown larvae (one day before spinning or cocooning) of the silk worm *Bombyx mori*(4). The second silk protein, sericin in liquid silk was removed by washing the silk gland thoroughly with deionized water. The sericin surrounds the bulk fiber fibroin as a separate layer and serves as a lubricant in the cocoon spinning. For stress-strain measurements, the silk gland was cut about 3cm in length(5). To examine the liquid crystal of fibroin, the specimen was prepared from the anterior division of the silk gland.

All experimental materials of spinning of silkworm used were full-grown larvae in spinning.

Method

Stress-strain curves were measured on a Tensilon tensile tester(UM-1-2) at 20°C. Strain rates were varied in a range of 10-1000 mm/min. The cross section of specimen was calculated from the length and weight of the specimen. X-ray diffraction patterns were recorded with a Rigaku D-3F x-ray apparatus. Raman spectra were taken on a JASCO Modle R-80 Raman spectrometer. Birefringence measurements were made by the optical method using a compensater, and optical diffraction patterns were obtained by laser light scattering. Viscosity of liquid silk was determined by penetration measurement using a penetrater made by Fudo Kogyo Ltd. The liquid crystal was observed under polarized light by a Nikon AMF microscope.

RESULTS

Crystallization of Silk Fibroin

Fibroin has a relatively simple structure and is easily available from natural products. Three types of conformation, the random coil, the α form(fibroin I, crank shaft pleated-sheet structure)(7, 8), and the β form(fibroin II, antiparallel-chain pleated-sheet structure)(7, 8) have been observed by x-ray diffraction and infrared spectroscopy. Three forms of fibroin film can be obtained by varying the preparation conditions such as concentration of fibroin and the casting and quenching temperatures of the aqueous solution. The α and β form conformations are very stable to heating and treatment with organic solvents. However, conformational changes occur easily from the random-coil to the α, β and well-oriented β forms by mechanical shearing, treatment with organic solvents, heat treatment, or by varying treating conditions and under several other conditions(4).

Crystallization of silk fibroin was studied in detail from dilute aqueous solution and from amorphous random-coil film. The effects of crystallization conditions on the conformation of fibroin are summarized in Figure 1.

Crystalline films in the α and β form conformations can be obtained by casting an aqueous solution at 0 to 40 ℃ and above 50 ℃, respectively, whereas by casting of a dilute solution at 0 to 45 ℃ results in amorphous films with a random-coil conformation (Figure 2)(6).

When an aqueous solution of fibroin is frozen by quenching prior to drying, β form crystals are obtained irrespective of the concentration of fibroin at quenching temperature from -2 to -20 ℃. On the other hand, when the solution is quenched to below -20 ℃ and then dried at 20 ℃, both α, β form crystals and an amorphous film with random-coil conformation are obtained depending on the concentration of fibroin in solution. Fibroin molecules in dilute aqueous solutions are in the random-coil conformation. Therefore, extremely rapid freezing of a very dilute solution does not permit a conformational change to occur to the β form, and the random-coil conformation in a dilute solution is maintained in fibroin films.

Silk gland

The silk glands are a pair of tubes lying one on each side of the larvae as shown in Figure 3. The schematic diagram and photograph show the silk gland from the silkworm *Bombyx mori*. Each gland is divided into three divisions: posterior, middle and anterior divisions. The epithelial wall of the gland consists of large hexagonal cells where the silk fibroin and silk sericin are synthesized. The silk proteins are stored inside the gland.

The posterior division, which is closed at one end, is narrow and very convoluted. This is where the main silk protein, fibroin, is synthesized. The fibroin moves forward into the wider, middle division, which serves as a

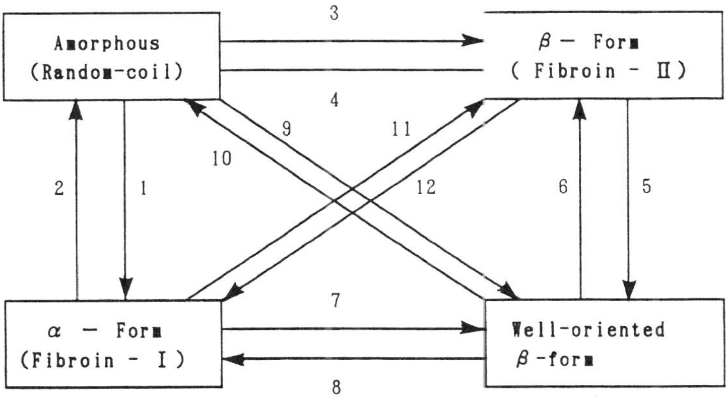

1) Casting from aqueous solution (>5%) at 0 to 45°C.
   Quenching aqueous solution (>5%) below −20°C prior to drying.
   Crystallization from aqueous solution at pH above 6.0.
   Epitaxial growth on Nylon 66.
   Treatment with water at 0 to 50°C.
2) Application of high pressure.
3) Casting from aqueous solution (>5%) above 45°C.
   Quenching aqueous solution (>5%) at 0 to −20°C prior to drying.
   Mechanical shearing.
   Application of an electric field to aqueous solution.
   Treatment with polar organic solvents.
   Crystal ization from aqueous solution at pH below 4.5.
   Heat treatment at 190 to 200°C.
   Treatment of film with hot water at above 60°C.
   Treatment of aqueous solution with enzyme (Chymotrypsin)[10].
4) Application of high pressure.
5) Drawing of liquid fibroin or film.
6) Application of high pressure.
   Treatment with salt.
7) Heating to 270°C[11].
   Mechanical shearing.
8) Treatment with salt solution.
9) Mechanical shearing.
   Epitaxial growth.
10) Application of high pressure.
11) Mechanical shearing.
12) Heat treatment by wet process.

Figure 1. Crystallization of silk fibroin induced under various conditions.

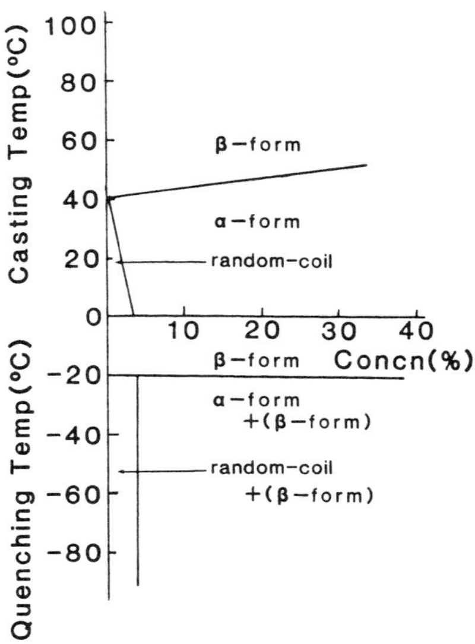

Figure 2. Effects of the casting temperature, freezing temperature and concentration on the confor-mation of silk fibroin.

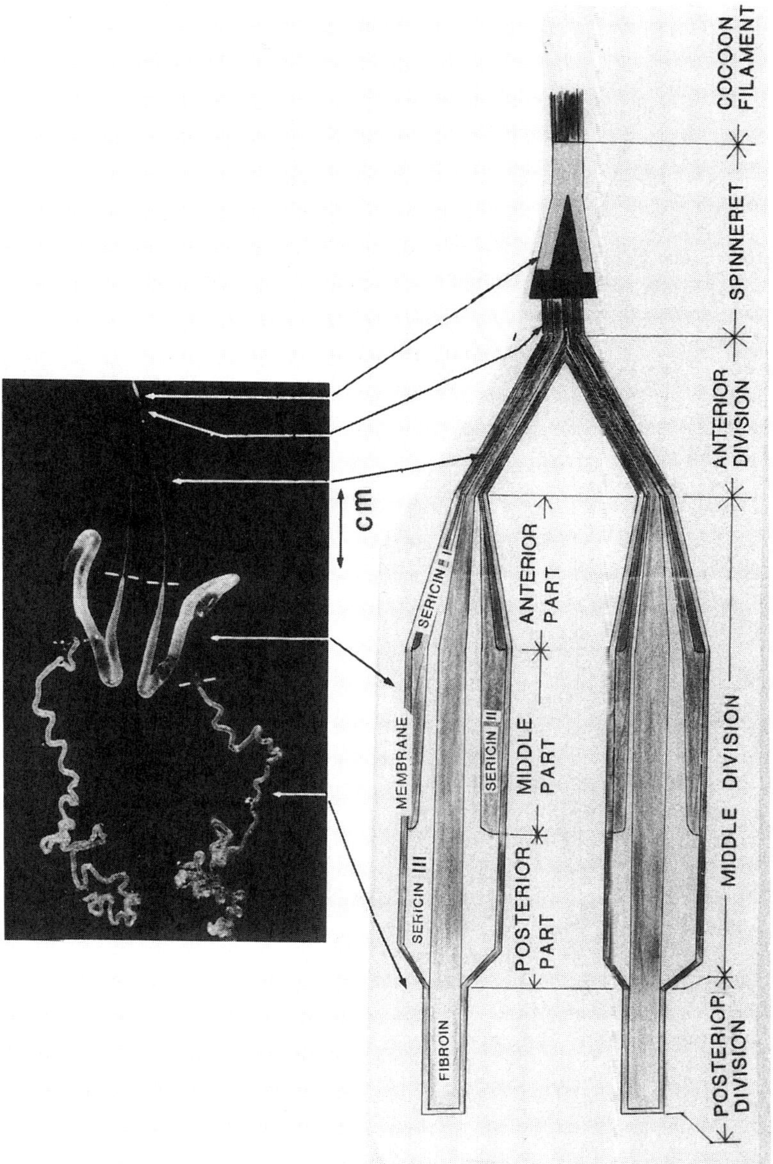

Figure 3. Photo micrograph and schematic diagram of the silk gland of the larvae of the silkworm, *Bombyx mori*.

reservoir. The second silk protein, sericin, is synthesized in this section and accumulates as a separate layer around the fibroin. The two proteins proceeds together without mixing into the anterior division, which becomes narrower as it approaches the spinneret. The two glands join together just before the spinneret. At the spinneret, twin fibroin cores are surrounded by a layer of sericin. The spinneret of silkworms consists of a silk press and an orifice, as seen in Figure 4.

The common gland, which contains two separate fibroin streams and an outer sericin stream, goes to the silk press without mixing. The silk press acts as a nozzle in the spinning of fibroin. In the cross section (Figure 4), the silk press is in the form of the two crescent shapes. In this part, the liquid silk is fixed for making the silk fiber. On extrusion through the spinneret into the air, the proteins are converted from a concentrated viscous solution into a thin, glossy, and strong filament, which cannot easily be reconverted into solution. A remarkable feature of this cocoon spinning is the stretching, brought about by the repeated drawing back of the silkworm's head. This stretching causes orientation of the long fibrous molecules in the direction of drawing. Water does not evaporate in the spinneret, but is subsequently lost from the spun filament in the air.

Nematic and Lyotropic Liquid Crystals(9)

The end of the anterior division of the silk gland is cut off immediately after being taken out from the cocooning silkworm. The liquid silk flows out spontaneously and continuously. The liquid silk flowing out is slightly birefringent under polarized light (Figure 5). In the same region of the anterior division of the silk gland, an order is detected in the direction perpendicular to the silk gland. Therefore, the fibroin molecules from the anterior region of the silk gland of *Bombyx mori* assume the nematic liquid crystal phase and the β form conformation.

On the other hand, when the silk gland is allowed to stand for 20min at 20 ℃ in the air and then the end of the anterior region is cut off, the liquid silk does not flow out spontaneously because the concentration of fibroin is increased due to evaporation of water. The liquid silk pushed out by hand is birefringent and oriented in the direction of the silk gland (in Figure 6). The force required to push out the liquid silk is very low. Therefore no crystallization is induced by this procedure.

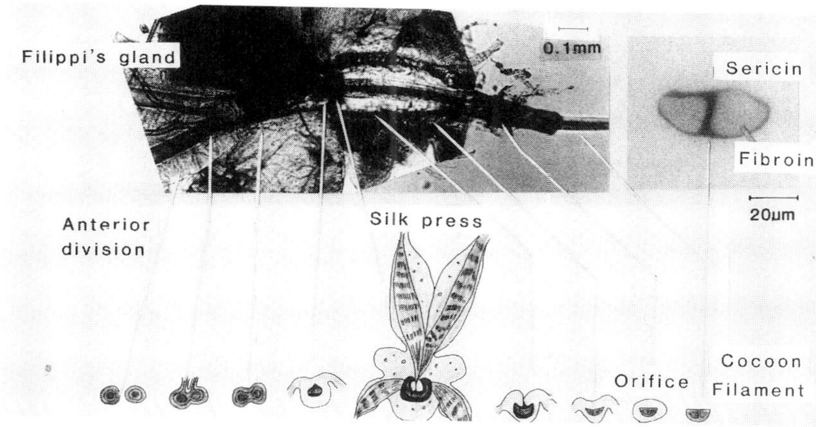

Figure 4. Photo micrograph of the spinneret and schematic cross sections of the spinneret of silkworm, *Bombyx mori*.

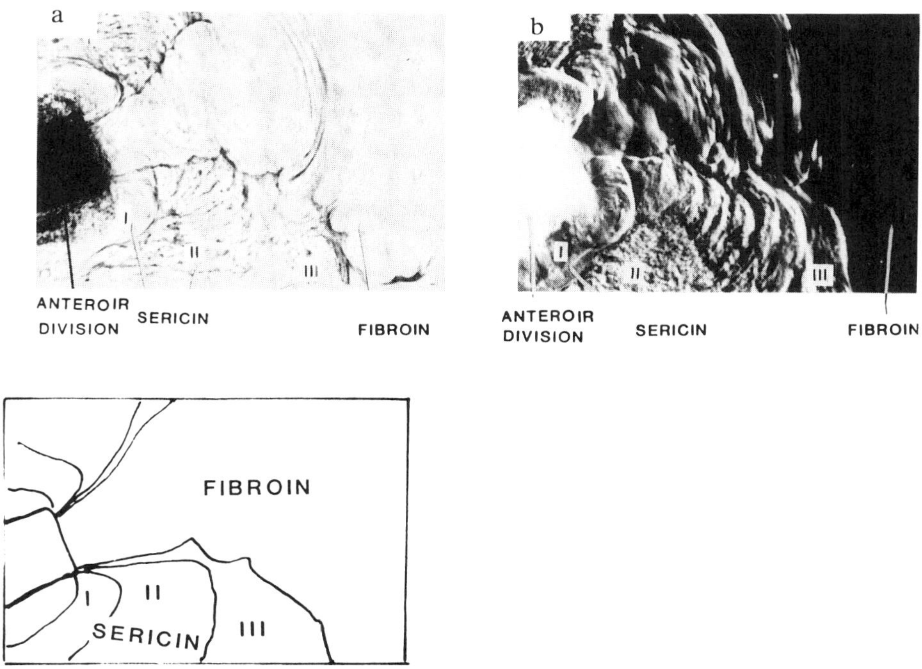

Figure 5. Photo micrographs of the liquid silk from the anterior division of silk gland; (a) in unpolarized light and (b) in polarized light. Observed immediately after the anterior division was cut off.

Moreover, liquid crystals are observed for wild silkworms, *Antheraea perni*, *Atheraea yamamai*, and *Dictyoploca japonica* (Figure 7).

From these results, the nematic liquid crystalline phase is produced in the liquid silk coming out from the anterior division of the silk gland. Microphotographs of the liquid crystal of silk fibroin immersed in water for 1min to 1hr are shown in Figure 8. The liquid crystal of fibroin is dispersed in immersing water. The gel of liquid crystal of silk fibroin is easily dissolved in water.

Dry Spinning

Figure 9 shows scanning electron micrographs of silkworm extruding filament through the spinneret into the air. The specimen in Figure 9a is freezed by quenching to -196 ℃. The specimen in Figure 9b is dried at room temperature after quenched to -196 ℃. Two fibroin filaments are observed in Figure 9a. No shrinkage occurs in the freezed specimen, even with loss of water during drying.

The water content of a cocoon fiber in spinning silkworm is 70% at the spinnert. When the fiber is got out in the air, the water is evaporated immediately from the fiber, many pores are formed in the fiber. Therefore, the spinning of silkworm is the dry spinning and the porous spinning.

Superdrawing

The anterior division of a spinning silkworm is taken out and cut off at its end and the liquid silk flowing out is drawn at the rate corresponding to that of cocoon spinning(6 ~ 8mm/sec). The liquid silk flows out very smoothly in the direction of drawing as shown in Figure 10. This is so-called elongational flow. The resulting filament is in the well-oriented β conformation from x-ray diffraction pattern. The cocoon spinning is the super drawing considering the tremendous decease in diameter of the silk gland.

Gel Spinning

Fibroin is synthesized in the posterior division and moves into the wider middle division as a 12% aqueous gel. The gelation induced by Ca^{2+}, Mg^{2+}, and k^{+} ions, which comes from mulberry leaves, the foodstuff of silkworm. The second protein, sericin is produced in the middle division. The

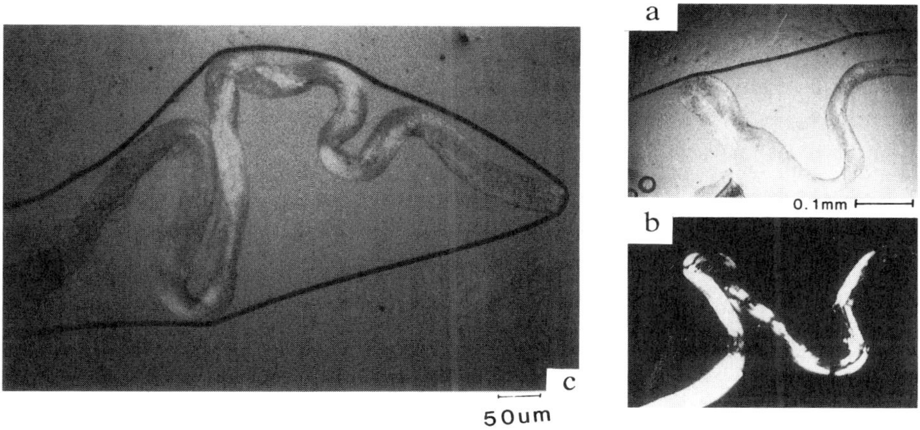

Figure 6. Photo micrographs of the nematic liquid crystal of liquid silk from the anterior division; (a) in unpolarized light, (b) in polarized light, and (c) in polarized light under a 1/4 λ sensitive color plate. Observed after standing for 20 min at 20 ℃.

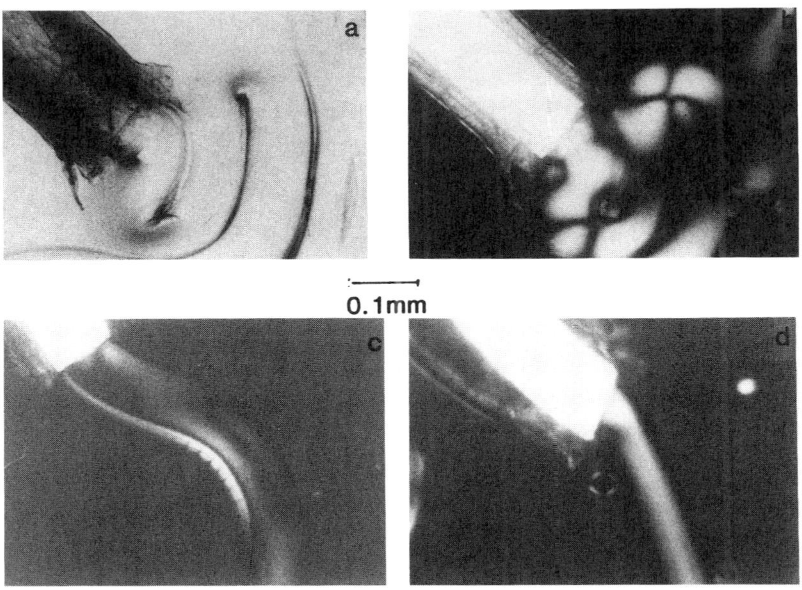

Figure 7. Photo micrographs in polarized light of the liquid silk from the anterior division of wild silkworms; (a) *Antheraea perni* in unpolarized light, (b) *Antheraea perni*, (c) *Antheraea yamamai*, and *Dictyoploca japonica*.

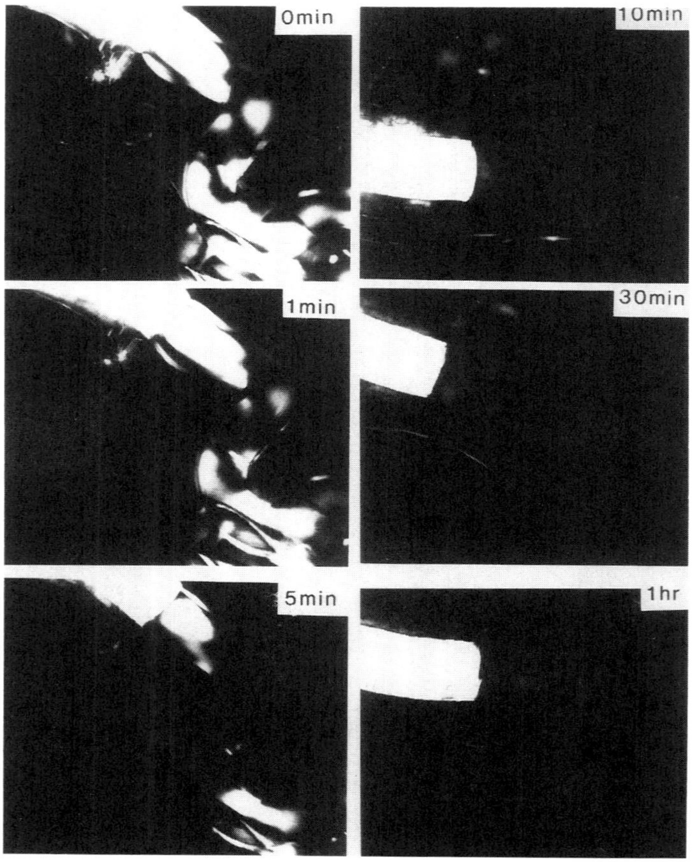

Figure 8. Lyotropic liquid crystal of the liquid silk from the anterior division immersed in water.

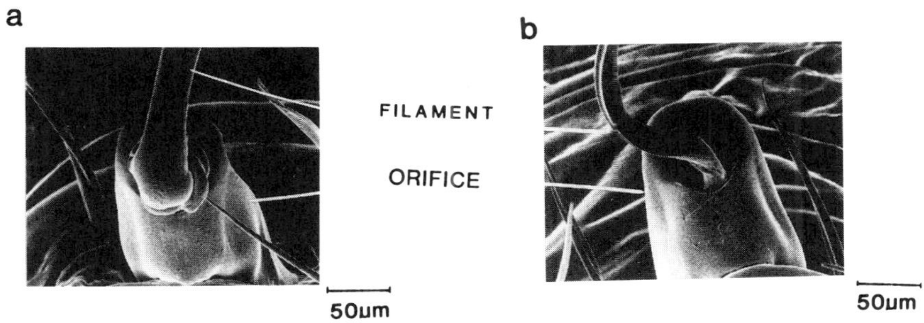

Figure 9. Scanning electron micrographs of the silkworm which is extruding filaments into the air; (a) quenched to -196 ℃ and (b) quenched to -196 ℃ and dried at room temperature.

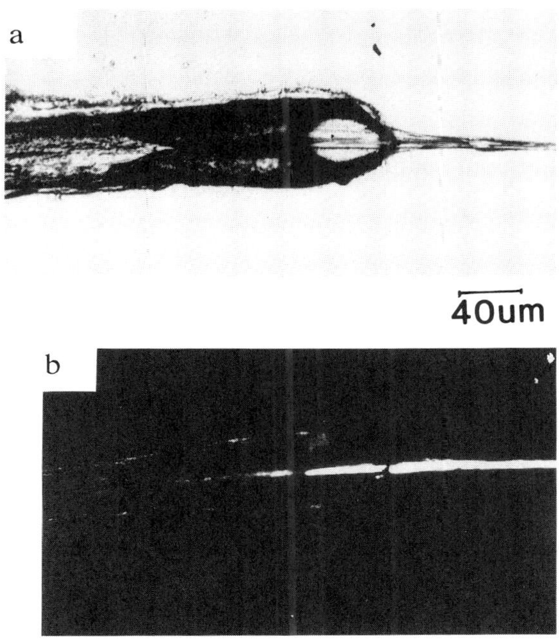

Figure 10. Photo micrographs of the drawn liquid silk from the anterior division; (a) in unpolarized light and (b) in polarized light.

water in the fibroin gel goes out into the sericin layer, and the fibroin gel is concentrated to about 25% by the further increase in metal cations. Moreover, the water in the sericin and fibroin solutions goes out through the cell wall of the gland. In the middle division consisting of three parts, anterior, middle, and posterior parts where sericin I, II, and III are synthesized, respectively (Figure 3). The total concentration of sericin is about 7%. The four proteins move forward together without mixing into the anterior division, which becomes narrower as it approaches the spinneret, and the twin cores of fibroin are surrounded by a matrix of sericin.

Viscosity of liquid silk is determined by measuring the stress required for a probe to penetrate into the specimen at predetermined rates. The viscosity of gel in the middle division of the liquid silk decreases as it approached the anterior division . Therefore,the viscosity of liquid silk may be low in the anterior division even though the liquid silk of the anterior division is too thin to allow measurement. The decrease in viscosity suggests the formation of an ordered structure in solution (Figure 11).

pH of Silk Gland

The pH of fibroin solution at middle division decreases as it approaches the anterior division, and become 4.9 at the anterior end of the middle division. The pH may be acidic in the anterior division, which is probably favorable for the β conformation, because β form crystals are obtained at pH 4.5 from an aqueous solution(9).

The yield stress (strength of gel) of aqueous solutions of regenerated fibroin from silk filaments changes with pH. AT pH 5-6.5, the yield stress reaches a maximum(Figure 12). At lower pH values, the yield stress and viscosity are low. Therefore, in the cocooning of silkworm, the viscosity of the liquid silk in the anterior division become lower due to the lowering in the pH and by the formation of liquid crystals. Figure 13 shows the yield stress of gel as a function of temperature. The yield stress of fibroin occurs at higher pH with increasing temperature.

Ion Spinning (Gel-Sol Transition)

The concentration of metal ions in the liquid silk increases from the posterior division to the anterior division. The gel strength is decreased with increasing Ca^{2+} concentration

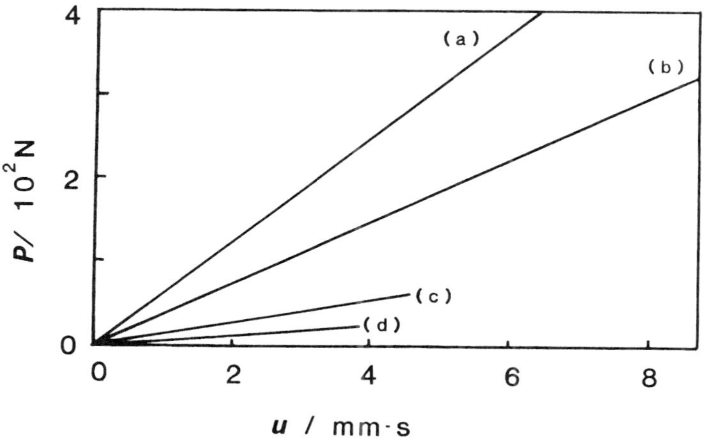

Figure 11. Stress required for penetration vs. rate of penetration at different parts of the middle division of liquid silk; (a) Middle portion, (b) posterior part of anterior portion, (c)middle part of anterior portion, and (d) anterior part of anterior portion of the middle division of liquid silk.

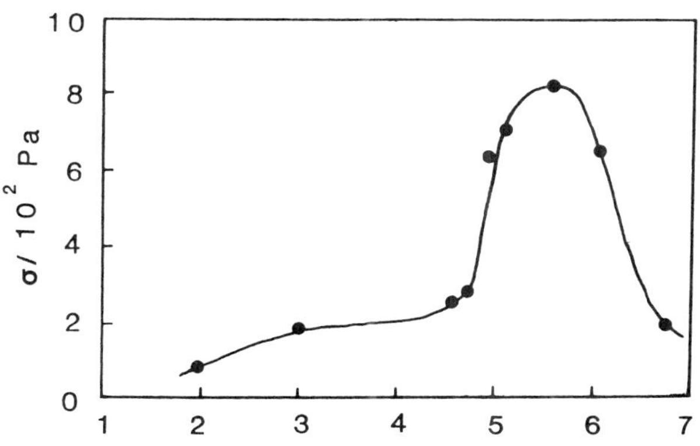

Figure 12. Change of yield stress (gel strength) of silk fibroin solution with pH at 20°C.

of in fibroin solutions (Figure 14). Therefore, the gel-to-sol transition occurs by Ca^{2+} in the anterior division, even though the mechanism is not yet clarified.

Effect of Draw Rate

When the liquid silk in the middle division of the silk gland is drawn, it behaves as a non-Newtonian fluid at the extension rates of 10-1000mm/min (Figure 15). As the extension rate is increased, the stress required to stretch the liquid silk increases. At lower rates of 10-75 mm/min, neither conformational change nor orientation of fibroin molecules was observed by x-ray diffraction and birefringence measurements. At rates higher than 500 mm/min, the stress-strain curves become irregular beyond the yield points and the specimens become opaque. Most of the water initially present is squeezed out by drawing. The yield point shifts to higher elongation with increasing extension rate. At lower extension rates the yield force is very low, whereas at 50-450 mm/min the value becomes almost constant (Figure 16). From x-ray diffraction patterns, laser Raman spectra, and optical diffraction pattern of undrawn and drawn specimens, the conformation is transformed to the β form by drawing (Figure 1).

Therefore, the conformational change does not occur until the extension rate of 500 mm/min is attained. At rates higher than 500mm/min, the conformation of fibroin in liquid silk undergo a transformation from the random coil to the β form, crystallization occurs to the well-oriented fibroin. Silk filament is produced by drawing liquid silk at rates higher than 500mm/min at 20℃. On the other hand, the spinning rate of silkworm is 360-480mm/min which is lower than that of in vitro. The gel-sol transition takes an important part in the lowering of spinning rate of silkworm in nature.

Zone Drawing and Self Drawing

A silkworm moves its head in a figure-eight pattern during cocoon spinning. Immediately after the liquid silk comes out from the spinnert, it is fixed at the outside walls by sericin, the outer layer of the filament. With this point fixed the liquid silk in the sol state is clamped by the silkpress of the nozzle and drawn out by the motion of silkworms head. Therefore this is zone drawing.

Cocoon filament coming out from the spinning part of silkworm is fixed at the outside walls. In the spinning of silkworm the spinnert is moved, while cocoon filament is fixed. Therefore, the spinning of silkworm is carried out by the force of silkworm itself.

If the movements of silkworm are restricted and its

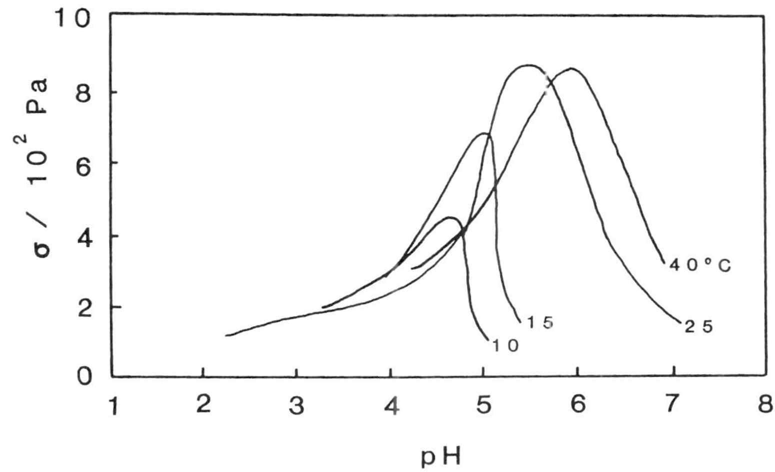

Figure 13. Change of yield stress of silk fibroin solution with pH at various temperatures.

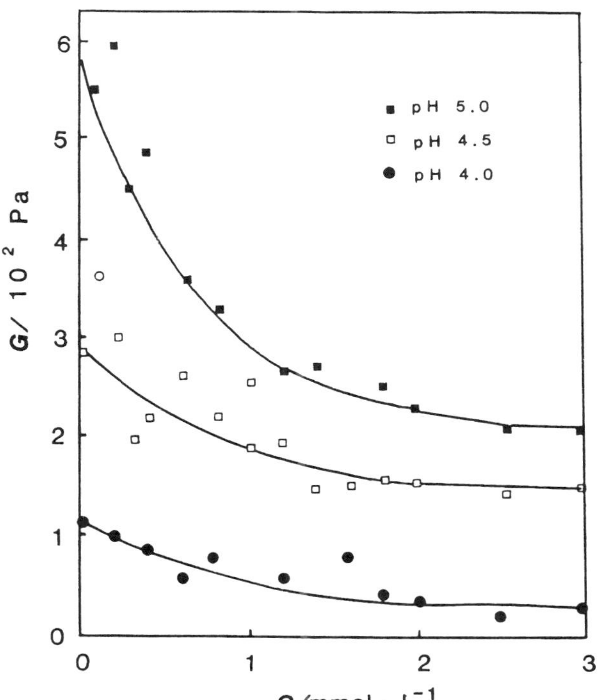

Figure 14. Change of yield stress of silk fibroin solution with Ca^{2+} concentration at various pH values.

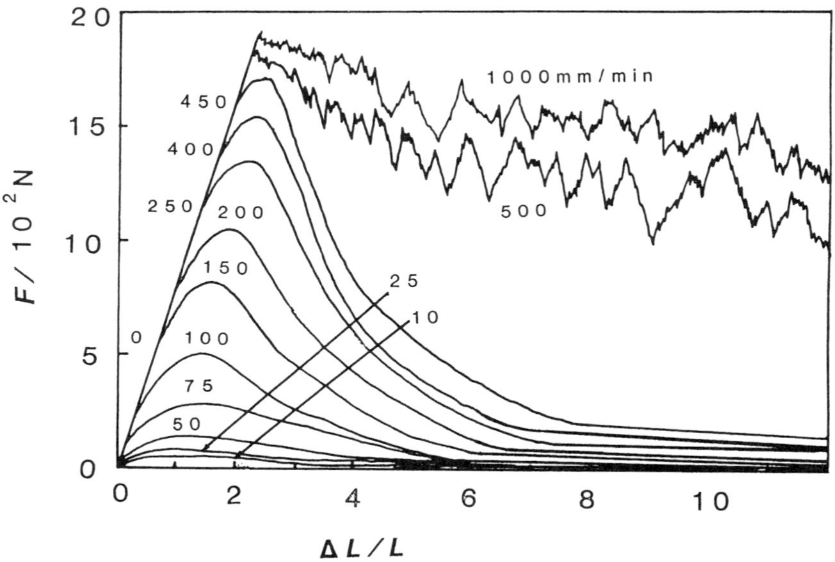

Figure 15. Stress-strain curves of liquid silk in the middle division of silk gland observed at extension rates of 10-1000 mm/min.

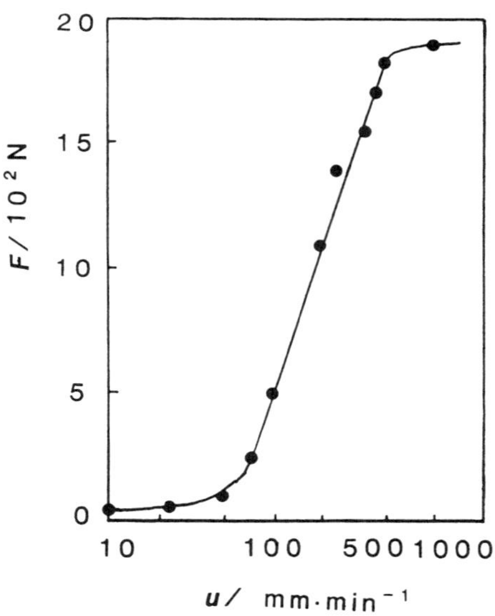

Figure 16. Yield stress-extension rate relationship for liquid silk.

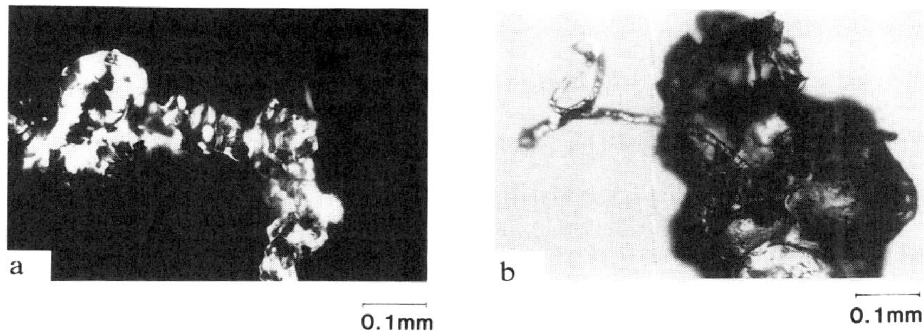

Figure 17. Liquid silk flowing out from the spinnert of silkworm when the movement of silkworm is restricted and it spinneret can not touch anywhere: (a) in unpolarized light and (b) in polarized light under a sensitive color plate.

spinneret can not touch anywhere during cocoon spinning. Then the silkworm could not produce filament (Figure 17).

Conclusion

Larvae of domestic silkworm, *Bombyx mori* and other wild silk worms, control the molecular orientation of their cocoon fiber with several sophisticated spinning techniques. Numerous physical and chemical variables can induce transitions among the four known phases of the silk fibroin. Fibroin solution becomes fibrous as it moves anteriorly through the silk gland. Cations mediate a gel-to-sol transition. Viscosity falls and strength increases as pH drops to about 4.9 and the incipient fibers assume a liquid crystal character. Strain-rate effects on yield stress are profound in lab-drawn fibers, the larva pulls fibers by segments in a figure-eight pattern and cannot spin if its movements are restrained. The resulting fiber is superdrawn. The silkworm performs molecular orientation control very accurately by methods involving numerous sophisticated spinning technologies such as gel spinning, liquid crystal spinning, high speed spinning, self-exerted spinning, zone elongation and porous spinning , ion spinning, dry spinning, crimp spinning, and low energy spinning which cannot yet be duplicated by advanced artificial spinning technologies.

Reference

1) O. Wada, Sen'i Gakkaishi, (J. Soc. Fiber Sci. Tech., Jpn.) **40**, P300 (1984).
2) M. Tani, Sen'i Gakkaishi, (J. Soc. Fiber Sci. Tech., Jpn.) **40**, P303 (1984).
3) S. Okamoto, Sen'i Gakkaishi, (J. Soc. Fiber Sci. Tech. Jpn.) **40**, P307 (1984).
4) J. Magoshi, Y. Magoshi, and S. Nakamura, J. Appl. Polym. Sci., Appl. Polym. Symp. **41**, 187 (1985).
5) J. Magoshi, Y. Magoshi, and S. Nakamura, Polym. Comm. **26**, 309 (1985).
6) J. Magoshi and S. Nakamura, Nat. Inst. Agro. Reso. **1**, 37 (1985).
7) M. Shimizu, J. Sericult. Sci. Jpn., **20**, 155 (1951)
8) O. Kratky and E.Schauenstein, Discuss, Faraday Soc., II, 171 (1951).
9) J. Magoshi and S. Nakamura, ACS Symp. Ser., vol **489**, 231 (1992).

RECEIVED July 27, 1993

Chapter 26

Spinning of Protein Polymer Fibers

J. Cappello[1] and Kevin P. McGrath[2]

[1]Protein Polymer Technologies, Inc., 10655 Sorrento Valley Road, San Diego, CA 92121
[2]Biotechnology Division, Natick Research, Development, and Engineering Center, U.S. Army, Natick, MA 01760–5020

Initial studies in the production of fibers from sequence controlled protein polymers have revealed that the properties of such fibers are extremely sensitive to processing conditions. We investigated methods for extrusion of high molecular weight protein polymers consisting of repeating blocks of amino acid sequence. The amino acid sequence of blocks used to design these protein polymers were modelled from the natural fibrous proteins, silk and elastin. These blocks were utilized in various lengths and dispersities such that a set of protein polymers was produced which could be systematically varied in composition from completely silk-like to primarily elastin-like. Additionally, an amino acid sequence block which confers mammalian cell adhesion was incorporated within the silk-like polymer. Several of these protein polymers were spun into fibers. The methods used for fiber spinning did not result in fibers with any appreciable molecular orientation. The properties of these fibers are discussed.

Commercially successful fibers have been produced from a variety of macromolecules either obtained from nature or synthesized chemically. Fiber properties are directly affected by the chemical nature of the monomers that make up the polymer, the composition and structure of the polymer chain, and the processing required to fashion the polymer into fibers. The more precisely one can control the composition of the polymer fibers, the more successfully one may produce fibers with properties ideally suited to a particular application.

Through our previous work, a set of high molecular weight protein polymers, consisting of a peptide sequence derived from the crystalline region of Bombyx mori silk fibroin have been designed for fiber production (1). One of these, SLP4 (silk-like polymer), consists of an extensive repetition of the hexapeptide sequence glycine-alanine-glycine-alanine-glycine-serine. The other polymers of this set contain discrete, periodic interruptions of the hexapeptide repeats with sequences designed to add chain flexibility, biological or chemical function. The silk-like polymer, SLP3, contains the pentapeptide block glycine-alanine-alanine-glycine-tyrosine included after

0097–6156/94/0544–0311$06.00/0

every nine repeats of the hexapeptide silk-like block. The SELP (silk-elastin-like polymer) block copolymers interrupt the repeated silk-like blocks with the inclusion of discrete numbers of the pentapeptide valine-proline-glycine-valine-glycine. This sequence block is derived from elastin, the major elastomeric fibrous protein in mammals (2). SELP1 contains four repeats of the elastin pentapeptide per interruption and SELP3 contains eight repeats. The polymer SLPF contains the insertion of a 17 amino acid sequence block containing the cell attachment epitope of fibronectin, a human plasma protein that mediates cell adhesion (3). Because we produce protein polymers by genetically controlled biosynthesis, the primary sequence of the chains is assured. Table I lists the sequence and theoretical molecular weights for each of these protein polymers.

Figure 1 graphically displays the sequences of these protein polymers as a function of the average hydrophobicity of the amino acids along the chain. Using a sliding window of 11 amino acids, the hydrophobicity was calculated using the Kyte-Doolittle amino acid hydrophobicity index. The hydrophobicity was then plotted for each polymer with respect to a zero value which is indicated by the horizontal lines. The silk-like blocks have a slight positive hydrophobicity as observed by values just above the zero line. Elastin-like blocks, because of their valine content, score a much greater hydrophobicity and are observed as plateaus above the silk blocks. Hydrophilic interruptions in the silk-like blocks, corresponding in SLPF to the positions of the cell attachment blocks, are observed as negative valleys below the zero line. This display clearly shows the polymeric nature of these protein polymers and how they compare in block length and dispersity.

The goal of the present study was to demonstrate the spinnability of these protein polymers and through the evaluation of the fibers obtained, begin to establish a relationship between the primary sequence of the chains and the ultimate physical properties of the fibers.

Methods and Results

Solvent Screening. The solubilities of all four polymers were determined in a number of potential dope solvents at room temperature. A threshold level of solubility of 50 mg/ml (5 weight percent) was used as a minimum criterion for suitability for fiber spinning. Solvents were chosen on the basis of previous knowledge or disclosures found in patents on the production of various protein fibers (4-23). Solubility was assessed visually on the basis of clarity and viscous homogeneity. The results are shown in Table II.

The polymers were soluble in strong acids such as sulfuric (35.6 M), dichloroacetic (12.1 M), formic (27 M), and phosphoric (44 M). However, in sulfuric and dichloroacetic acids the polymer solutions turned dark brown and purple indicating some reactivity. Gel permeation chromatography of the polymer dissolved in these solvents revealed that the polymer chains had undergone virtually complete degradation (data not shown). The polypeptide chains were relatively stable in formic and phosphoric acid for 16 hours; however, greater exposure also caused degradation (data not shown). From these results, two solvents consisting of 50% phosphoric acid (22 M) and 9M LiBr were chosen for further analysis based on their ability to solvate the polymers and their miscibility with a variety of potential coagulant systems.

Coagulation studies were initiated using polymer dopes of SLP3, SLP4, and SELP3 at several different polymer concentrations, as well as at different

Table I. Amino Acid Sequences of Protein Polymers

Polymer Name/Amino Acid Sequence
SLP4
fMDPVVLQRRDWENPGVTQLNRLAAHPPFASDPMGAGS
$[(GAGAGS)_6]_{27}$
$(GAGAGS)_5$
GAGAMKPGRYQLSAGRYHYQLVWCQK
Chain length=1398 amino acids   Molecular weight=76 kd
SLP3
fMDPVVLQRRDWENPGVTQLNRLAAHPPFASDPMGAGS
$(GAGAGS)_6$ GAAGY
$[(GAGAGS)_9\ GAAGY]_{18}$
$(GAGAGS)_2$
GAGAMDPGRYQLSAGRYHYQLVWCQK
Chain length=1171 amino acids   Molecular weight=83 kd
SELP1
fMDPVVLQRRDWENPGVTQLNRLAAHPPFASDPMGAGS
(GAGAGS) $[GAA(VPGVG)_4\ VAAGY\ (GAGAGS)_9]_{12}$
$GAA(VPGVG)_4$ VAAGY $(GAGAGS)_2$
GAGAMKPGRYQLSAGRYHYQLVWCQK
Chain length=1205 amino acids   Molecular weight=89 kd
SELP3
fMDPVVLQRRDWENPGVTQLNRLAAHPPFASDPMGAGS
$(GAGAGS)_2\ [(GVGVP)_8\ (GAGAGS)_8]_{11}\ (GVGVP)_8\ (GAGAGS)_5$
GAGAMKPGRYQLSAGRYHYQLVWCQK
Chain length=1113 amino acids   Molecular weight=89 kd
SLPF
fMDPVVLQRRDWENPGVTQLNRLAAHPPFASDPMGAGS
$(GAGAGS)_6$ GAAVTGRGDSPASAAGY
$[(GAGAGS)_9\ GAAVTGRGDSPASAAGY]_{12}\ (GAGAGS)_2$
GAGAMKPGRYQLSAGRYHYQLVWCQK
Chain length=980 amino acids   Molecular weight=73 kd

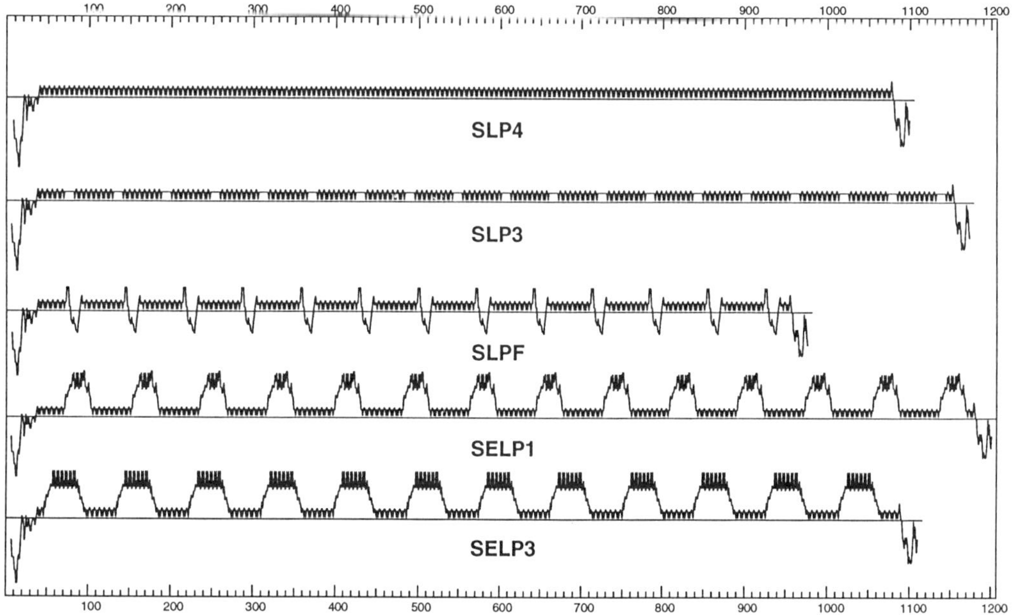

Figure 1: Hydropathy plot of protein polymers. The amino acid sequences of five protein polymers (Table I) are plotted on the basis of hydrophobicity using the Kyte and Doolittle scale and a sliding window length of 11 residues. Numbers along the top and bottom of the figure indicate amino acid sequence position. Each polymer has its own x-axis. Units in the y-direction are arbitrary but identical scales are used for all polymers. Points above the x-axes are hydrophobic and those below are hydrophilic. Stretches of silk-like blocks can be seen as short plateaus just above the x-axes. Elastin-like blocks are elevated points above the silk-like blocks, and the peaks below the x-axis in the polymer, SLPF, mark the positions of the cell attachment blocks. The head and tail peptides, which are not part of the repeating polymer sequence, are observed as peaks below the x-axes of all polymers.

concentrations of the coagulating species. SELP1 was excluded from these screening studies because of its limited supply. Because its composition is intermediate to the other polymers, its behavior in the various coagulants might be predicted to fall between SLP3 and SELP3 (see Table I and Figure 1).

Table II. Solubility Characteristics of Protein Polymers (wt%)

SOLVENT	*SLP3*	*SLP4*	*SELP1*	*SELP3*
Li salts:				
9M LiBr	10	10	10	10
6M LiBr	10	<5	ND	10
4.5M LiSCN	10	5	ND	10
Acids:				
88% HCOOH	10	10	10	15
85% H_3PO_4	>20	>10	20	>20
50% H_3PO_4	20	10	ND	>20
DCA[a]	10	10	10	10
H_2SO_4[b]	10	10	10	10

ND-not done. [a]turned purple. [b]turned dark brown.

The coagulants were chosen based on disclosures in the literature, their ability to precipitate proteins, their ready availability, and their low cost (4-23). They consisted of water at high temperature (>60° C), ammonium sulfate solutions, acetic acid, isopropanol, and acetone. The coagulants were tested individually and at several concentrations with polymer dopes of SLP3, SLP4, and SELP3 at several different polymer concentrations by extruding a stream of dope solution through a 32 gauge needle into a bath containing 150-200 ml of the coagulant and visually assessing the degree of precipitation, the speed of precipitation, and the morphology of the precipitate (amorphous, globular or fibrous).

The initial screen of all coagulants showed that none of the polymer solutions would precipitate in hot water regardless of temperature. In acetone, all the polymers rapidly precipitated upon contact, but the solidified precipitate showed no fibrous quality. In the case of isopropanol, the coagulation was too slow based on a reasonable residence time projected for the spinning of fibers (approximately one minute). Aqueous solutions of acetic acid and ammonium sulfate gave good coagulation results in this initial screen and were chosen for more detailed evaluation. Although all combinations of dopes and coagulants were tried, the more compatible were 9 M LiBr polymer dope with acetic acid coagulant and phosphoric acid polymer dope with ammonium sulfate coagulant. The results of these studies are listed (Tables III and IV).

Based on these results, spinnability was predicted for SLP3 and SLP4 using the LiBr polymer dope/acetic acid coagulant system. For SELP3, the phosphoric acid polymer dope/ammonium sulfate coagulant system gave the best results. Under these conditions, the precipitates formed were fibrous enough to be directly pulled from the coagulant vessel as apparent monofilaments or multifilaments.

Table III. Coagulation Results Using Phosphoric Acid Polymer Dope and Ammonium Sulfate Coagulant Solution

	POLYMER DOPE COAGULANT CONCENTRATION (% of saturated aqueous solution at 25° C)					
Wt% polymer	*50%*	*60%*	*70%*	*80%*	*90%*	*100%*
10% SLP3	no ppt	no ppt	v. slow no spin	PPT no spin	PPT no spin	PPT no spin
7.5% SLP3	no ppt	no ppt	no ppt	v. slow no spin	PPT no spin	PPT no spin
5% SLP3	no ppt	no ppt	no ppt	no ppt	v. slow no spin	PPT no spin
10% SLP4	no ppt	no ppt	v. slow no spin	PPT no spin	PPT no spin	PPT no spin
7.5% SLP4	no ppt	no ppt	no ppt	no ppt	v. slow no spin	PPT no spin
5% SLP4	no ppt	no ppt	no ppt	no ppt	no ppt	v. slow no spin
10% SELP3	PPT no spin	PPT no spin	PPT **fibers**	PPT **fibers**	PPT **fibers**	PPT **fibers**
7.5% SELP3	PPT no spin	PPT no spin	PPT no spin	PPT no spin	PPT **fibers**	PPT **fibers**
5% SELP3	PPT no spin	PPT no spin	PPT no spin	PPT no spin	PPT no spin	PPT no spin

Table IV. Coagulation Results Using LiBr Polymer Dope and Acetic Acid Coagulant Solution

	POLYMER DOPE COAGULANT CONCENTRATION (Molarity of acetic acid)					
Wt% polymer	*2 M*	*4 M*	*6 M*	*8 M*	*10 M*	*17.4 M*
10% SLP3	no ppt	no ppt	no ppt	no ppt	no ppt	PPT **fibers**
10% SLP4	no ppt	no ppt	no ppt	no ppt	no ppt	PPT **fibers**
10% SELP3	no ppt	no ppt	no ppt	no ppt	no ppt	PPT no spin

Explanation of terms used in Tables III and IV:

no ppt - no precipitate formed in less than 5 minutes;
v. slow - precipitate formed between 1 minute and 5 minutes;
PPT - precipitate formed in less than 1 minute;
no spin - precipitate morphology was amorphous or globular, probably not spinnable;
fibers - fibrous precipitate observed upon extrusion from needle.

Fiber Spinning.

Phosphoric Acid:Ammonium Sulfate Fiber Spinning System. Although each of the dope/coagulant systems was shown to be successful for hand-pulling syringe extruded polymer dopes (20 weight percent protein or greater), the phosphoric acid polymer dopes yielded less viscous solutions at polymer concentrations greater than 20%. Suitable conditions, therefore, consisted of dissolving the polymer at 25 weight percent in 50% aqueous phosphoric acid and extruding directly into a coagulant bath containing 75% ammonium sulfate (3 M). This system was mechanized to produce continuously extruded fibers by using a motor driven syringe pump and reeler. Fifty and 200 meters of fiber were produced from SELP3 and SELP1 dopes, respectively, but SLP3 and SLP4 fibers could not be produced in this system.

In the cases of SELP3 and SELP1, the threadlines were coagulated immediately upon emerging from the tip of the needle (32 gauge, id=0.01 cm). Microscopic analysis of these fibers indicated that the fiber diameter varied from 0.004 cm for a reeling rate of 72 cm/min to 0.003 cm for 360 cm/min. In contrast, the extrusion of SLP3 and SLP4 dopes resulted in a liquid polymer droplet forming at the tip of the needle. An ultrathin fiber could be drawn from this droplet but with insufficient strength to allow reeling at any speed. A 50% blend of SLP3 and SELP3 was successfully spun using this system, although breakage occurred occasionally during reeling.

The fibers produced using this method were washed in an attempt to remove residual salt remaining from the coagulant. Both the SELP1 and SELP3 fibers had very poor stability in water. Upon wetting, the fibers swelled, became somewhat translucent, lost their rigidity, and in some cases, disintegrated (Figure 2, only SELP3 fibers are shown). Without washing, the dried fibers were mechanically stable but had no elasticity and were quite brittle. The poor fiber characteristics in both dry and wet states could be attributed to the presence of salt within the fibers. This was supported by determining that the protein content of the SELP3 fibers was approximately 37% by weight (data not shown). The exception to this behavior was the 50% SLP3:SELP3 blend fibers which did not swell appreciably in water but were still quite brittle in their dry state even after washing (Figure 3).

LiBr/Acetic Acid/Acetone Fiber Spinning System. Our experience with the phosphoric acid/ammonium sulfate spinning system taught us that a salt free coagulant system would be necessary for obtaining better quality fibers. If the new systems were to succeed in the spinning of the SLPs, it would be necessary to achieve a much more rapid coagulation of SLP3 and SLP4. According to our coagulant screening studies, the fastest precipitation of all the polymers occurred in acetone. However, such rapid precipitation was not conducive to spinnability. Mixtures of acetic acid and acetone were more effective.

Two of the four polymers, SLP3 and SELP3, were used to define spinning conditions by hand pulling fibers using 9M LiBr as the solvent and acetic acid/acetone as the coagulant. For SLP3 and SELP3, coagulant volume to volume mixtures of acetic acid to acetone of 60/40 and 80/20, respectively, were determined to be adequate. The viscosities of the dopes were so high that a concentration maximum of only 20 weight percent polymer could be pushed through the 32 gauge needle using the syringe pump. Since extrusion was considered to be more ideal with 40 weight

A

B

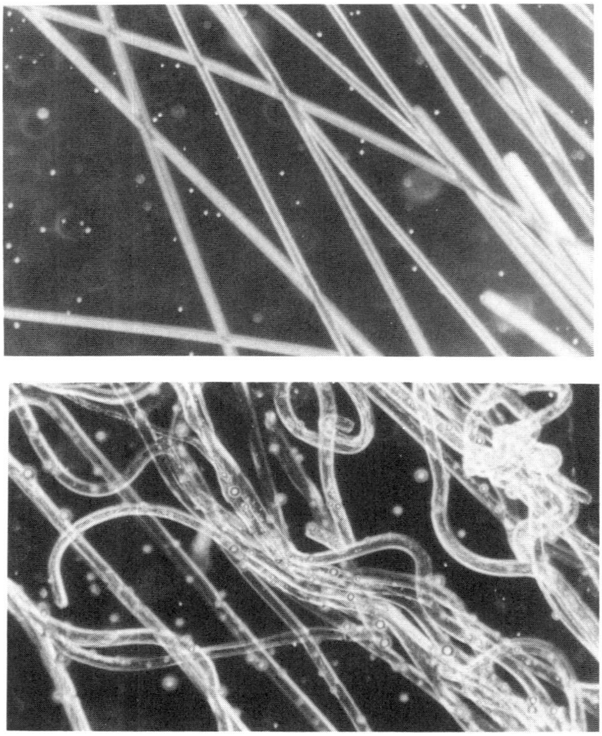

Figure 2: Photomicrographs of SELP3 protein polymer fibers. Bright field photographs at 160x magnification of SELP3 fibers spun continuously from phosphoric acid dope solution into a coagulant of ammonium sulfate solution at 360 cm/min. A: Dry fibers. B: Wet fibers in water.

percent polymer dopes, we elected to conduct the spinning through a 26 gauge needle (id=0.02 cm). Even so, with the syringe pump at its highest dispensing setting (2.7 ml/min) the flow rate from the needle was no greater than 5-10 ul/min.

Using these conditions, greater than 60 meters of SELP3 fibers were spun and reeled. The diameter of the fibers were extremely small which forced the reeling rate below 15 cm/min. The fiber bundle was washed in fresh coagulant for 2 hours, then in water for 16 hours, and dried at 100° C. This fiber bundle went proceeded to molecular analysis as described below. A photomicrograph of this fiber bundle is shown in Figure 4.

These same conditions were used in attempts to spin SLP3 and SLP4 dopes. To improve the speed of coagulation, various coagulant mixtures were tried, but again, similar to the experience with the extrusion of SLP3 and SLP4 into ammonium sulfate, liquid polymer dope emerged from the needle and dropped to the bottom of the coagulant vessel. Fibers could be hand drawn carefully from the pool, but they were so fragile that they could not be reeled. The mechanism used for continuous extrusion was inadequate for the difficulties encountered with this spinning system. We were not able to translate the hand pulled fiber conditions using the syringe pump and reeler into a continuous process. The syringe pump could not develop sufficient

A

B

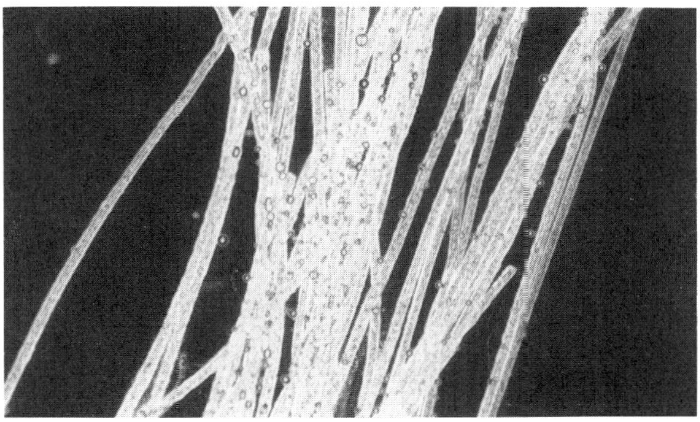

Figure 3: Photomicrographs of SLP3 and SELP3 blended protein polymer fibers. Bright field photographs at 160x magnification of SLP3/SELP3 fibers spun continuously from phosphoric acid dope solution into a coagulant of ammonium sulfate solution at 360 cm/min. A: Dry fibers. B: Wet fibers in water.

uniform pressure to extrude 40 weight percent solutions through the 32 gauge needle. It seemed again that the rate of coagulation of SLP3 and SLP4 dopes in acetic acid/acetone mixtures containing up to 40% acetone was still too slow. However, further acceleration of the coagulant by increasing the acetone content produced precipitated droplets. It is plausible that an increased coagulation rate must be complemented with an increase in the flow rate of extrusion. This was not possible using our laboratory apparatus.

Although we were still unable to mechanize the spinning and reeling of SLP3 and SLP4 using this system, this salt free coagulant allowed at least the production of fibers by hand pulling methods for all four polymers. This gave us a single process by which fibers of all four polymer compositions could be analyzed and compared.

Fiber preparation for molecular analysis. We produced fiber samples of all four polymers for molecular analysis by hand pulling methods. Fibers were produced by extruding 40 weight percent polymer solutions in 9 M LiBr from a 26 gauge needle through a coagulant bath containing acetic acid or acetic acid/acetone mixtures. The fibers were fixed under slight tension in the bath and allowed to cure for at least 30 minutes. These fibers were then washed with distilled, deionized water and air dried

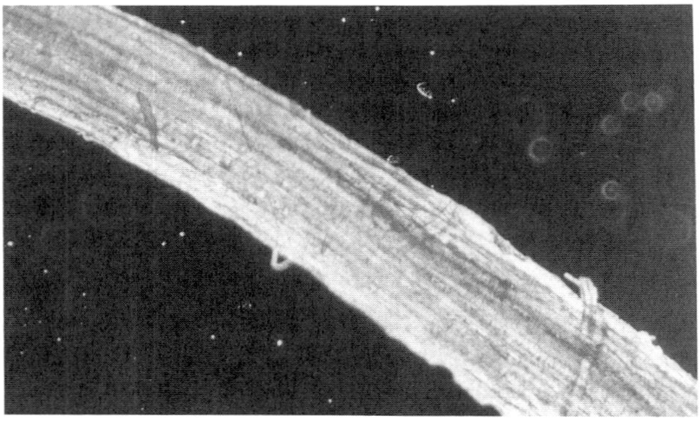

Figure 4: Photomicrograph of SELP3 protein polymer fiber bundle. Bright field photograph at 160x magnification of SELP3 fibers spun continuously from LiBr dope solution into a coagulant containing acetic acid and acetone and reeled on a spool.

A

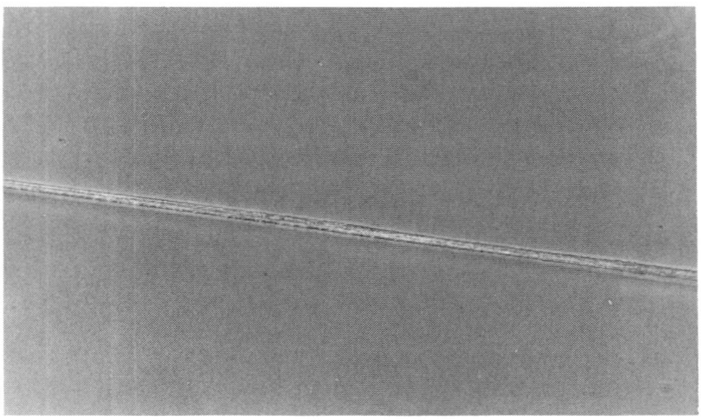

B

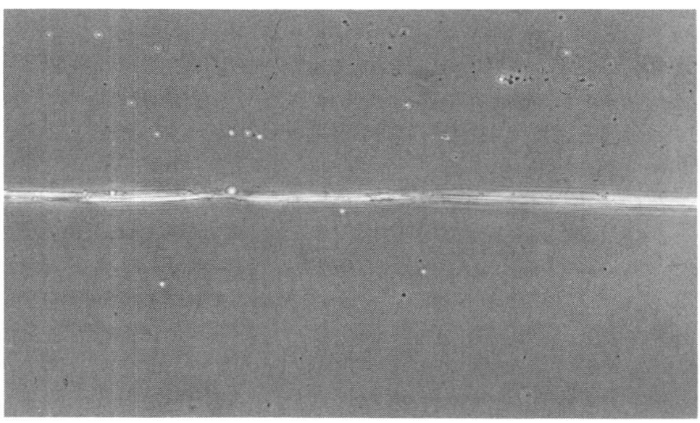

Figure 5: Photomicrographs of protein polymer fibers. Bright field photograph at 160x magnification of hand spun fiber from LiBr dope solution into a coagulant containing acetic acid and acetone. A: SLP3. B: SLP4.

A

B

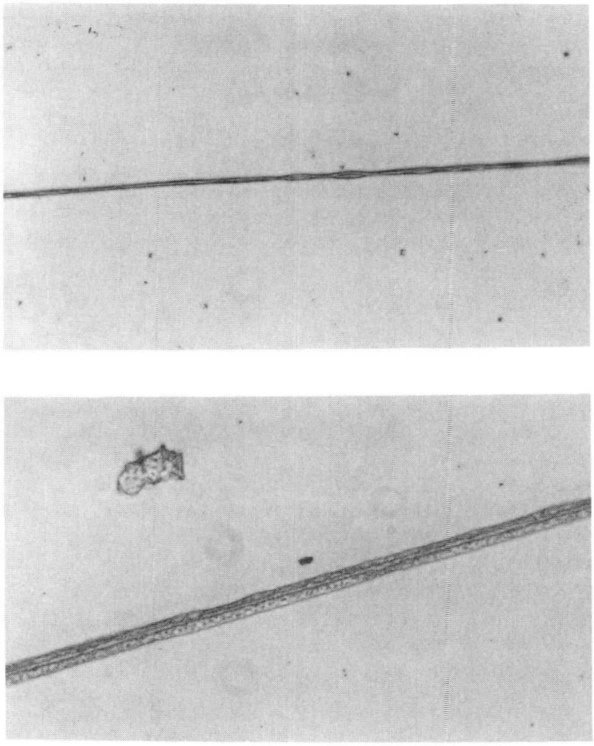

Figure 6: Photomicrographs of protein polymer fibers. Bright field photograph at 160x magnification of hand spun fiber from LiBr dope solution into a coagulant containing acetic acid and acetone. A: SELP1. B: SELP3.

at 80°C. Microscopic inspection of these fibers revealed reasonably smooth morphologies, although some variation in fiber diameter was observed. Photographs of representatives of these fibers are shown in Figures 5 and 6.

Additionally, the continuous production SELP3 fibers previously produced and described above were mechanically stretched as a fiber bundle in order to affect changes in their molecular alignment by suspending a 32 gram weight from them. The fibers were suspended within an enclosed chamber at saturated humidity and ambient temperature. Over a period of 14 days the fibers elongated approximately 400%. The fiber bundle was then removed from the chamber and allowed to dry in air under tension. During drying the fibers shrank in length by approximately 50%. Both the hand-pulled fibers and the stretched SELP3 fiber bundle were analyzed by x-ray diffraction.

Fiber molecular analysis. All fibers produced and listed in Table V were examined

Table V. X-ray Diffraction Analysis of Protein Polymer Fibers

1. Polymer: **SLP4** — Method of Preparation: Hand-pulled, LiBr/Acetic acid:acetone.

Exposure time (hr): 30
Camera distance, in mm: 53.1

Film distances, in mm:		Intensity	Resolution	Unit cell dimension*
1st ring	24	moderate	sharp	1.87 Å
2nd ring	19	strong	sharp	2.29 Å
3rd ring	18	moderate	sharp	2.40 Å
4th ring	9	v. strong	diffuse	4.61 Å

2. Polymer: **SLP3** — Method of Preparation: Hand-pulled, LiBr/Acetic acid:acetone.

Exposure time (hr): 31
Camera distance, in mm: 53.1

Film distances, in mm:		Intensity	Resolution	Unit cell dimension*
1st ring	24	moderate	sharp	1.87 Å
2nd ring	19	v. strong	diffuse	2.29 Å
3rd ring	9	moderate	diffuse	4.61 Å

3. Polymer: **SELP1** — Method of Preparation: Hand-pulled, LiBr/Acetic acid:acetone.

Exposure time (hr): 32
Camera distance, in mm: 53.1

Film distances, in mm:		Intensity	Resolution	Unit cell dimension*
1st ring	24	weak	sharp	1.87 Å
2nd ring	18.5	moderate	diffuse	2.34 Å
3rd ring	9	v. strong	diffuse	4.61 Å

4. Polymer: **SELP3** — Method of Preparation: Hand-pulled, LiBr/Acetic acid:acetone.

Exposure time (hr): 36
Camera distance, in mm: 74.7

Film distances, in mm:		Intensity	Resolution	Unit cell dimension*
1st ring	34	weak	sharp	1.86 Å
2nd ring	26.5	strong	moderate	2.31 Å
3rd ring	12.5	moderate	diffuse	4.67 Å

Table V.—Continued

5. Polymer: **SELP3**		Method of Preparation: Continuous syringe pump extrusion, LiBr/acetic acid. Fiber bundle not stretched.		
Exposure time (hr): 36				
Camera distance, in mm:		75.2		
Film distances, in mm:		Intensity	Resolution	Unit cell dimension*
1st ring	33.5	weak	moderate	1.89 Å
4th ring	27	strong	diffuse	2.28 Å
5th ring	12.5	moderate	diffuse	4.69 Å
6. Polymer: **SELP3**		Method of Preparation: Continuous syringe pump extrusion, LiBr/acetic acid. Fiber bundle stretched.		
Exposure time (hr): 32				
Camera distance, in mm:		53.1		
Film distances, in mm:		Intensity	Resolution	Unit cell dimension*
1st ring	34	v. weak	sharp	1.43 Å
2nd ring	24	weak	sharp	1.87 Å
3rd ring	19	strong	moderate	2.29 Å
4th ring	14	moderate	sharp	3.02 Å
5th ring	9	v. strong	diffuse	4.61 Å

* Error in film distance measurements= +/- 0.3 mm which allows variance in unit cell dimension= 6%.

by light microscopy in order to evaluate fiber morphology, surface texture, diameter, and uniformity. The fibers were examined both in the dry state and after wetting in water. These examinations showed that although extruded through the same size needle (26 gauge), the hand pulled fibers varied in diameter from 10 um for SELP1, 20 um for SLP3 and SLP4, to 40 um for SELP3. The SLP3 fibers appeared the most uniform in diameter while SLP4 and SELP1 were more inconsistent. The hand pulled fibers showed obvious noncontinuous internal morphology (rough appearance) when observed by light microscopy. Especially in the cases of SELP3 and SLP3, as seen in the fiber photographs in Figures 5 and 6, this roughness could represent microcrystalline defects in the fiber due to incomplete or nonuniform coagulation.

The SELP3 fibers extruded continuously from phosphoric acid dopes into ammonium sulfate coagulants swelled to a great extent when immersed in water. In some cases the swelling caused them to disintegrate when viewed under the microscope. The SELP3 fibers extruded continuously using the LiBr/acetic acid system were stable in water. The wetting of the hand-pulled fibers of all four compositions also was of little consequence to their appearance or internal morphology. In the dry state, all fibers were white and opaque.

Hand-pulled fibers of all four polymers, along with the SELP3 fiber bundle before and after stretching, were analyzed by wide angle x-ray scattering in order to determine their crystallinity, unit cell dimensions, and degree of orientation. Fibers were prepared for x-ray analysis by cutting into short lengths (approximately 1-2 cm), dipping briefly in distilled water, and mounting across two pieces of adhesive tape. Periodically, a drop of water was placed on the bundle of fibers, retained briefly and aspirated away. This allowed the formation of tight bundles. After all fibers were mounted, the bundles were secured with a second piece of adhesive tape. The samples were dried at 65° C for one hour. All samples were analyzed on a Statton 2 x-ray camera at 53.1 mm camera distance using an x-ray source with a wavelength of 1.54178 Å. The exceptions were samples #4 and #5 (see Table V) which were analyzed at camera lengths of 74.7 and 75.2 mm, respectively. The diffraction patterns for the four hand-pulled fiber samples and the SELP3 fiber bundle were analyzed and the corresponding unit cell dimensions are presented in Table V.

The strong reflection at approximately 4.6 Å in all samples is clear evidence for the presence of crystalline portions containing substantial amounts of beta sheet structures (24). This reflection is present in all samples, including those with high elastin content (SELP1 and SELP3). Additional reflections may be tentatively assigned to higher order reflections of the interchain distance within the beta sheet (those at 2.29-2.31 Å, n = 2). An additional reflection present in all samples, at 1.86 Å, has yet to be satisfactorily assigned.

X-ray diffraction was performed on the fiber bundle of SELP3 under slight tension to determine its crystallinity, unit cell dimensions, and degree of orientation. A strong reflection at 4.6-4.7 Å was observed, characteristic of a hydrogen-bonded ß-sheet. An additional reflection at 10.2 Å was observed, but has not yet been assigned. The fibers exhibited little or no orientation; this is not entirely unexpected, as the material is flexible and might require considerable tension in order to fully align the polymer chains.

The molecular effects of mechanical stretching of the SELP3 fiber bundle are appraised by comparing the patterns generated before and after stretching of the same sample (Table V, patterns #5 and #6). Two additional reflections appear upon stretching of the SELP3 fiber bundle corresponding to unit cell dimensions of 1.43 Å and 3.02 Å. These reflections are also absent from all hand-pulled fibers. Their assignments cannot be made at this time.

No evidence for orientation is seen in any of the samples, including the SELP3 fiber bundle that had been mechanically stretched. If sufficient orientation of the crystallites within the fibers had been achieved, the x-ray scattering rings observed for these samples would have appeared as discontinuous arcs. The lack of evidence for orientation indicates that in relation to the angle of irradiation (perpendicular to the longitudinal fiber axis) crystalline domains are as likely to be aligned with the fiber axis as opposed to it.

Discussion

These studies indicate that sequence controlled protein polymers consisting of silk-like and elastin-like amino acid sequence blocks can be spun into fibers using solution extrusion techniques. The choices of solvents for production of the dopes and for coagulation of the extruded filaments are important, affecting the overall quality and properties of the fibers produced. Because protein polymers consisting of silk-like blocks have a high tendency to form interchain hydrogen bonded complexes, solvents of high chaotropic character are required for producing dopes. Formic acid, phosphoric acid, high concentrations of lithium halide salts in water, and hexafluoroisopropanol have all been useful in dissolving these polymers. In these studies we chose to investigate dopes produced in phosphoric acid and aqueous LiBr. The choice of coagulants was made on the basis of speed of coagulation. The quicker the extruded liquid stream solidified in the coagulant bath, the less frequently the filament would break during reeling. However, the quicker the coagulation, the greater the tendency for the proteins to crystallize prior to attaining an oriented alignment within the fiber. None of our attempts to date have yielded fibers which display any measurable molecular orientation. Until spinning parameters are developed which can achieve such fibers, ultimate physical properties of protein polymer fibers can not be assessed.

Recently, fiber spinning and mechanical properties of fibers produced from protein polymers were reported (25). Fibers were produced as continuous filaments from protein polymers matching our compositions, SLP4, SELP1, SELP3, and SLPF (see Table I and Figure 1). The dopes were produced containing from 5 to 30% by weight protein in hexafluoroisopropanol and in formic acid containing either LiCl or LiBr at 5 to 15% by weight. The coagulant used was methanol. The filaments were drawn while wet at various draw ratios to improve mechanical properties. The properties obtained for these protein polymer fibers are displayed in Table VI.

These results confirm that a range of fiber properties can be obtained from a single protein polymer composition depending on a number of processing variables including solvent, protein weight percent in the dope, time of coagulation, draw ratio, drying parameters, etc. As expected, tenacity and initial modulus of these fibers is improved by drawing. Trapping the drawn configuration of the fiber by drying it at its fixed elongated length gives the greatest improvement in strength. Drawing and allowing the fiber to shrink during drying improves elongation.

Once processing variables which yield optimal properties can be set for spinning of fibers from silk-like and silk-elastin-like protein polymers, the physical properties of the fibers can be related to the compositional changes between protein chains. With this information, protein polymer chains may be designed with a range of properties suited to specific materials applications.

Table VI. Properties of Protein Polymer Fibers

Polymer	*Draw Ratio*	*Size*[a]	*Tenacity*[b]	*Elong.*[c]	*Init. Mod.*[d]	*Solvent*
SLP4	1x	96.0	0.5	2.1	33.4	HFIP
SLP4[e]	2.5x	42.0	1.0	13.0	40.0	FA+10% LiCl
SLP4	3x	34.0	1.9	26.0	55.0	HFIP
SELP1[e]	4x	36.0	2.0	8.0	62.0	HFIP
SELP1[f]	4x	48.0	1.3	45.0	57.0	HFIP
SELP3	2x	6.3	1.2	17.0	33.0	HFIP
SELP3[e]	4x	38.0	1.0	16.0	47.0	FA+10% LiCl
SELP3[e]	6x	5.0	3.0	11.0	73.0	HFIP
SLPF	1x	95.0	0.5	2.1	31.0	HFIP
SLPF[e]	4x	29.0	1.8	19.0	65.0	HFIP
SLPF[f]	4x	31.0	1.7	40.0	52.0	HFIP

Abbreviations: FA, formic acid; LiCl, lithium chloride; HFIP, hexafluoroisopopanol
[a]Fiber size in denier, [b]grams/denier, [c]Elongation in percent of resting length, [d]grams/denier
[e]The ends of these filaments were fixed while air drying after drawing to prevent shrinkage.
[f]These filaments were allowed to shrink while air drying after drawing.
Reference: Lock, R.L. (25).

Acknowledgments

This work was supported by an SBIR contract sponsored by the Department of the Army (contract number DAAK60-90-C-0056). This publication represents the views of the authors and does not purport to reflect the positions of the Department of the Army or the Department of Defense. We appreciate the contributions of H. Creel and Dr. D. Tirrell for molecular analysis. We thank Dr. M. Goodman for providing equipment, discussion and advice.

Literature Cited

1. Cappello, J.; Crissman, J.; Dorman, M.; Mikolajczak, M.; Textor, G.; Marquet, M.; Ferrari, F. *Biotechnol. Prog* **1990**, 6, pp. 198.
2. Urry, D.W.; Okamoto, K.; Harris, R.D.; Hendrix, C.F.; Long, M.M. *Biochemistry* **1976,** 15, p. 4083.
3. Cappello, J. and Crissman, J.W. *Polymer Preprints* **1990,** 31, p. 193 .
4. Ferretti, A., BP 482 731, **1936.**
5. Whittier, E. O., and Gould, S. P., USP 2 140 274, **1938.**
6. Whittier, E. O., and Gould, S. P., USP 2 167 202, **1939.**
7. Balmaceda, E., and Rha, C., J. Food Sci., 39, 226, **1974.**
8. Lundgren, H.P. *Adv. Prot. Chem.* **1949**, 5, pp. 305.
9. Magoshi, J., Magoshi, Y., and Nakamura, S., Biorheology, 18, 142, **1981.**

10. Khanraev, A.L., Geller, B.E., and Ibragimova, A.N. *Viniti* **1978,** pp. 3178-79, (Chem. Abst. 92,7716, **1980**).
11. Xu, M.; Lewis, R.V. *Proc. Natl. Acad. Sci. USA* **1990,** 87, pp. 7120.
12. Pritchard, W. USP 2 516 145, **1950.**
13. MacDonald, R.N. USP 2 534 283, **1950.**
14. MacDonald, R.N. USP 2 560 584, **1951.**
15. MacDonald, R.N. USP 2 572 843, **1951.**
16. MacDonald, R.N. USP 2 650 214, **1953.**
17. MacDonald, R.N. USP 2 789 973, **1957.**
18. Baird, W.; Parry, E.G.; Robinson, S. BP 646 033, **1950.**
19. Courtalds, LTD. Belg. Pat. 577 609, **1959.**
20. Courtalds, LTD. Belg. Pat. 591 015, **1960.**
21. Graham, B. USP 2 692 247, **1954.**
22. Noguchi, J.; Tokura, S.; and Nichi, N.; *Angew. Makromol. Chemie* **1972**, 22, pp. 107.
23. Asahi Kasei Kogyo Kabushiki Kaisha BP 1 030 801, **1966 .**
24. Fraser, R.D.B.; MacRae, T.P.; Stewart, F.H.C.; Suzuki, E. *J. Mol. Biol.* **1965**, 11, pp. 706.
25. Lock, R.L. European Patent Office, Publication number 0 488 687 A2, **1992.**

RECEIVED May 4, 1993

Chapter 27

Elastomeric Network Models for the Frame and Viscid Silks from the Orb Web of the Spider *Araneus diadematus*

J. M. Gosline[1], C. C. Pollak[1], P. A. Guerette[1], A. Cheng[1], M. E. DeMont[1], and M. W. Denny[2]

[1]Department of Zoology, University of British Columbia, Vancouver, British Columbia V6T 1Z4, Canada
[2]Hopkins Marine Station, Stanford University, Pacific Grove, CA 93950

Models for the molecular networks present in the frame and viscid silks of the spider, *Araneus diadematus*, have been developed from the non-Gaussian mechanical behaviour of these silks when tested in their elastomeric state. Frame silk contains a short chain network, crosslinked by β-sheet crystals, where the network chains behave as 2 equivalent random links and contain approximately 15 to 20 amino acid residues. The β-sheet crystals act as reinforcing fillers, and estimates based on the degree of reinforcement suggest a crystal aspect ratio of around 5 and a crystal volume fraction of 0.25 in the dry fiber. Viscid silk appears to have fewer crosslinking crystals and longer chains. The network chains in viscid silk contain 5 to 6 equivalent random links, on average, but values for this parameter can vary between 2 and 14. β-sheet crystals crosslink the network, but usually are not present in sufficient quantity to reinforce the network.

Silk fibres from the cocoon of the silk moth, *Bombyx mori*, have formed the basis for a centuries old textile industry, but with the advent of genetic engineering it is possible to envision the design and synthesis of novel silks based on the vast diversity found in nature. Arthropods produce many different silks that exhibit a broad range of mechanical properties, and spiders are unique in that individual species are able produce multiple silks, often with dramatically different mechanical properties. Thus, spiders provide an ideal model system for studying the relationship between molecular structure and mechanical properties.

A key feature in virtually all silks is that they contain extensive regions of crystalline structure, and much that is known about the structure of silks is based on the analysis of these crystals by X-ray diffraction. The crystals are most often β-sheet structures with the crystal chain axis running parallel to the fiber axis, but silks with β-sheet crystals running normal to the fiber axis, and with α-helical or

0097–6156/94/0544–0328$06.00/0

collagen triple-helical crystal are also known (*1*). In general, however, the majority of the protein in silks is not crystalline but appears "amorphous" in X-ray patterns, and silks are probably best described as elastomeric networks that are crosslinked by and in some instances strongly reinforced by these crystals (*2-4*). Thus, in order to understand structure-property relationships in silks it is essential that we have information on the organization of the amorphous portions of these networks.

This paper presents an analysis of frame silk and viscid silk from the orb web of the spider, *Araneus diadematus*, components that exhibit dramatically different mechanical properties. Frame silk is produced by the major ampullate gland and is used to form the frame of the orb web, as well as the radii and the guys that attach the web to external structures. Viscid silk forms the catching spiral and contains a pair of core fibers and a glue coating. In the web these two silks exhibit very different properties, as shown in Figure 1. The frame silk is rigid, with an initial tensile modulus of about 10 GPa, strong and modestly extensible, failing at extensions of about 25 - 30%. The viscid silk, on the other hand, is elastomeric, with an initial tensile modulus of about 3 MPa, and very extensible, failing at extensions of 150% or more (*5*). Both silks are capable of absorbing enormous amounts of energy before breaking, which reflects their major function of absorbing the kinetic energy of flying insects that impact the web (*3,5*). Because these two silks exhibit large differences in mechanical properties, it is likely that a knowledge of their network structure will provide important insights into the molecular mechanisms for modulating the mechanical properties of silks.

Experimental Approach.

Most silks function in air and have very low water contents. In these situations amorphous protein chains will be glassy (*6*), and it is difficult to distinguish the contribution of the rigid, glassy portions of the network from the rigid, crystalline portions. Fortunately, frame silk super-contracts when immersed in water (*7*). In this process the fiber contracts to about 60% of its dry length and absorbs water, swelling to about 2.2 times its dry volume. As shown in Figure 2, super-contraction causes a dramatic change in properties. X-ray diffraction (*8*) indicates that this hydration does not alter the β-sheet crystals in the silk, and thermoelastic studies indicate that the water-swollen network exhibits rubber-like or conformational entropy elasticity (*2*). Thus, when hydrated, the rigid frame silk fiber is converted into an elastomeric material with properties very similar to those of viscid silk. Because the amorphous chains are kinetically free in these hydrated networks, they play a dominant role in determining mechanical properties, and this makes it possible to use the Kinetic Theory of Rubber Elasticity (*9-11*) to evaluate mechanical properties and create molecular models for the structure of these silks.

Silk Test Apparatus. To create network models from mechanical information it is necessary to have accurate records of the equilibrium elastic behaviour of the elastomeric silk fibers, and because the silk fibers are small (1-4 μm in diameter) we designed a microscope-based test system to measure mechanical properties (Figure 3). The system uses the deflection of thin glass rods for force measurement,

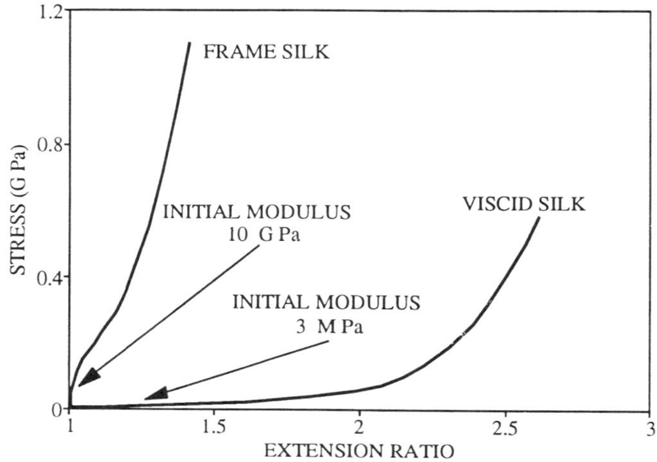

Figure 1. The stress-extension behaviour of frame and viscid silks.

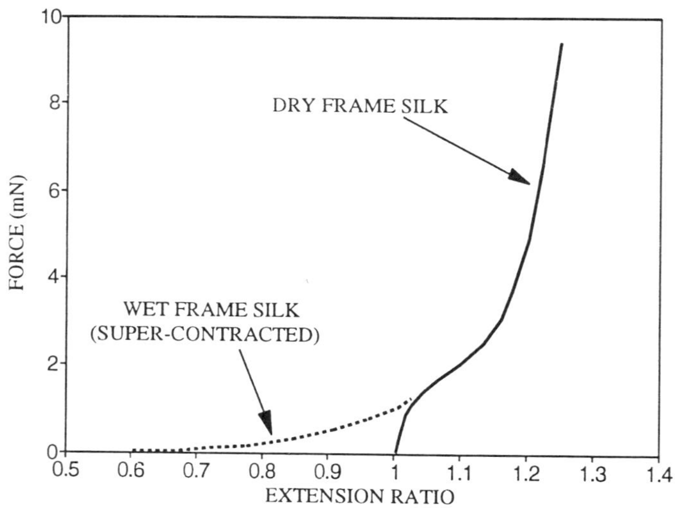

Figure 2. Force-extension data for a single sample of frame silk in its dry, native state and when immersed in water. In this plot extension ratio is defined as sample length divided by the initial length of the dry fiber to illustrate the changes in sample length that accompany hydration.

where the deflection of the rod is followed by a video dimension analyzer (VDA). Our system employs a fixed, reference rod and a moving rod to which the silk fiber is glued. The moving rod was typically about 15 mm long and 50 to 100 μm in diameter, and the setup can detect nano-Newton forces if necessary. For the studies reported here we used glass rods with diameters close to 100 μm, giving excellent accuracy in the micro-Newton force range. The glass rods were assembled on a large, glass, microscope slide, and one end of the silk fiber was glued to the moving rod. The other end of the silk fiber was glued to a glass cover slip, which in turn was glued to a micrometer (M). We selected silk fibers from the webs of large spiders to obtain samples of large diameter. Fiber diameters were measured with a polarizing microscope at 100X under oil immersion, and we estimate that stress values calculated from the force and diameter data are accurate to about ± 10%. With initial silk fiber lengths of 10 to 15 mm and extension measured by a micrometer, we estimate that the extension values are accurate to about ± 2%.

In order to ensure that test data represent the true equilibrium mechanical behaviour we extended the samples in small steps (ca. 2 - 4% extension) and allowed the fibers to relax to an equilibrium force. This generally required that the samples be extended at intervals of 10 minutes or so. Samples were typically tested through several cycles to obtain consistent behaviour before data were collected for further analysis.

Non-Gaussian Network Models. The essence of the analytical procedure used in this study is shown in Figure 4 (*10*). Random networks create rubber-like elasticity because the coiled, flexible chains between network junction points can be extended with little force and will recoil elastically. When the network chains are extended, their conformational entropy is reduced (i.e. they become "ordered"), and elastic recoil occurs because random thermal motion causes the chains to return to their initial, coiled, high entropy state. Simple models, based on the Gaussian statistics of random-walks, predict that random networks can be extended without limit at very low stiffness, where the stiffness (specifically, G, the shear modulus) provides a measure of the number and size of the network chains in the material. Real networks, however, become stiff as the network chains approach their fully extended lengths, and real networks must be modelled with non-Gaussian statistics that represent the network chains as random-walks with a finite number of "steps" (N). Thus, in Figure 4, networks constructed from chains with N = 5, N = 25 or N = 100 "steps" depart from the Gaussian behaviour at increasingly large extensions.

The connection between the statistics of random-walks and the mechanics of real polymer networks is provided by the concept of an *equivalent random link*. Although individual bonds in the backbone of a polymer do not provide sufficient rotational mobility to mimic the complete rotational freedom assumed for each step in an ideal random-walk, a length of polymer with several bonds, each with limited mobility, can approach the rotational freedom assumed for the ideal random-walk. Thus, the shear modulus, G, provides information on the number of effective, random chains in a rubber-like material, and non-Gaussian analysis of the shape of force-elongation curves provides an estimate of the number of equivalent random links in these chains.

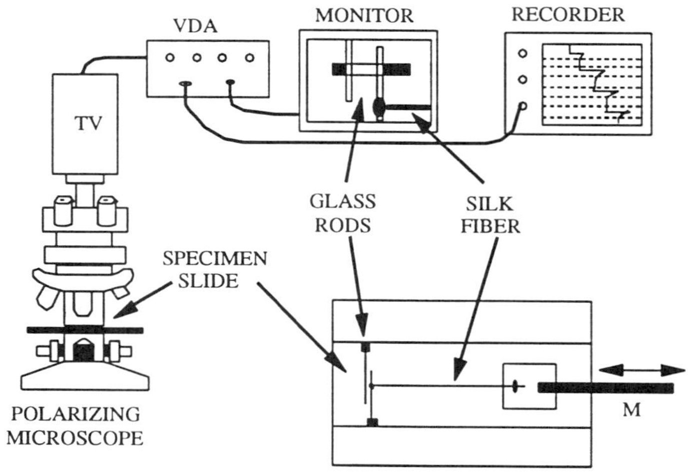

Figure 3. Apparatus for testing elastomeric silk samples.

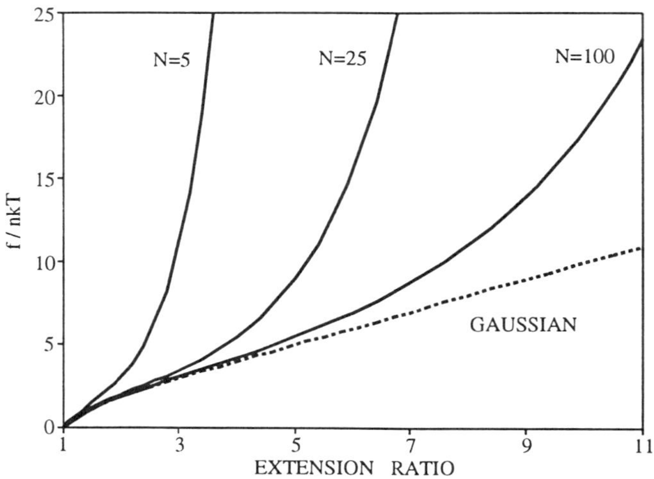

Figure 4. Non-Gaussian network properties of short-chain polymer networks (*11*).

Non-Gaussian Analysis of Frame Silk.

Figure 5 shows typical stress-extension behaviour for a single sample of hydrated, super-contracted frame silk. This sample had super-contracted to 0.62 of the dry slack length, a value that is typical of all frame silk samples from this spider, with little variation. The initial tensile modulus of 10 MPa is also typical of super-contracted frame silk samples tested, again with little variation. The modulus changes dramatically with extension, however. The initial modulus is maintained for only the first 10 to 20% of extension, and by 60 to 70% extension modulus has risen almost 100 fold. Thus, the network in super-contracted frame silk must contain short chains with few equivalent random links.

As shown in Figure 6, data of this type can be fit to non-Gaussian network models to give estimates of the relevant network parameters (G, the shear modulus and N, the number of equivalent random links). In fitting these data to the non-Gaussian model the extension values from Figure 5 have been transformed, with a series expansion approximation to the inverse Langevin function for a tetra-functional network (*10*), into a non-Gaussian extension term, and then linear regression analysis has been used to establish the quality of the fit to the model. In this analysis N, the number of equivalent random links per chain, is adjusted to obtain the best linear fit between stress and non-Gaussian extension (i.e. maximize the coefficient of determination, R^2). In all cases very high values of R^2 were obtained, indicating that the data are well described by the non-Gaussian model.

Table I. Non-Gaussian Analysis for an Individual Frame Silk Sample

STRAIN AMPLIFICATION	LINKS per CHAIN	MODULUS (MPa)	M_c (Daltons)	A.A. per LINK
1.0	0.40	0.007	190,000	5650
1.4	1.16	0.55	2,400	25
1.6	1.46	0.76	1,740	14
1.8	1.74	0.90	1,470	10
2.0	2.02	0.98	1,320	8
2.2	2.28	1.04	1,270	7
2.4	2.58	1.90	1,200	5.5
2.6	2.86	2.12	1,180	4.9

Table I summarizes the results of this analysis for a single sample. Note, however, that the table includes two new parameters, Strain Amplification and the number of amino acid residues per equivalent random link (A.A. per LINK). Strain amplification in a polymer network arises from the presence of reinforcing filler particles. If we assume that the crosslinks in the frame silk network occupy negligible volume, then there is no reinforcement and the strain amplification will be 1.0. If the crystals occupy significant volume, and in particular if the crystals are elongated, then they will reinforce the network and increase its stiffness. One way to account for the reinforcing effect of fillers in a rubber network is to assume

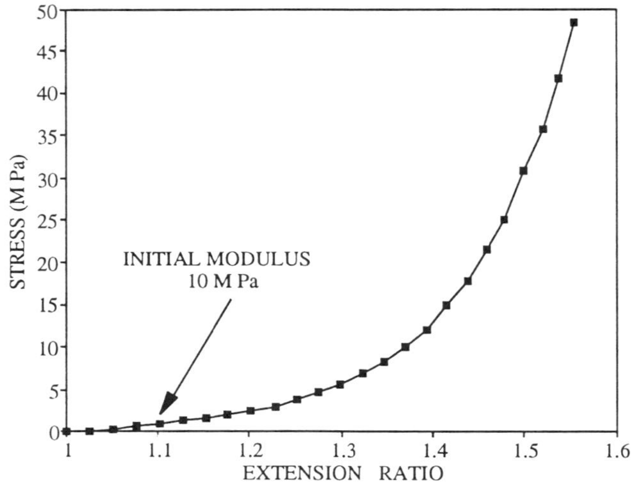

Figure 5. Stress-extension data for super-contracted frame silk.

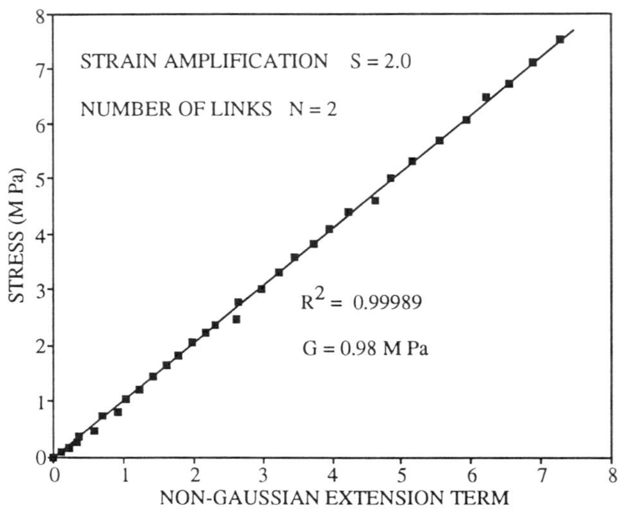

Figure 6. Regression of stress-extension data for super-contracted frame silk, showing the fit to a non-Gaussian network model.

that the fillers magnify the local strain on the network chains, creating strain amplification (*11*). Because frame silk contains a significant fraction of rigid crystals, it is reinforced by these crystals, but the extent of this reinforcement, and hence the magnitude of the strain amplification, is unknown. For this reason, the network analysis has been carried out with strain amplification as an adjustable parameter, and evaluation of this parameter is based on the ability of the model to predict the properties of the equivalent random links.

Because the model provides an estimate of the shear modulus (G), it is possible to calculate the average molecular weight (M_c) of the network chains between crosslinks (*9-11*). The average size of the amino acids in *Araneus* frame silk is 84 Daltons (*12*). With this information and the number of equivalent random links per chain from Table I, it is possible to estimate the number of amino acids per equivalent random link. This estimate is listed in the final column of Table I for a range of values of strain amplification. The importance of this last parameter is that it provides a measure of the inherent flexibility of the polymer backbone, and the magnitude of this parameter for protein is known to range from 7 to 12 amino acids per link, with a value of 7 to 9 being most appropriate for glycine-rich proteins like silk (*6*). Thus, on the basis of this criterion it is possible to infer a final model for the network structure of frame silk.

Network Model. Inspection of Table I indicates that at a Strain Amplification of 2.0 the number of A.A. per LINK = 8, suggesting that the most appropriate network parameters are: shear modulus, G = 0.98 MPa, molecular weight of network chains, M_c = 1,320 Daltons and links per chain, N = 2. Then, from M_c and the average amino acid mass, we calculate that network chains are about 16 amino acids long.

The analysis above was based on data from a single frame silk sample, but similar analyses for other samples from webs produced by different spiders gave essentially the same result, but with a small amount of variation. Figure 7 shows stress extension data for 4 additional samples, which when subjected to the same analysis revealed the following: the best-fit strain amplification values ranged from 1.6 to 2.2, the values for the number of links per chain fell between 1.8 and 2.2, and the number of amino acid residues per network chain fell between 15 and 20. Thus, the model described above seems quite robust and likely holds to a good first approximation for frame silk in general.

Crystal Shape and Volume Fraction. The strain amplification factor of 1.6 to 2.2 inferred in this analysis can also be used to make some estimates on the shape and the volume fraction of the crystalline regions in the frame silk network. The curves in Figure 8 predict the strain amplifying effect of rigid, rod-like particles as a function of filler volume fraction (v) for a series of different aspect ratios (f) from 3 to 7. The equation in the figure shows the relationship used to calculate strain amplification (*11*). The horizontal arrows indicate the range of strain amplification inferred above for super-contracted frame silk, and the intersection of these arrows with the lines for f = 7 and f = 3 indicate a plausible range for the crystal volume fraction in the hydrated silk network. Accordingly, we conclude that crystals with aspect ratios between 3 and 7, at volume fractions between 0.09 and 0.14 are present

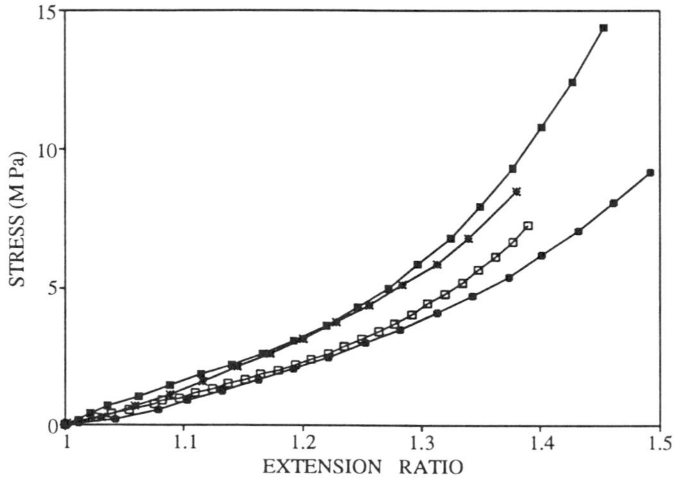

Figure 7. Stress-extension data for 4 additional super-contracted frame silk samples.

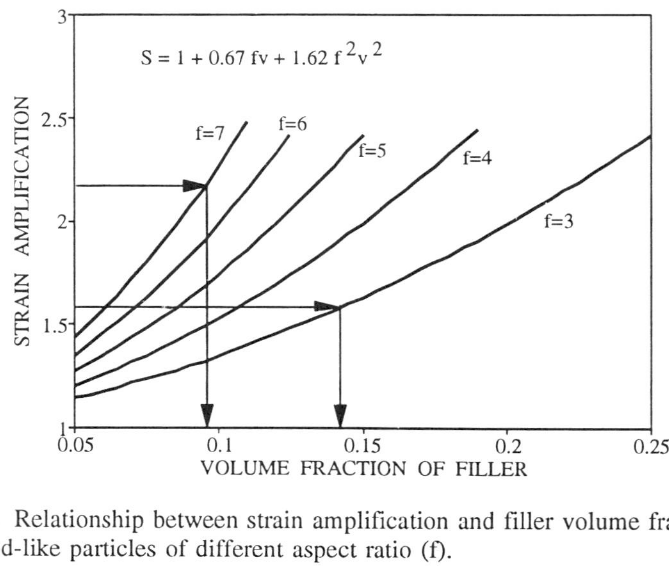

Figure 8. Relationship between strain amplification and filler volume fraction (v) for rod-like particles of different aspect ratio (f).

in the frame silk network. This calculation is based on the hydrated network volume, and the volume of the dry fiber is 0.45 that of the hydrated silk. If the volume of crystals remains unchanged during super-contraction, then the volume fraction of crystals in the dry fiber probably falls between about 0.2 and 0.31.

Non-Gaussian Analysis of Viscid Silk.

The non-Gaussian analysis of viscid silk samples revealed interesting differences in network properties from those observed for frame silk. First, viscid silk exists in an elastomeric state when taken from the web in its "dry" condition. Although it is in equilibrium with air, it is not strictly dry because the hygroscopic glue on the viscid silk draws water from air (*13*). Also, the long-range elasticity of viscid silk indicates that the amorphous chains are not glassy but are mobile and therefore must be plasticized by water or some of the other small molecules present in the glue. (Gosline, et. al, In *Design and Processing of Materials by Biomimicking,* Aksay, I. Ed., Am. Institute of Physics, New York, in press)

Second, even though the "dry" viscid silk is elastomeric, it super-contracts in water. However, the magnitude of the super-contraction and the effect of water on mechanical behaviour are much more variable for viscid silk than for frame silk. The super-contraction ratio was observed to vary from 0.3 to 0.85 for a series of fibers from webs spun by different animals, but the variation can be almost as large for samples from a single web. As shown in Figure 9, the stiffness of the wet fiber can vary dramatically from sample to sample. This figure shows "dry" - wet force extension properties measured for two samples that illustrate the full range of behaviours observed. In Figure 9A we see a sample that super-contracted to about 0.60, and when stretched it exhibited a mechanical behaviour quite similar to that of the "dry" fiber. Figure 9B shows another fiber that contracted to about the same extent, but when stretched this fiber showed dramatically reduced stiffness. This change in properties reflects the loss of network crosslinks with hydration, and obviously the extent of this loss in crosslinks is quite variable. At present we know little about the network crosslinks, but we believe that β-sheet crystals are involved. Certainly, non-covalent, physical linkages play a dominant role because these fibers will dissolve when immersed in protein denaturing agents.

Figure 10 shows typical stress extension properties for viscid silk. The initial tensile modulus of the "dry" fiber is around 3 MPa, but variation is large. The modulus of the wet fiber is always less than that of the "dry" fiber, in part because of the loss of network crosslinks with hydration, but also because the cross-sectional area increases dramatically with the swelling and contraction caused by immersion in water. The properties of the wet silk are therefore even more variable than those for the "dry" viscid silk.

Non-Gaussian analysis of "dry" and wet viscid silk samples confirms the variable nature of this material. Figure 11 shows the properties of three fibers that illustrate the full range of network properties observed for a series of 15 samples collected from 5 webs spun by different animals, and Table II shows average (± one standard deviation) values for the number of links per chain and the shear modulus (G). On average, viscid silk contain longer network chains than frame silk, with 5

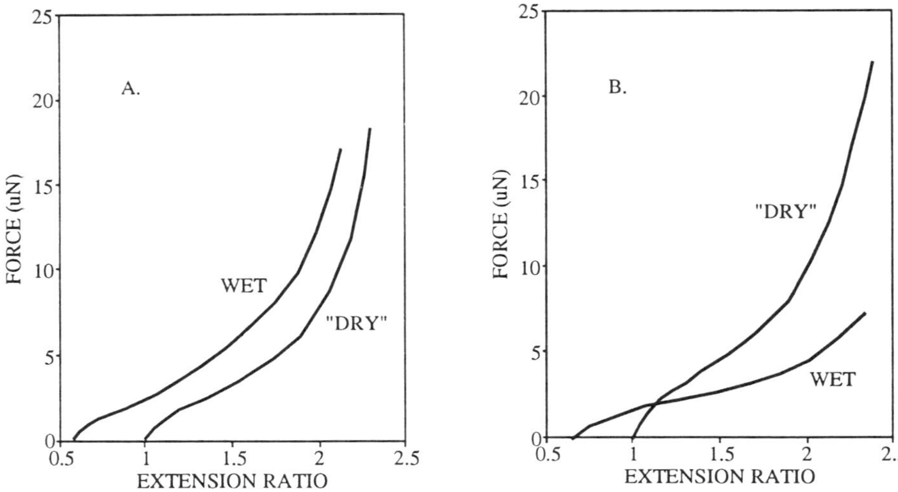

Figure 9. Force-extension data for two viscid silk samples, showing the range of behaviours seen when "dry" fibers super-contract in water. Extension ratio is defined as sample length divided by the initial length of the "dry" fiber to illustrate the changes in sample length that accompany hydration.

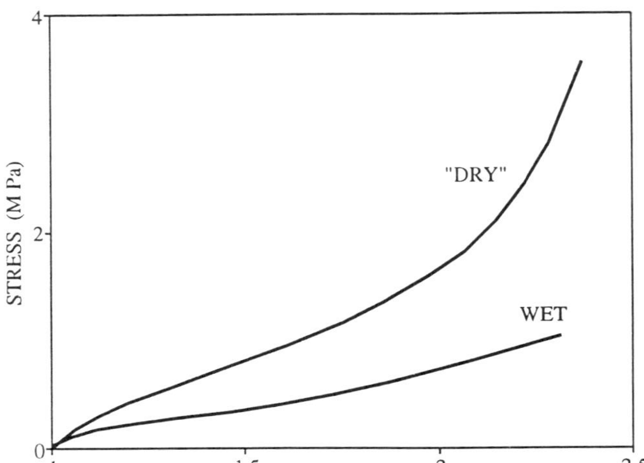

Figure 10. Typical stress-extension data for "dry" and wet viscid silk.

to 6 links per chain, compared to the two-link chains in frame silk. In addition, the number of links per chain increases and the shear modulus decreases when viscid silk is immersed in water, changes that likely reflect the tendency of water to disrupt

Table II. Average Non-Gaussian Network Parameters for Viscid Silk

SAMPLE TYPE	LINKS per CHAIN	SHEAR MODULUS (MPa)
"DRY"	4.8 ± 2.2	0.67 ± 0.32
WET	6.6 ± 2.8	0.32 ± 0.18

some of the physical interactions that form the network crosslinks. Curve B in Figure 11 shows the properties of one fiber that closely matches the averages listed in Table II. It is clear from curves A and C, however, that there is great variation. Curve A shows data for a single, "dry" fiber that exhibited the shortest chains and the highest modulus. Curve C shows data for a single, wet fiber that represents the other extreme. Thus, chain length, expressed as the number of links per chain, varies seven fold, and at 8 amino acids per random link, the network chains in viscid silk contain from 15 to 110 amino acids, with an average of about 40 residues.

Finally, it remains to consider the reinforcement of these elastomeric networks by crystalline filler particles. With the exception of the single fiber, plotted as curve A in Figure 11, all "dry" and wet viscid silk fibers produced reasonable values for the number of A.A. per LINK without strain amplification. This indicates that the crystalline crosslinks are small enough or are present at sufficiently low volume fraction that they do not significantly reinforce the viscid silk network. The single exception (curve A, Figure 11) required a strain amplification of about 1.7 and thus had network properties very similar to those of super-contracted frame silk fibers. At present we do not know why this sample showed such an exceptional network structure.

Conclusions.

The use of non-Gaussian network analysis provides interesting insights into the molecular organization of spider's silks. As shown in Figure 12, frame and viscid silk both contain crystal crosslinked networks, but they differ in the length of random chains and the crystal volume fraction. The diagrams represent these networks as they likely exist in the dry state, and they show the following.

1. Frame Silk: Network chains, approximately 15 to 20 amino acid residues in length (ca. 2 equivalent random links), are nearly fully extended in the direction of the fiber axis. They were locked into this extended conformation when the wet (i.e. elastomeric) fiber was drawn from the spinneret and dried. That is, drawing, followed by drying has created a polymer glass with a high degree of preferred molecular orientation parallel to the fiber axis. The reinforcing crystal filler-particles, which occupy about 25% of the volume of the dry fiber, have been aligned

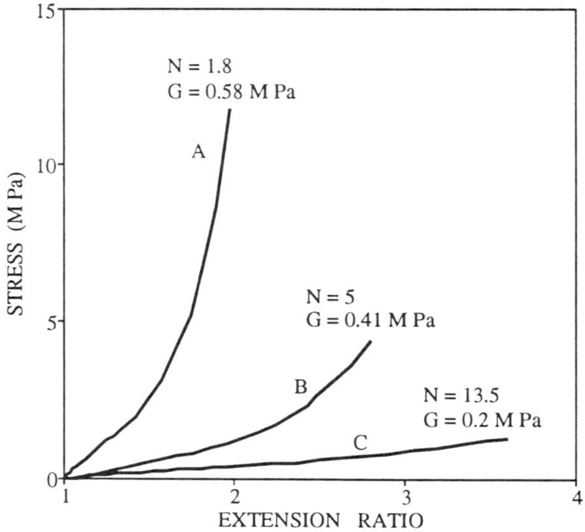

Figure 11. Network properties of three viscid silk samples.

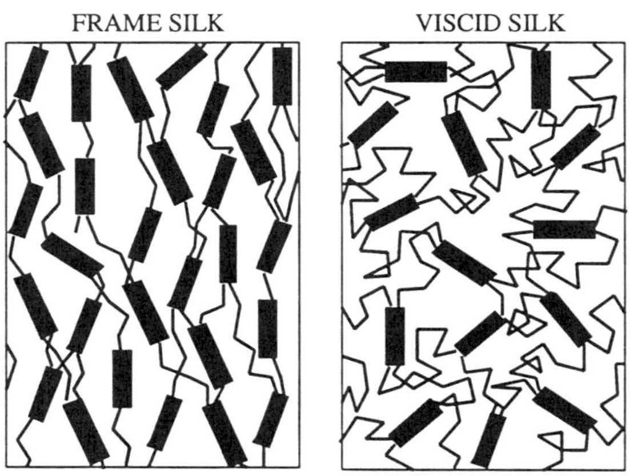

Figure 12. Models for the network structure of frame and viscid silks.

in this drawing process and remain aligned with the fiber axis because they are embedded within the rigid, glassy network.

2. Viscid Silk: The network chains in both the "dry" and the wet state remain coiled, flexible and mobile, allowing this silk to exhibit rubber-like elasticity. The chains, on average, contain about 40 amino acids residues (5 to 6 equivalent random links), and show little or no preferred orientation. The crystals that crosslink the network are present in small quantities and are also quite randomly oriented. However, the super-contraction seen for viscid silk indicates at least some degree of preferred orientation in "dry" fibers, and it is possible that some residual orientation remains in wet fibers.

Acknowledgments.

This research was supported by grants from N.S.E.R.C of Canada and from E.I. DuPont de Nemours & Co., Inc. to J.M.G.

Literature Cited.

1. Lucas, F. & Rudall, K.M. In *Comprehensive Biochemistry*; Florkin, M. & Stotz, E.H., Eds.; Elsevier, Amsterdam, **1968**, Vol 26B; pp 475-558.
2. Gosline, J.M., Denny, M.W. & DeMont, M.E., *Nature, Lond.*, **1984**, *309*, 551-552.
3. Gosline, J.M., DeMont, M.E. & Denny, M.W., *Endeavour*, **1986**, *10*, 37-43.
4. Gosline, J.M., Shadwick, R.E., DeMont, M.E. & Denny, M.W. In *Biological and Synthetic Polymer Networks*; Kramer, O., Ed.; Elsevier Applied Science, London, **1988**, pp. 57-77.
5. Denny, M.W. *J. exp. Biol.*, **1976**, *65*, 486-506.
6. Gosline, J.M. *Rubber Chem. Tech.*, **1987**, *60,* 417-438.
7 Work, R. W. *Textile Res.*, **1977**, *47*, 650-662.
8. Work, R.W. & Morosoff, N. *Text. Res. J.* **1982**, *52*, 349-356.
9. Flory, P.J. *Statistical Mechanics of Chain Molecules*, John Wiley & Sons, New York, **1969**.
10. Treloar, L.R.G. *The Physics of Rubber Elasticity*; Oxford Univ. Press, Oxford, **1975.**
11. Mullins, L. In *The Mechanical Properties of Biological Materials*; Vincent, J.F.V. & Currey, J.D., Eds.; Society for Experimental Biology, Cambridge, **1980**, pp. 273-288.
12. Andersen, S.O. *Comp. Biochem. Physiol.*, **1970,** *35*, 705-714.
13. Volrath, F., Fairbrother, W.J., Williams, R.J.P., Tillinghast, E.K., Bernstein, D.T., Gallagher, K.S. & Townley, M.A., *Nature, Lond.* **1990**, *345*, 526-527.

RECEIVED June 29, 1993

Chapter 28

Formation and Properties of Silk Thin Films

Wayne S. Muller, Lynne A. Samuelson, Stephen A. Fossey, and David Kaplan

Biotechnology Division, Natick Research, Development, and Engineering Center, U.S. Army, Natick, MA 01760–5020

As an immobilization matrix cast fibroin silk films exhibit useful properties including stability to most solvents, biological compatibility, phase transition for the physical entrapment of reactive molecules, and the capablity to retain high activity of the entrapped molecules. However, in the casting process there is limited control over the density, thickness and the orientation of the polymer chains. The Langmuir-Blodgett (LB) technique can enhance the control of the membrane structure and allow improved control over membrane properties. We have formed natural silk films using the Langmuir technique. Silk fibroin, regenerated from *Bombyx mori* cocoons, formed stable LB thin films as indicated from pressure/area isotherms. Multiple layers of the silk fibroin were deposited on a number of substrates and basic information about the physical properties of the LB films were obtained with transmission electron microscopy (TEM) and ellipsometry data. Preliminary analysis of electron diffraction data from the film indicates a polycrystalline structure consistent with the known structure of silk. Infrared spectrometric analysis of these silk films using attenuated total reflectance (ATR) gave wavenumbers for amide I, II, III and V bands, which are in agreement with the silk II conformation reported for cast silk membranes.

Over the centuries, silk has been valued as a textile fiber because of its strength, elasticity, softness, lustre, absorbency and affinity for dyes. *Bombyx mori* silk consists of two types of proteins, fibroin and sericin. Fibroin is the protein that forms the filaments of silkworm silk and gives silk its unique physical and chemical properties. Sericins are the group of gummy proteins which bind the fibroin filaments. Silk fibroin can be used in various forms, such as gels, powders, fibers, or membranes, depending on application. Recently *(1-5)* silk fibroin has been used as an immobilization matrix for enzymes. As a biomaterial it has many advantages over both natural and synthetic materials used in biosensor systems. These attributes include its biological compatibility, stability to most solvents including water and good tensile strength and elasticity properties. Our interests lie in the membrane properties of silk and the utilization of the phase transition (silk I to silk II) in the processing of water-soluble fibroin polymer to water-insoluble films.

Fibroin contains a unique amino acid composition and primary structure *(6,7)*. Its major advantage as an enzyme immobilization matrix is that it entraps the enzyme. This entrapment of the enzyme without the usual cross-linking chemicals alleviates the problems of residual cross-linking chemicals in the matrix, which can deactivate the enzyme and the chemical cross-linking, which can negatively impact enzyme activity. The entrapment process is accomplished by physical, chemical or mechanical treatment of the membrane (e.g., change in temperature, pH, solvent, mechanical shear or stretch), which induces a phase transition. The phase transition is a conformational change of the protein from mostly a random coil to a ß sheet conformation entrapping the reactive molecule. *Bombyx mori* silk fibroin has been used as an immobilization matrix for enzymes such as glucose oxidase *(1-4)*, alkaline phosphatase*(8)*, peroxidase *(5)*, and invertase *(9)*.

Silk fibroin exhibits at least three conformations: random coil, silk I, and silk II. All three conformations can be formed by the appropriate preparation conditions and each is interchangeable under certain conditions *(10)*. Silk structures studied for phase transitions as immobilization matrices have in the past been in the form of cast membranes. The cast silk membranes have good properties as membrane materials; however, the casting process has limitations. There is limited control over the thickness of the membrane or the density and the orientation of the polymer chains. Since the functionality of these membranes, including permeability, the activity of entrapped enzymes, and mechanical integrity, is dependent in part on conformation, density, and orientation of the polymer chains, new processing techniques to control these properties would be useful. The Langmuir-Blodgett (LB) technique is used in this study in an attempt to enhance the control of the physical processing of silk fibroin protein. This enhanced control could therefore provide new opportunity for these membranes as immobilization matrices and in other barrier and biomaterial applications.

We describe the formation and characterization of natural silk fibroin films using the LB technique. Basic information regarding the physical characteristics of the thin films is obtained from pressure-area isotherms, electron micrographs, and ellipsometry. Analysis of the silk fibroin LB films with infrared spectroscopy and electron diffraction provides insight into the silk conformation favored at ambient temperature and the expected polycrystalline order of the silk film.

Experimental

B. mori cocoon silk was regenerated using 9.3M LiBr solution and dialyzed for three days in distilled water. A Pasteur pipet was used to apply the solubilized silk to a Lauda Filmbalance FW2 (Brinkmann Instr. Inc., Westbury, NY). To generate pressure/area isotherms, the silk fibroin was added to the surface of a Milli Q water subphase at 24° C and the barrier compressed at a rate of 46 cm^2/min (approximately 125 mg of silk was added to form isotherm in Figure 1). Transmission Electron Microscopy (TEM) was performed on the films using a Hitachi H 600 (Rockville, MD) with samples collected from the surface of the trough using T 1000 grids. Fourier Transform Infrared Reflectance (FTIR) Attenuated Total Reflection (ATR) analysis was performed on a Nicolet 20SXB Infrared Spectrometer (Madison, WI) with an accessory holder for ATR (Harrick Scientific Co., Ossining, NY). Thin film samples for FTIR analysis were transferred and collected on germanium prisms at a pressure of 16.7 mN/m, a dipping speed of 0.2 cm/min and a temperature of 20° C. The silk films were deposited on frosted glass slides under the same conditions used for the FTIR samples. These films were analyzed on a Thin Film Ellipsometer Type 43603-200E (Rudolph Research, Flanders, NJ) to determine the thickness of the transferred material.

Results & Discussion

Figure 1 is a typical pressure/area isotherm of the soluble silk. The steep slope and smooth appearance of the curve indicate the formation of a stable film. It should be noted that the x-axis of the isotherm is in arbitrary units because the silk fibroin has an estimated molecular weight of 350 kDa to 415 kDa *(11)*, which exceeds the limits of the Lauda software program. A molecular weight of 75.53 Da was used as the basis of standardizing the calculation for the x-axis units, derived from the amino acid composition of fibroin *(12,13)* and is based on the average weight of each amino acid monomer in the silk fibroin polymer. It remains difficult to obtain an accurate quantitation of the area per molecule in the stablized film due to the complex secondary structure of the silk fibroin protein.

Silk fibroin does not exhibit the typical amphipathic character of LB materials suitable for monolayer formation. Also, silk fibroin exhibits unique solubility characteristics *(7,14)*, which can make the material difficult to work with in an LB system. Silk fibroin is insoluble in volatile nonpolar solvents often used in the application of surfactants to an aqueous subphase. Silk fibroin is also insoluble in dilute acids and alkali, and resistant to most proteolytic enzymes *(11,15)*, but soluble in 9.3M LiBr aqueous solution. After dialysis, the solubilized silk fibroin remains in solution if undisturbed. After application to the trough, a portion of the polymer enters the aqueous subphase and the remainder sits at the air-water interface. This event is evident from the protein residue observed when cleaning the trough. This

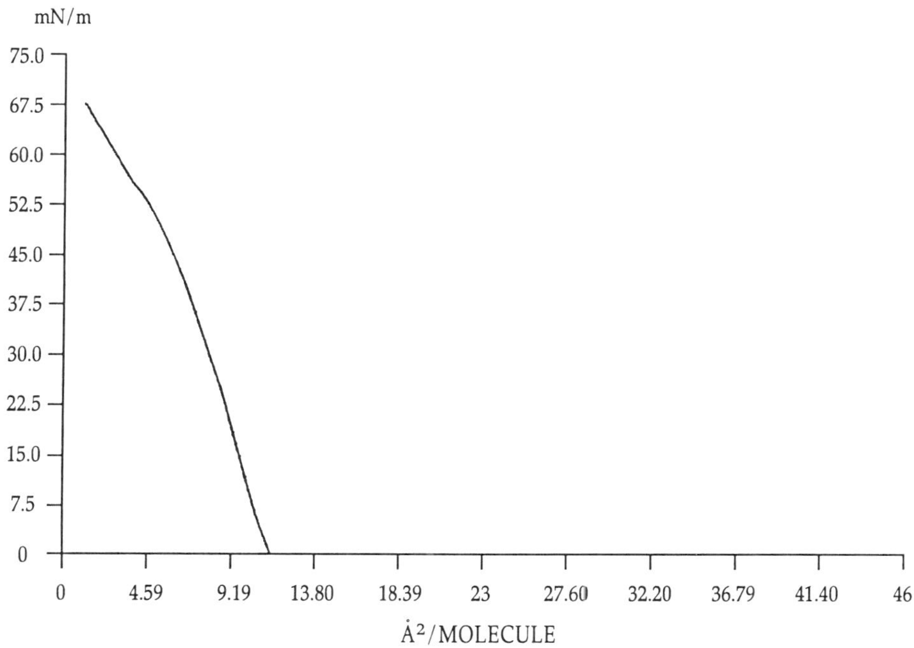

Figure 1. Typical pressure/area isotherm of solubilized silk fibroin at 24° C.

unusual solubility behavior adds to the difficulty in obtaining an accurate determination of the area per polymer chain.

The silk fibroin films demonstrate excellent stability and transfer properties, indicative of a well-behaved Langmuir-Blodgett system. The compression barrier has been stabilized for half hour to hour time periods at various pressures (10, 15, 20, 25, 30, 35 mN/m) with no indication of film collapse. Films remained stable overnight (16 h) without a change in area when studied at our usual working pressure for transfer at 16.7 mN/m. In Figure 1, a typical isotherm for silk, there is no sudden drop in surface pressure with compression, which would be indicative of a collapsed film. The steady rise in surface pressure of the isotherm indicates the increasing resistance to dense molecular packing. Upon complete collapse of the film, it is possible to remove very long fibers with the tip of a pipet.

Table I presents the ellipsometry data for silk films indicating the relative thickness of a transferred LB silk film. Many biomolecular materials exhibit Y or Z type deposition *(16)*. For Y deposition on a hydrophilic surface the first monolayer is transferred as the substrate is raised through the subphase. Subsequently a monolayer is deposited on each traversal of the surface. With Z deposition material transfer occurs only on movement of substrate up through the monolayer. We believe that Y type deposition is characteristic of the silk fibroin thin film based on our observations of the change in area and shape of the meniscus at the air/water interface during vertical deposition. The average thickness of a monolayer determined from the data in Table I was approximately 11.6 to 11.9 Å. The ellipsometry data indicated that the average thickness increased with the number of layers deposited. The increased average thickness maybe due to a combination of factors, the inefficient packing of layers with increasing numbers and/or the nonuniformity of the layers. Therefore, by extrapolation back to a single layer, a average value of 11.6 to 11.9 Å was established.

Silk from *B. mori* consists of antiparallel ß sheets as first described by Marsh et al. *(17)*. The fibroin consists of both crystalline (short side-chain amino acid monomers glycine, alanine, serine) and amorphous (amino acids with bulkier side chains) domains. There have been two types of crystalline structures proposed for silk, silk I and silk II. For silk II, the insoluble and more stable form of silk, the reported unit cell based on X-ray diffraction data has an interchain distance of 9.4 Å, a fiber axis distance of 6.97 Å, and an intersheet distance of 9.2 Å *(17)*. The silk does not exhibit the typical amphipathic character of LB materials thus the polymer chains at the air/water interface are probably parallel to the surface. A single chain of silk crystalline domain is 4.5 to 4.7 Å thick while the amorphous chains vary in their thickness. It maybe in forming the silk LB film the nonaqueous soluble silk II (crystalline domains) are at the air/water interface and the soluble amophous domains are buckled into the subphase. Such a model of the film structure on the surface would be consistent with the ellipsometry value of 11.6 to 11.9 Å and the increase in average thickness of the layers. The amorphous domains would not be uniform in their arrangement underneath the crystalline domains creating an irregular surface with the ellipsometry value representing the average thickness of the silk film deposited on the substrate.

Silk I is also of interest due to its metastable nature, its role in the natural processing of silk, and its unresolved structure. Silk I is the conformation of the soluble form of silk fibroin that rapidly undergoes a phase transition to the insoluble silk II conformation *(11,18)*. This transition can be activated by mechanical agitation, exposure to hydrophilic organic solvents, or temperature changes *(19-23)*. Due to this instability, experimental studies with silk I have been primarily on samples of low orientation (attempts to improve orientation result in conversion to silk II). Therefore, elucidation of the structure of silk I has depended on molecular modeling and comparisons with the limited available experimental evidence *(11)*. A number of models have been proposed, including the Lotz and Keith [24] crankshaft model based on poly(L Ala-Gly) lamaellar crystals (intersheet distance of 14.4 Å) and a recent model of Fossey et al. *(18)* based on conformational energy calculations with copolymers of Gly-Ala and stacked sheets (intersheet distance of 11.3 Å).

Infrared spectroscopy was used to partially characterize the structure of silk in the LB film. Yoshimizu and Asakura *(25)*, in a study on cast films with a thickness of 100-250 μm, employed FTIR (ATR) to determine the conformational transition of the silk membrane surface treated with methanol. The absorption bands observed for membranes treated with methanol had wavenumbers of 1625 (amide I), 1528 (amide II), and 1260 cm^{-1} (amide III), characteristic of a silk II structure. Membranes without methanol treatment showed absorption bands at 1650 (amide I), 1535 (amide II), and 1235 cm^{-1} (amide II), which were assigned the random coil conformation. In addition, Asakura et al. [26] observed that the amide V band had a frequency of 700 cm^{-1} for silk II compared to a 650 cm^{-1} for the random coil conformation.

Table II compares the FTIR wavenumbers for cast membranes (silk I, silk II) with results obtained on silk fibroin LB films. A total of eleven silk layers were deposited on a germanium prism. Absorption bands were observed at 1624 (amide I), 1522 (amide II), 1258 cm^{-1} (amide III), and 700 (amide V). A pronounced shoulder at 1260 cm^{-1} is a critical feature to distinguish silk II from random coil/silk I in cast silk films treated with methanol. In the FTIR spectra for silk fibroin LB films we have identified this shoulder at 1258 cm^{-1}. This confirms that at least part of the silk fibroin has a silk II conformation.

The explanation for the dominance of a silk II structure in these thin films may be in the mechanical forces present during the application of the solubilized fibroin to the surface of the trough, during compression of the surface of the trough, and/or during transfer of the silk fibroin. During application of the solubilized silk to the surface of the trough with the Pasteur pipet, some shear may induce, in part, a silk II conformation. The silk II conformation, due to its insolubility in the aqueous subphase, would therefore form the thin film. Another possible source of mechanical shear is the stretching of silk fibroin during transfer and deposition on solid supports. Once a silk fibroin film is formed there may be resistance to transfer and deposition by interchain forces between the polymer chains. The energy needed to break these interchain forces may contribute to the phase transition of the deposited material.

Figure 2 is a TEM of a single layer of the LB silk fibroin film formed at 24°C. The edge of the TEM grid appears in the micrograph as the black areas at the upper and lower corners, while the dark areas in the field are assumed to be thicker regions of the film. The clear or spherical areas in the micrograph are holes in the film. These holes appear irregularly throughout the film as do the thicker regions, and may arise during film drying on the TEM grid. This conclusion is supported by the ellipsometry data which would not be consistent if holes were uniform throughout the film. The physical appearance of the film is affected by the drying and processing conditions. Figure 3 is a micrograph of a silk fibroin LB film where the temperature of the subphase was elevated to 45°C. The physical appearance of the film is different from that seen in Figure 2. The film has striations throughout. The small dark cubic shapes in the film we believe are crystals of LiBr, not removed during the dialysis process, precipitating out of the solution at the elevated temperature. Altering the temperature of the subphase imparts significant changes in the physical appearance of the silk fibroin film. Further studies are underway to correlate the physical environment of the subphase and drying conditions of transferred films with the structure of the LB silk film formed.

Figure 4 shows a micrograph of an electron diffraction pattern of an LB silk fibroin film. The electron diffraction pattern is typical of a polycrystalline material which is characteristic of silk. Minoura et al. *(22)* and Magoshi et al. *(23)* have published X-ray diffraction data on cast silk thick films which have similar patterns as those observed here for the LB silk fibroin film. Asakura et al. *(26)* and Minoura et al. *(22)*, using X-ray diffraction, observed that the preparation and physical treatment of silk films are critical factors in determining whether silk I or silk II structures form. We have yet to confirm with electron diffraction whether the LB silk film is silk I or silk II. Further work is underway to utilize electron diffracion to determine the structure of silk in the films.

Based on the IR data discussed earlier, we expect a silk II conformation to be confirmed for the conditions under which these films were prepared. to determine the conformation characteristic of the LB silk film individual reflections within the electron diffraction pattern would be critical for structure identification. Reflections with d spacings of 9.70, 4.69, and 4.30 Å would indicate a silk II conformation while reflections at 7.25 and 4.50 Å would be indicative of a silk I conformation. Under the appropriate conditions, using the Langmuir method, it may be possible to obtain a well oriented silk I film. This would provide a unique opportunity to experimentally characterize the silk I structure since this structure has eluded definitive structural characterization due to its metastable state.

Table I. Ellipsometry data on silk fibroin LB films deposited on frosted glass slides

Film Characteristic	Number of Layers		
	3	5	7
Multi-Layer Thickness Å	37	65	94
Calculated Monolayer Thickness Å	12.3	13	13.4

Reproduced from *Mater. Res. Soc. Symp. Proc.* **1993** *292,* 181.

Table II. Comparison of FTIR wavenumbers reported for cast films [17,18] (silk I, silk II) versus those obtained for silk fibroin LB films

Absorption Bands	Cast Membranes		Silk Fibroin LB Films
	Silk I	Silk II	
Amide I	1650	1625	1624
Amide II	1535	1528	1522
Amide III	1235	1260	1258
Amide V	650	700	700

Reproduced from *Mater. Res. Soc. Symp. Proc.* **1993** *292,* 181.

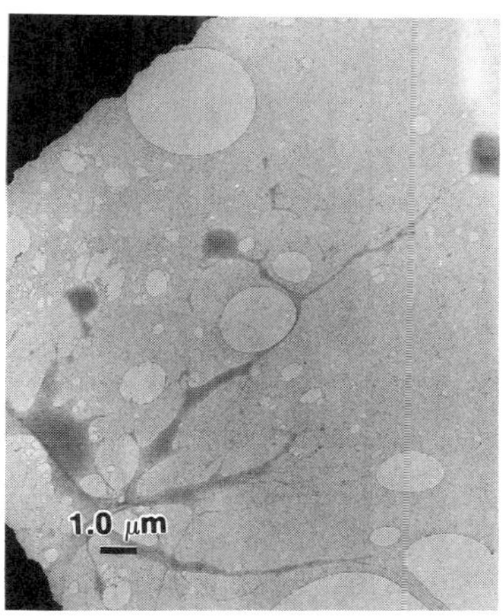

Figure 2. A transmission electron micrograph of the deposited silk fibroin LB film at 24°C. Reproduced from *Mater. Res. Soc. Symp. Proc.* **1993** *292,* 181.

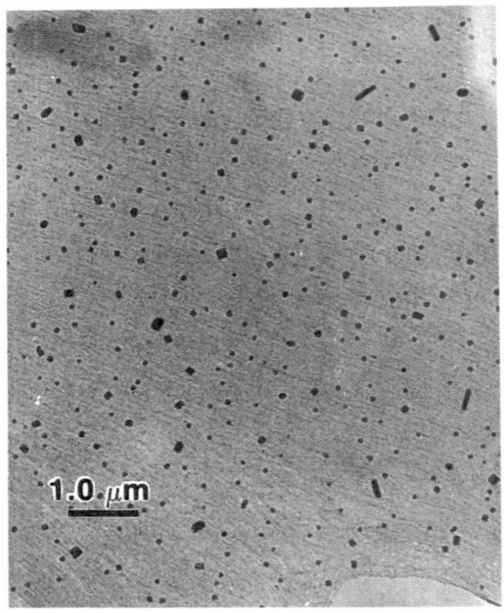

Figure 3. A transmission electron micrograph of the deposited silk fibroin LB film at 45°C. Reproduced from *Mater. Res. Soc. Symp. Proc.* **1993** *292,* 181.

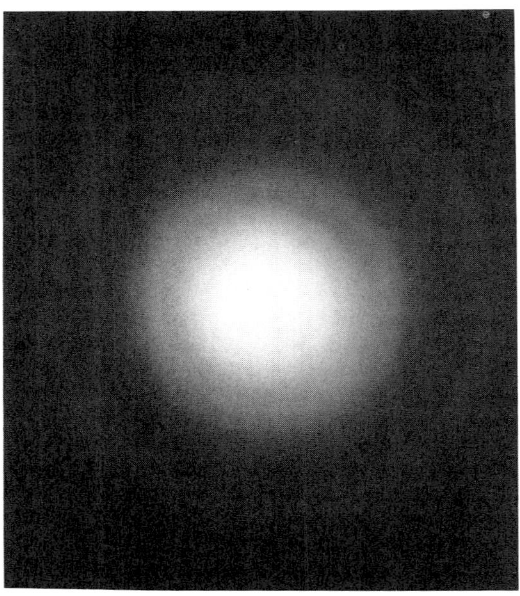

Figure 4. Electron diffraction pattern of silk fibroin LB film. Reproduced from *Mater. Res. Soc. Symp. Proc.* **1993** *292,* 181.

Acknowledgments

We thank Marian Goldsmith (University of Rhode Island, Kingston, R.I.) for supplying the silkworm cocoons, Abe King and Sam Cohen (Natick) for the TEM analysis and helpful discussions, Aaron Bluhm (Natick) for the FTIR analysis, and Tony Chen (University of Massachusetts, Lowell, MA) for the ellipsometry analysis.

Literature Cited

1. Kuzuhara, A.; Asakura, T., Tomoda, R., Matsunaga, T. *J. Biotechnol.* **1987,** *5,* 199.
2. Asakura, T.; Yoshimizu, H., Kuzuhara, A., Matsunaga, T.J., *J. Seric. Sci. Jpn.***1988,***57,* 203.
3. Demura, M.; Asakura, T. *Biotechnol. Bioeng.* **1989,** *33,* 598.
4. Demura, M.; Asakura, T., Kurso, T. *Biosensors* **1989**, *4,* 361.
5. Demura, M.; Asakura, T.,Nakamura, E., Tamura, H.*J. Biotechnol.* **1989,***10,* 113.
6. Ilzuka, E. *J. Appl. Poly.Sci.* **1985,***41,* 163.
7. Robson,R.M. *Fiber Chemistry Handbook of Science and Technology,* Marcel Dekker, Inc: New York,N Y, 1985; Vol. IV, pp 647-699.
8. Asakura, T.; Kanetake, J., Demura, M. *Poly-Plast. Technol. Eng.* **1989,** *28,* 453.
9. Yoshimizu, H.; Asakura, T. *J. Appl. Poly. Sci.* **1990,** *40,* 127.
10. Magoshi, J.; Magoshi, Y., Nakamura, S. *J. Poly. Sci., Appl. Polym. Symp.* **1985,***41,* 187.
11. Kaplan, D.L.; Lombardi, S.J., Muller, W.S., Fossey, S. In *Biomaterials Novel Materials from Biological Scources,* Byrom D., Ed.; Stockton Press, New York, NY, 1991, pp 3-53.
12. Lucas, F.; Shaw, J.T.B., Smith, S.G., *Adv. Prot. Chem.*, **1958,** *13,* 107.
13. Lucas, F. *Nature* **1966,** *210,* 952.
14. Otterburn, M.S. In *Chemistry of Natural Protein Fibers,* Asquith R.S., Plenum Press, New York, NY, 1977, pp 53-80.
15. Minoura, N.; Tsukada, M., Masanobu, N. *Biomaterials* **1990,***11,* 430.
16. Swart, R.M., *Langmuir-Blodgett Films,* Roberts, G., Plenum Press, New York, NY, 1990, pp 273-307.
17. Marsh R.E.; Corey R.B., Pauling, L. *Biochim. Biophysics Acta* **1955,***16,* 1.
18. Fossey, S.A.; Nemethy, G., Gibson, K.D., Scheraga, H.A. *Biopolymers* **1991,***31,*1529.
19. Magoshi, J. *Kobunshi Ronbunshu* **1974,** *31,* 765.
20. Ishida, M.; Asakura, T., Yokoi, M., Saito, H. *Macromolecules* **1990,** *23,* 88.
21. Magoshi J.; Mogoshi, Y., Nakamura, *S. J. Appl. Poly. Sci., Appl Poly. Sym.* **1985,***41,*187.
22. Minoura, N.; Masuhiro, T., Masanobu, N. *Polymer* **1990,** *31,* 265.

23. Magoshi, J.; Kamiyama, S., Nakamura, S. *Proceedings of th 7th International Wool Textile Research Conference*, **1985**, *1*, 337.
24. Lotz, B.; Keith, H.D. *J. Mol. Biol.* **1971,** *61*, 201.
25. Yoshimizu, H.; Asakura, T. *J. Appl. Poly. Sci.* **1990,** *40*, 1745.
26. Asakura, T.; Kuzuhara, A., Tabeta, R., Saito, H.*Macromolecules* **1985,***18*, 1841.

RECEIVED June 29, 1993

Chapter 29

Applications of Silk

T. J. Bunning[1], H. Jiang[1], W. Wade Adams[1], Robert L. Crane[1], Barry Farmer[2], and David Kaplan[3]

[1]Materials Directorate, Wright Laboratory, WL/MLPJ, Wright-Patterson AFB, OH 45433–7702
[2]Department of Materials Science, Thornton Hall, University of Virginia, Charlottesville, VA 22903–2442
[3]Biotechnology Division, Natick Research, Development, and Engineering Center, U.S. Army, Natick, MA 01760–5020

Modern interest in silk started in the late 1960's with the first extensive mechanical property examination by Zemlin in 1968 *(1)*. Today, the same unique balance of material properties, good tensile strength, elasticity, and resistance to fracture is the primary driving mechanism behind the continuing interest in silks. Supercontraction, the mechanical response of some silks to moisture, has also been identified as being of potential technological value *(2-4)*. It is well known that the diverse primary and secondary structure of silk fibers is the basis from which some of their physical properties arise. Through interactions between and modifications of these structures to respond to specific chemical and mechanical environments, the spider tailors the properties of a particular silk for each specific application. It is this control of properties that we hope to better understand. Gosline stated it well in a recent interview *(5)*: "We do not aim to copy nature directly, but to adapt her designs and processes to our own purposes".

Using the spider's processing machinery directly to produce commercial silk is not viable. However, the possibility of using water-soluble precursors in a very energy efficient process that yields high performance fibers with controlled variability in their properties is an attractive goal. In addition, the final material could be environmentally friendly. This is in marked contrast to conventional fiber processes which typically employ environmentally undesirable solvents, (e.g. sulfuric acid for poly(para-phenyleneterephthalamide) [PPTA] and polyphosphoric acid for poly(para-phenylenebenzobisoxazole) [PBO]), require expensive precursors and large energy consumption and result in materials that are difficult to recycle/reuse. The potential cost savings from an industrial point of view is enormous. However, before such lofty goals can be achieved, we must first understand in some detail how nature utilizes the molecular control of chemical and physical chain properties to formulate protein polymers into materials with extraordinary mechanical properties *(6, 7)*. This chapter briefly summarizes applications for both fiber- and film-based silk materials. Most of the applications discussed take advantage of silk's unusual mechanical properties. Other more esoteric applications, including the use of silk gels as a cure for hangovers *(8)*, have also been proposed.

0097–6156/94/0544–0353$06.00/0

Fiber-based Materials

Silk fibers have been primarily used in the textile industry due to their superior properties *(9)*. Starting in the 1800's in Madagascar, spider silk has been woven into a variety of clothing apparel including socks, shirts, ties, and dresses. The only other commercial use of spider silk fiber has been as reticle lines in optical devices *(10)*. Many uses for silk among native cultures have been documented including fishing lines and nets, as well as a number of sinister applications *(11)*. These include the dooming bag, a close knitted silk bag said to have a hypnotic effect, and the smothering cap, similar to a dunce's cap except made by matting spider silk, used for putting adulteresses to death.

Silkworm silk has been utilized for perhaps 6000 years, primarily as a textile fiber. An entire industry, sericulture, has been built upon raising, harvesting, and processing silkworms and their silk for this application. Silks are desirable for textile applications due to the mechanical properties and dyeability in addition to surface luster or sheen. In several museums examples are found of body armor from the times of Ghengis Khan, where leather and silk were used in combination to protect a warrior's body, especially from arrows *(12)*. The storied history of using silks in many ceremonial activities has also resulted in a need to understand how to best preserve these textiles *(13)*. Many "silk-like" synthetic fibers and textiles have also been produced in the past in order to provide lower cost facsimiles.

Recently, the excellent mechanical properties of spider silk fibers have led to a number of possible applications based on the property of large energy absorption before failure. This is a consequence of the fact that unlike most fibrous materials, silk elongation increases at faster loading rates. This gives rise to a number of shock absorber applications.

Research leading to these types of applications is ongoing in a number of laboratories as shown in this book. There is interest in using the fibers for personnel protection from ballistic projectiles *(14)*. From an orb-weaving spider's perspective, the sole purpose of the orb-silks is to absorb kinetic energy. To preclude web failure, the kinetic energy of the incoming prey must be dissipated. Through its viscoelastic response mechanism, the web dissipates 70% of this energy in the form of heat. The chemically harsh processing conditions necessary to process fibers currently used in body armor could be avoided if a silk-like material having similar or better properties were developed. The incorporation of high extensibility and high compressive strength silk fibers into lightweight cloth could have a significant effect on the way we design energy absorbing clothing.

The high tensile strength of silks would also be useful in the design of cables and ropes. This could be advantageous over conventional nylon rope in marine applications where high stretching ultimately leads to sudden and drastic failure. The high extensibility and the possibility of encoding saltwater protection into a silk-based fibrous material is one example. There is precedence in nature for this as several insects spin silks in an underwater environment *(15)*. These fibers possess their own unique set of properties. Cables or fish nets (even bungee cords) could potentially be made with superior properties if the costs were reasonable. The superior strength of lightweight fibers may also find use in the next generation of parachutes and high performance sails for sailboats and hang gliders. The ability to dye silk, already demonstrated by the conventional textile silk industry, is likewise advantageous.

The use of these materials in impact sensitive composite systems is also being considered for future applications *(16)*. Researchers hope that silk's potential high elasticity and tensile strengths will benefit applications requiring strong reinforcing fibers. The absence of kink-band formation, possibly an indication of good compressive strength *(17)*, could lead to applications where high compressive strength is required, such as structural components on vehicles with flexing wings or where impact damage from tools, ice, and stones is likely to occur *(16)*. Large recoverable deformations have been observed for both compression and extension for dragline silk from *Nephila clavipes*.

Since spiders produce a variety of silks, each tailored to a specific functional job, a wide range of additional applications may be realized when these other protein-based materials are understood. For example, adhesives which bond to a wide range of environmental substrates, even in wet conditions, and protective capsules derived from egg case silk are just two examples. Similarly, novel displays and responses in the electromagnetic spectrum *(18)* may prove useful in labeling, pattern recognition, and related scenarios.

The incorporation of silks into medical textiles is attractive because its protein-based molecular structure should yield excellent biocompatibility *in vivo*. Artificial tendons, blood vessels, and skin grafts could greatly benefit from the mechanical response of silks. Currently, silk fibroin is widely used as surgical thread due to its excellent mechanical and physical properties and good microbial resistance *(19)*.

Another potential benefit is the production of submicron diameter fibers which is in the realm of possibility using the spider's processing conditions. Man-made fibers of that size are difficult to make using conventional processes. These small diameter fibers possess superior properties, including high filament stiffness and strength*(20)*. which allows waterproof and windproof clothing that is permeable to water vapor to be manufactured. One could also speculate about applications such as corneal implants, where the submicron silk fibers would take the place of 30nm diameter collagen fibers *(21)*. The ability to control the molecular and microstructure will eventually lead to custom engineered mechanical properties.

Supercontraction of some silks has been postulated as having the possibility of doing molecular work. With the advent of micromachinery, one could conceive of a complex series of micromachines linked by ultra-fine spider silk fibers. The activation switch could be moisture causing a physical change in the amount of interaction through supercontraction. On this small scale, the high tensile strength would also undoubtedly be an added benefit. This phenomena could also possibly be utilized as moisture/humidity sensors.

Film/Membrane Applications

The complex secondary molecular structure of silks can be used to control specific interactions in different chemical and mechanical environments. The presence and amounts of particular secondary structure (silk I, silk II, α-helical, or random coil) of silk fibroin can be modified and controlled through stretching, compression, or with chemical and annealing treatments. These conformational changes can be utilized in the formation of membranes of stable, thin films for a variety of barrier applications*(22)*.

One area of current research interest is the use of silk in enzyme immobilization technology *(23-25)*. This approach can be used in the production of pharmaceuticals, cosmetics, specialty chemicals, fuels, and foodstuffs. Problems associated with immobilization of enzymes by conventional covalent coupling techniques can be reduced using silk-entrapped enzymatic systems. Advantages offered include the continuous operation of a microbial or enzymatic reactor, stabilization and higher activity of catalysts, and higher cell densities than those in conventional fermentation methods.

Although there are many different methods of immobilization available, their commercial applicability has been limited to date. Commercially produced chemicals using immobilization technology include the enzyme b-lactamase *(26)*, bulk chemicals ethylene oxide *(27)* and butadiol *(28)*, and the anti-inflammatory pharmaceutical shikonin *(29)*. Other promising applications include the immobilization of mammalian cells *(30)*, waste water treatment *(31-33)*, production and transformation of steroids *(34-36)*, and the production of antibodies *(37)*. To date, disadvantages encountered with proposed current immobilization methods include the mechanical instability of the support matrix, loss of enzymatic activity, harmful effects on the enzyme or cell due to harsh immobilization processes, and diffusional limitations.

Recently, researchers have examined a number of entrapment immobilization schemes to try to minimize these disadvantages. Gel systems such as calcium alginate and k-carraggenan have received the most attention due to successful ethanol production *(38)*. However, the main problem has been the weak mechanical stability of these systems, which leads to failure of the reactor systems due to increased pressure. Silk systems have recently shown promise as an immobilization matrix . A number of enzymes including peroxidase, glucose oxidase, and invertase have been immobilized successfully. One advantage of silk is the variability of its structure with slight changes in the environment. This allows the simultaneous insolubilization of the carrier and the immobilization of the enzyme without the use of chemicals. Tests indicate no disruption of enzyme activity, no inhibition to oxidation, good mechanical stability, and no diffusional problems in the incorporation of glucose oxidase for the development of a glucose sensor *(24)*. Biophotosensors have also been developed using an immobilized silk-based system *(25)*. The issue of embrittlement may be a factor if nonaqueous environments or ambient air exposures are encountered.

The degree of control which is available for silk systems leads to the possibility of their use in other permeable or semi-permeable membrane structures. Controlled delivery of drug dosages through capsules or membranes placed on the skin is one example. They could also be used as sensors for humidity or stress (especially compression). The large changes that the natural protein undergoes under a variety of environmental conditions may lead to devices where changes in this structure are used as a probe.

Conclusions

This paper has outlined a few of the potential technological uses for silk fiber and film technology. In this brief discussion we could not give proper attention to the long, difficult, and expensive research and development necessary to identify commercial applications or products for any new material. Indeed, just to list the properties that would have to be optimized for any successful application would significantly increase the length of this article. Therefore, this discourse only presents an abbreviated list of potential uses/products that current silk research could yield. In addition it is essential to recognize the need to achieve reasonable economic silk-production systems for these applications to be realized. This may occur through synthetic analogs, genetics with the native genes *(39)*, or genetics with synthetic genes *(40)*. Some of the most interesting applications for silk may in the long run not have anything to do with silk. For example, by developing a thorough understanding of the structure of the silk fibers we should be able to design new synthetic analogs tailored to achieve specific functional properties. Similarly, as we elucidate the processing steps and controls used in spinning silk fibers, an expanded field of environmentally-compatible approaches to fiber/film manufacturing may be realized.

Literature Cited

(1) Zemlin, J.C., *Technical report 69-29-CM (AD 684333)*, U.S. Army Natick Laboratories, **1968**.
(2) Work, R.W., *J. Exp. Biol.* **1985**, *118*, 379-404.
(3) Work, R.W. and Morosoff, N., *Texas Research J.*, **1982**, *52*, 349-356.
(4) Work, R.W. and Morosoff, N., *J. Arach.*, **1981**, *9*, 299-308.
(5) Nash, J.M., *Time*, **1993**, *March 8*, 58-59.
(6) O'Brien, J. et. al., in this volume.
(7) Cantor et. al., in this volume
(8) Hirabayashi, K., Ayub, Z.H., and Kume, Y., *Sen-I Gakkaishi*, **1990**, *46(11)*, 521-524.
(9) Kaplan, D.L., Lombardi, S.J., Muller, W.S., and Fossey, S.A.in *Silks* ; Editor., D. Byron, Biomaterials: Novel Materials from Biological Sources, Stockton Press, N.Y., 2-51, **1991**.
(10) Lucas, F., Shaw, J.T.B., and Smith, S.G., *J. Mol. Biol.*, **1960**, *2*, 339-349.
(11) McKeown, K.C., *Spider Wonders of Australia*; Angus and Robertson Limited; Sydney, Australia, **1936**.
(12) Hyde, N., *National Geographic*, **1984**, *165(1)*, 2-50.
(13) Becker, M. and Tuross, N., in this volume.
(14) Cunniff, P. et. al., in this volume.
(15) Case, S. et. al., in this volume.
(16) Mahoney, D. et. al., in this volume.
(17) Pachter, R., in this volume.
(18) Craig, C. et. al., in this volume.
(19) Asakura, T., Yoshimizu, H., and Kakizaki, M., *Biotech. Bioeng.*, **1990**, *35*, 511-517.
(20) Heidenreich, I and Ninow, H., *Melliand Textilberichte*, **1991**, *72(12)*, 971.
(21) McCally, R.L. and Farrell, R.A. in *Noninvasive Diagnostic Techniques in Ophthalmology*, B. R. Masters, Ed., Springer-Verlag; New York, pp 189-210, **1990**.
(22) Muller, W.S., Samuelson, L.A., Fossey, S.A., and Kaplan, D.L., in this volume.
(23) Yoshimizu, H. and Asakura, T., *J. Appl. Polym. Sci.*, **1990**, *40*, 127-134.
(24) Demura, M., Asakura, T., and Kuroo, T., *Biosensors*, **1989**, *4*, 361-372.
(25) Demura, M., Asakura, T., Nakamura, E., and Tamura, H., *J. Biotech.* **1989**, *10*, 113-120.
(26) Indoes, D.S., Smith, N.J., Taylor, D.P., Cohen, S.N., Michaels, A.S., and Robertson, C.R., *Biotech. Bioeng.*, **1983**, *25*, 2653-2681.
(27) de Bont, J.A.M. and van Ginkel, C.G., *Enz. Micro. Tech.*, **1983**, *5*, 55-59.
(28) Kautola, H., Linko, Y-Y., and Linko, P., *Ann. N. Y. Aca. Sci.*, **1984**, *434*, 454-458.
(29) Sahai, O. and Knuth, M., *Biotech. Prog.* **1985**, *1*, 1-9.
(30) Hu, W.S., Giard, D.J., and Wang, D.I.C., *Biotech. Bioeng.*, **1985**, *27*, 1466-1476.
(31) Duss, I.J., Tanaka, H., Suheyla, U., and Denac, M., *Ann. N. Y. Aca. Sci.*, **1983**, *413*, 168-183.
(32) van Ginkel, C.G., Tramper, J., Luyben, K., Ch., A.M., and Klapwijk, A., *Enz. Micro. Tech.*, **1983**, *5*, 297-303.
(33) Narjari, N.K., Khilar, K.C., and Mahajan, S.P., *Biotech. Bioeng.*, **1983**, *26*, 1445-1448.
(34) Koshcheyenko, K.A., Turkina, M.V., and Skryabin, G.K., *Enz. Micro. Tech.* **1983**, *5*, 14-21.

(35) Mozes, N. and Rouxhet, P.G.., *Enz. Micro.Tech.,* **1984**, *6*, 497-503.
(36) Lian-Wan, Y. and Li-Chan, Z., *Ann. N.Y. Aca. Sci.*, **1984**, *434,* 459-460.
(37) Vandamme, E.J., *Enz. Micro.Tech.*, **1983**, *5*, 403-416.
(38) Kuu, W.Y. and Polack, J.A., *Biotech. Bioeng.,* **1983**, *25*, 1995-2006.
(39) Hinman, M. et. al., in this volume.
(40) Cappello, J., Crissman, J., Dorman, M., Mikolajczak, M., Textor, G., Marquet, M., and Ferrari, F., *Biotech.Prog.* **1990**, *6*, 198-202.

RECEIVED July 15, 1993

Author Index

Affiliation Index

Subject Index

Production: C. Buzzell-Martin
Indexing: Deborah H. Steiner
Acquisition: Anne Wilson
Cover design: Alan Kahan

Printed and bound by Maple Press, York, PA